Atlas of Descriptive Embryology

Second Edition

Cover: Photomicrograph of sea urchin embryos from a commercially prepared whole mount of fixed and stained specimens; 16-mm achromatic objective; 5 X Huygenian ocular, Rheinberg illumination.

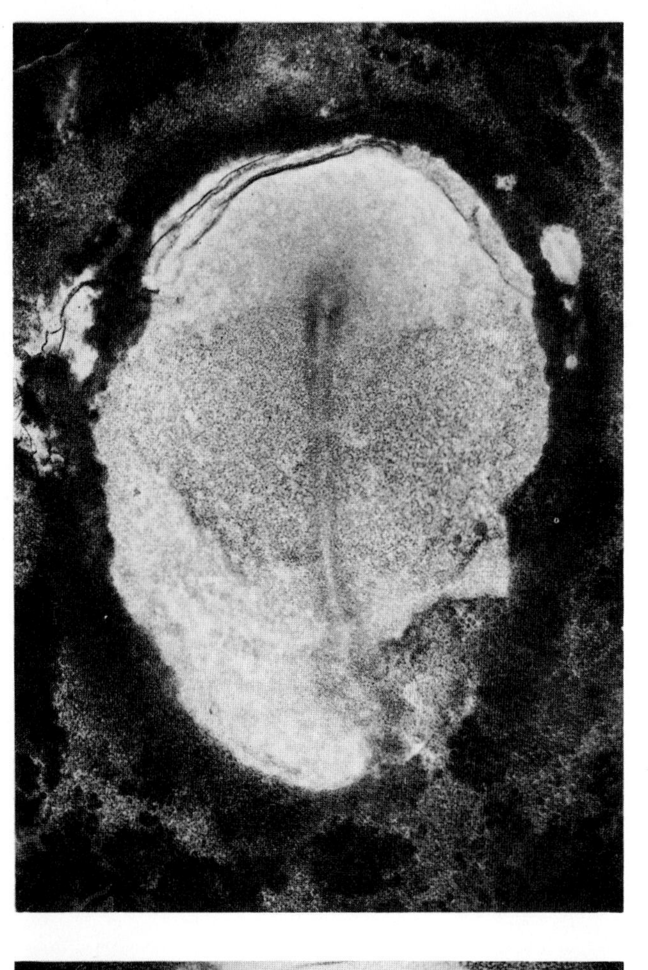
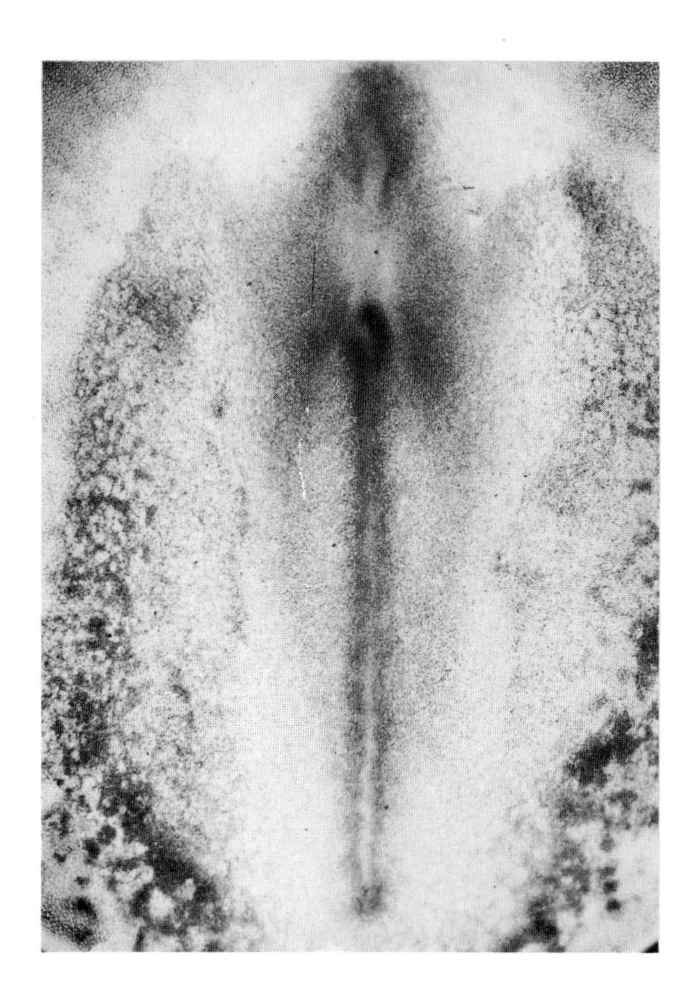
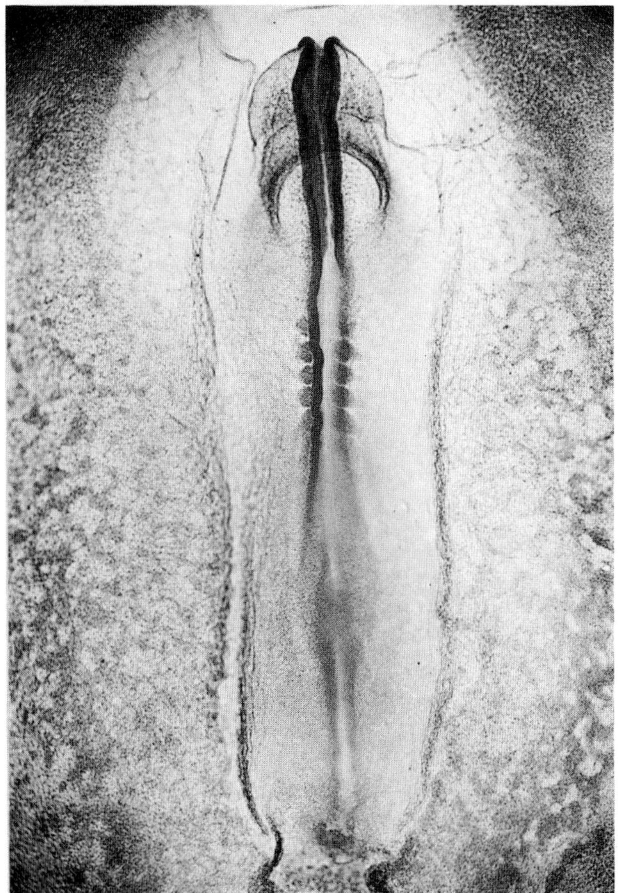
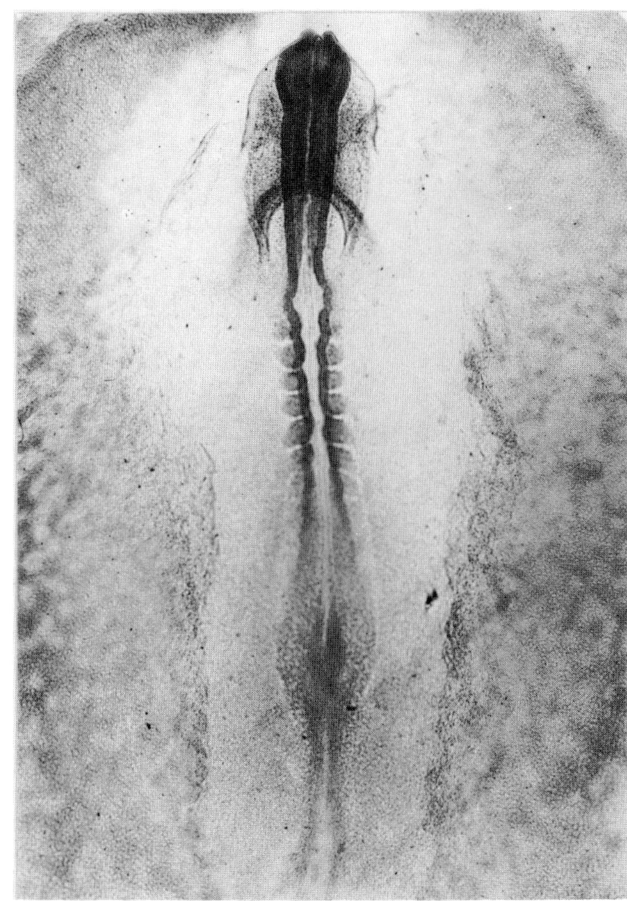

Atlas of Descriptive Embryology

Second Edition

Willis W. Mathews
Wayne State University

Macmillan Publishing Co., Inc.
New York
Collier Macmillan Publishers
London

Printed in the United States of America

Macmillan Publishing Co., Inc.
866 Third Avenue, New York, New York 10022

Collier Macmillan Canada, Ltd.

Library of Congress Cataloging in Publication Data

Mathews, Willis W (date)

Atlas of descriptive embryology.

Includes index.

1. Embryology—Atlases. I. Title.

QL956.M38 1976 599'.03'30222 75-22465

ISBN 0-02-377110-0

Printing: 1 2 3 4 5 6 7 8 Year: 6 7 8 9 0 1 2

Preface to the Second Edition

The favorable reception of the first edition of this atlas has encouraged the author to try to augment its usefulness with new material. A total of 16 drawings and 27 photomicrographs have been added. These figures have illustrated 69 additional structural features and these have been added to the "Glossary and Synopsis of Development."

The section on gametogenesis now includes spermatogenesis in the rat, grasshopper, frog, and chicken, and oogenesis in the cat, frog, sea urchin, and chicken. Frog development has been covered with ten additional drawings. The chick development section now includes five new drawings and six new photomicrographs. Five new transverse sections of the 10-mm pig embryo were added together with a drawing of the circulatory system of the pig. Finally, a section on the human placenta was included.

I hope that these additions and the corrections which were also made will adapt the work to a greater variety of courses in embryology and developmental biology.

W. W. M.

Preface to the First Edition

Descriptive embryology still constitutes a body of knowledge fundamental to modern developmental biology. Yet, the growth of experimental and biochemical embryology and other curricular demands allows less time for the course study of descriptive embryology. As a consequence, students have generally felt the need for detailed, accurate pictures of their laboratory materials which are fully labeled. This atlas was prepared, hopefully, to satisfy such a need. With its help, together with that of his text and laboratory manual, a student should be able to carry forward his studies quickly, efficiently and mainly by his own efforts. With descriptive embryology in hand, the student and his lecturer will then be free to devote more time to comparative, physiological and experimental studies of development.

The slide preparations which were photographed for this atlas were virtually all obtained from biological supply houses. The figures should, therefore, closely resemble the slides the student will receive for his laboratory work. They will provide a useful supplement to the reconstructions, diagrams and incompletely labeled photomicrographs found in most texts and manuals.

The problem of what to include in the atlas was easily resolved in its main features. The chick embryo has been standard laboratory material for the study of development since classical times. Amphibian and sea urchin embryos have been used for many important experimental investigations of development. Some acquaintance with the development of these groups prepares the student to understand the intricacies of the experiments. The pig embryo is certainly the most widely used example of mammalian development, so it was included without hesitation. Materials for the study of gametogenesis and fertilization are available in greater variety, but the cat, the rat and *Ascaris* serve very well.

The Glossary will likely prove to be valuable. It contains all terms used in the figures. Following each term is a list of figures in which it occurs. Then, related structures are given and the term defined. Next, the development of most structures including their origin and developmental fate is summarized. Important synonyms are also listed. The definitions of terms and summaries of development are occasionally incomplete where they were necessarily limited to the animal groups included in the atlas.

If a word of advice may be offered to the student, you will do well to study your slides and other preparations thoroughly. The illustrations in this atlas are not an adequate substitute for firsthand observations of embryos. No book of practical size could show all microscopic structures and their interrelations that are important. Use the atlas to check your identifications of structures, then fully explore the morphology of the parts with your microscope. The atlas will again be helpful in reviewing for exams. The figures will provide a quick recapitulation of your slides and the Glossary will supply a summary of the development of each part.

Many hollow organs and structures have been identified on the plates of this atlas by extending a label line to the cavity or lumen of the part. This practice helps to clarify the labels, but it should be recognized by the student that the organ so identified is actually the wall or tissue surrounding the space. Examples of this kind of labeling are: blood vessel, fig. 2; Graafian follicles, fig. 6; archenteron, figs. 30, 45.

My wife prepared the drawings inserted in many of the figures. I gratefully acknowledge this and much other help in preparing this atlas.

W. W. M.

Contents

1. Gametogenesis

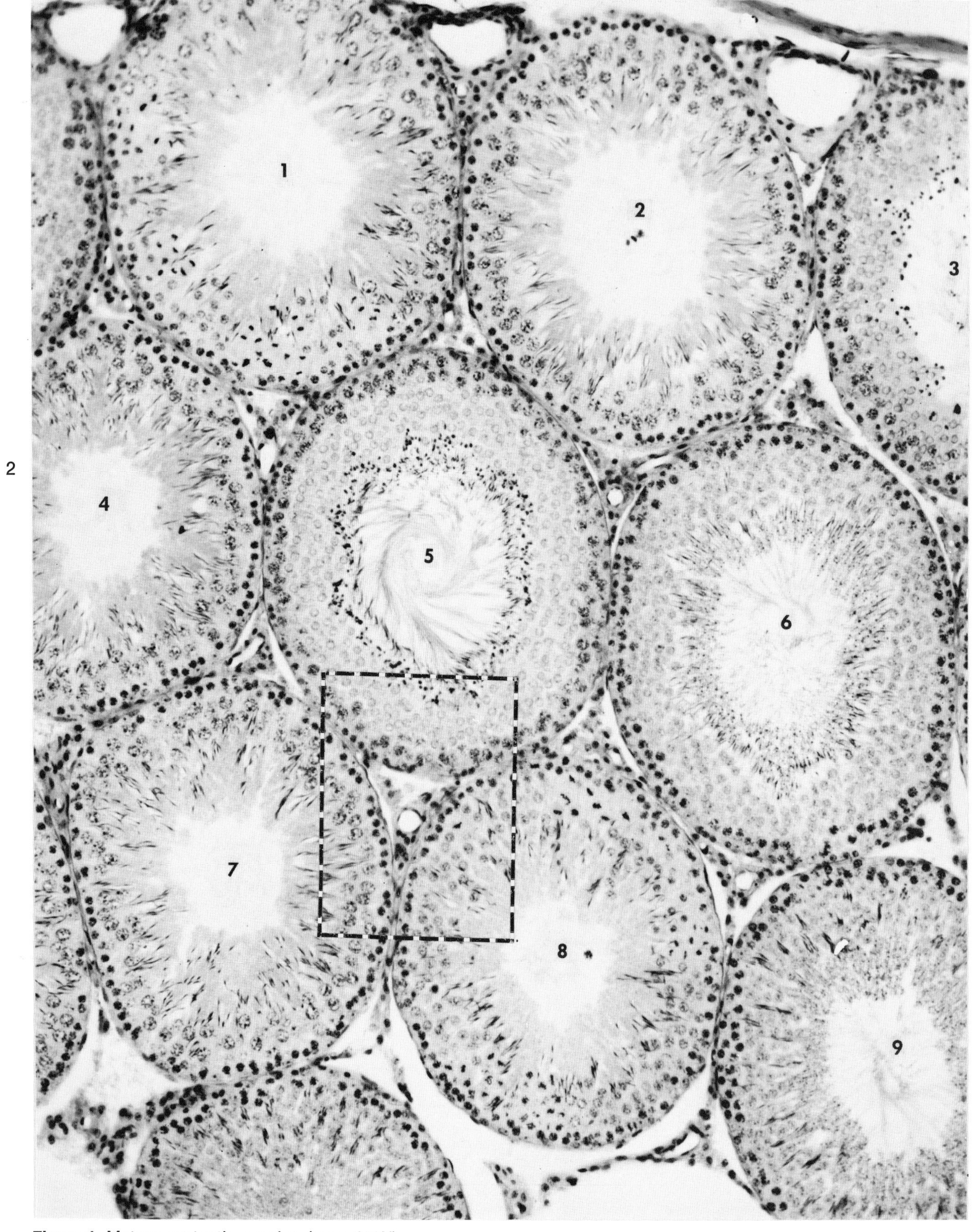

Figure 1 Mature rat testis, section (mag. 200X)

The dashed rectangle indicates the area shown in fig. 2. The numbered seminiferous tubules show various stages in spermatogenesis. The stages and cell associations* are as follows:

Tubules	Stages	Cells
1 and 8	XIII, XIV	spermatogonia and Sertoli cells (the layers of cells from basement membrane to lumen) immature primary spermatocytes mature primary spermatocytes (some in first maturation division) secondary spermatocytes differentiating spermatids
2 and 7	XII	spermatogonia and Sertoli cells immature primary spermatocytes mature primary spermatocytes differentiating spermatids
3	VIII	spermatogonia and Sertoli cells resting primary spermatocytes growing primary spermatocytes spermatids sperm
4	XI	spermatogonia and Sertoli cells immature primary spermatocytes growing primary spermatocytes differentiating spermatids
5	VII	spermatogonia and Sertoli cells resting primary spermatocytes growing primary spermatocytes spermatids sperm
6	VI	spermatogonia and Sertoli cells growing primary spermatocytes spermatids differentiating spermatids
9	VI	spermatogonia and Sertoli cells immature primary spermatocytes spermatids differentiating spermatids

*Based on Clermont, Leblond, and Messier in Roy O. Greep (ed)., *Histology,* 2nd ed., McGraw-Hill Book Company, New York, 1966.

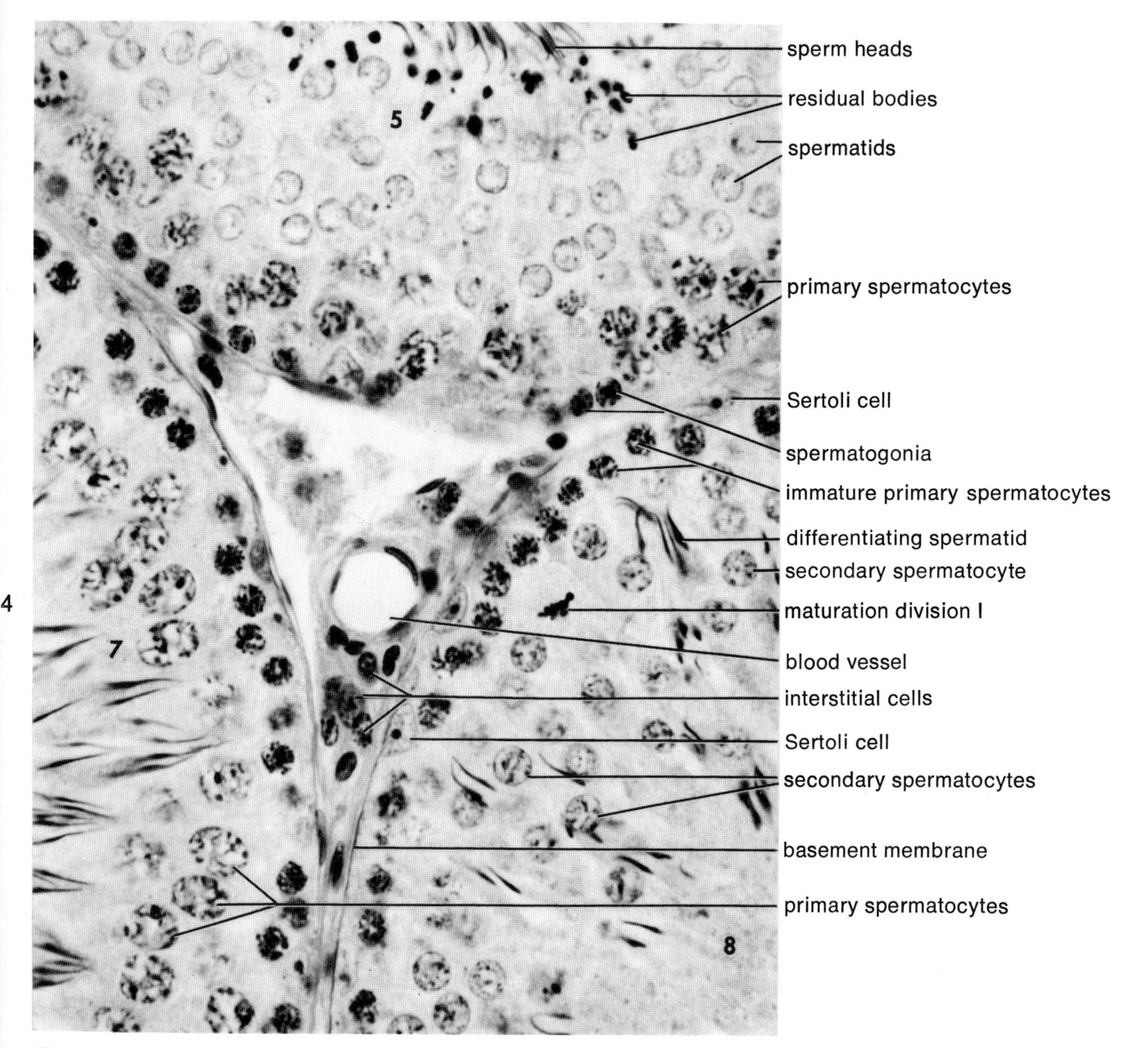

Figure 2 Mature rat testis, section (mag. 650X)

Label lines indicate cell nuclei. Outlines of surrounding cytoplasm are indistinct. The stages of spermatogenesis and cell associations are as follows:

Tubules	Stages	Cells
5	VII	spermatogonia and Sertoli cells resting primary spermatocytes growing primary spermatocytes spermatids sperm
7	XII	spermatogonia and Sertoli cells immature primary spermatocytes mature primary spermatocytes differentiating spermatids
8	XIII, XIV	spermatogonia and Sertoli cells immature primary spermatocytes mature spermatocytes (some in first maturation division) secondary spermatocytes differentiating spermatids

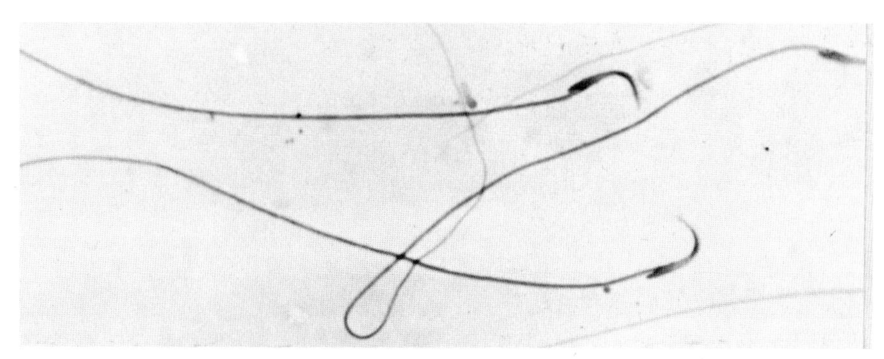

Figure 3 Rat sperm smear (mag. 710X)

Figure 4 Mature cat ovary, section through cortex (mag. 125X)

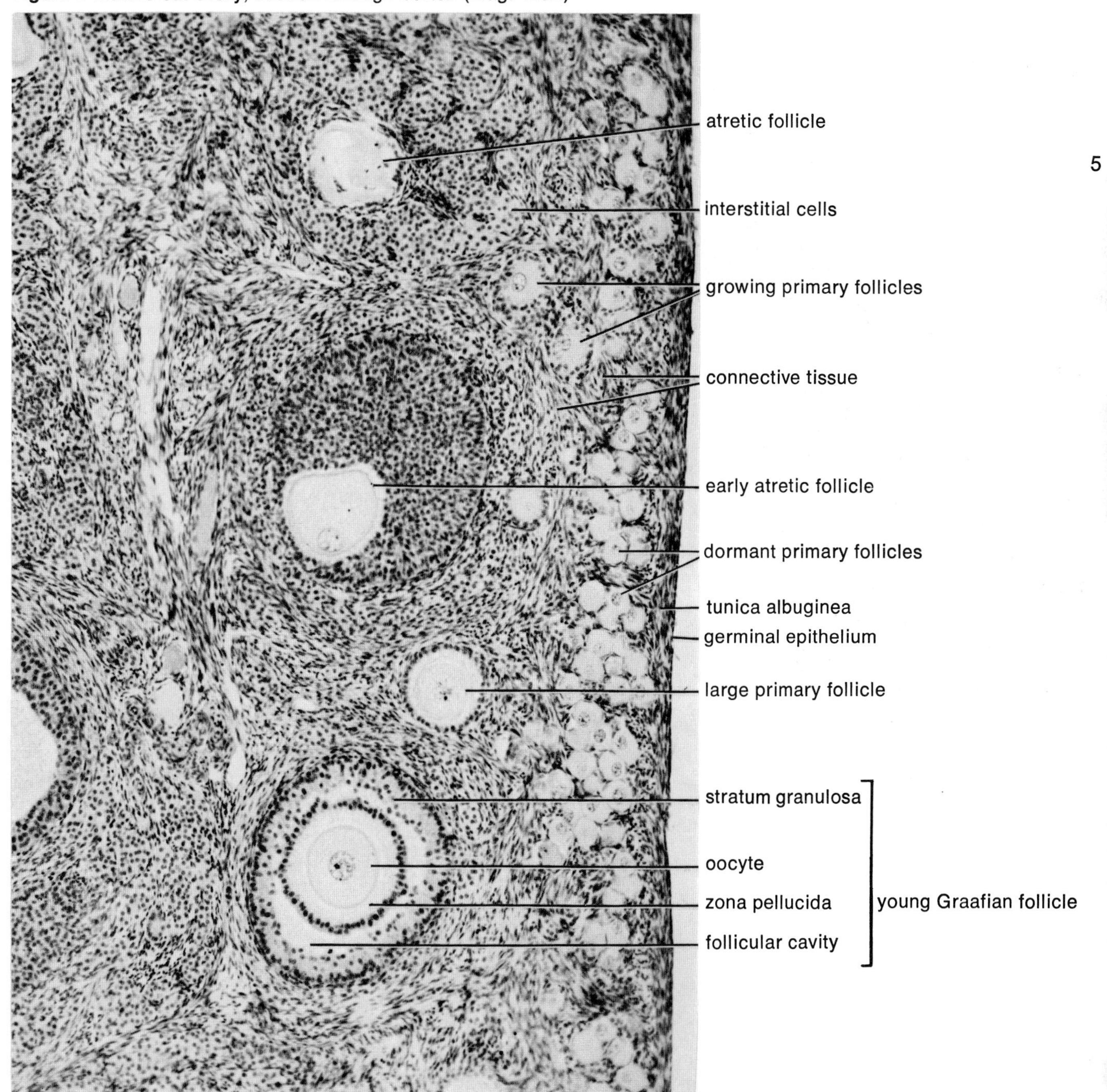

primary follicles
tunica albuginea
atretic follicle
theca externa
theca interna
stratum granulosa
zona pellucida
oocyte
cumulus oophorus
follicular cavity
blood vessels

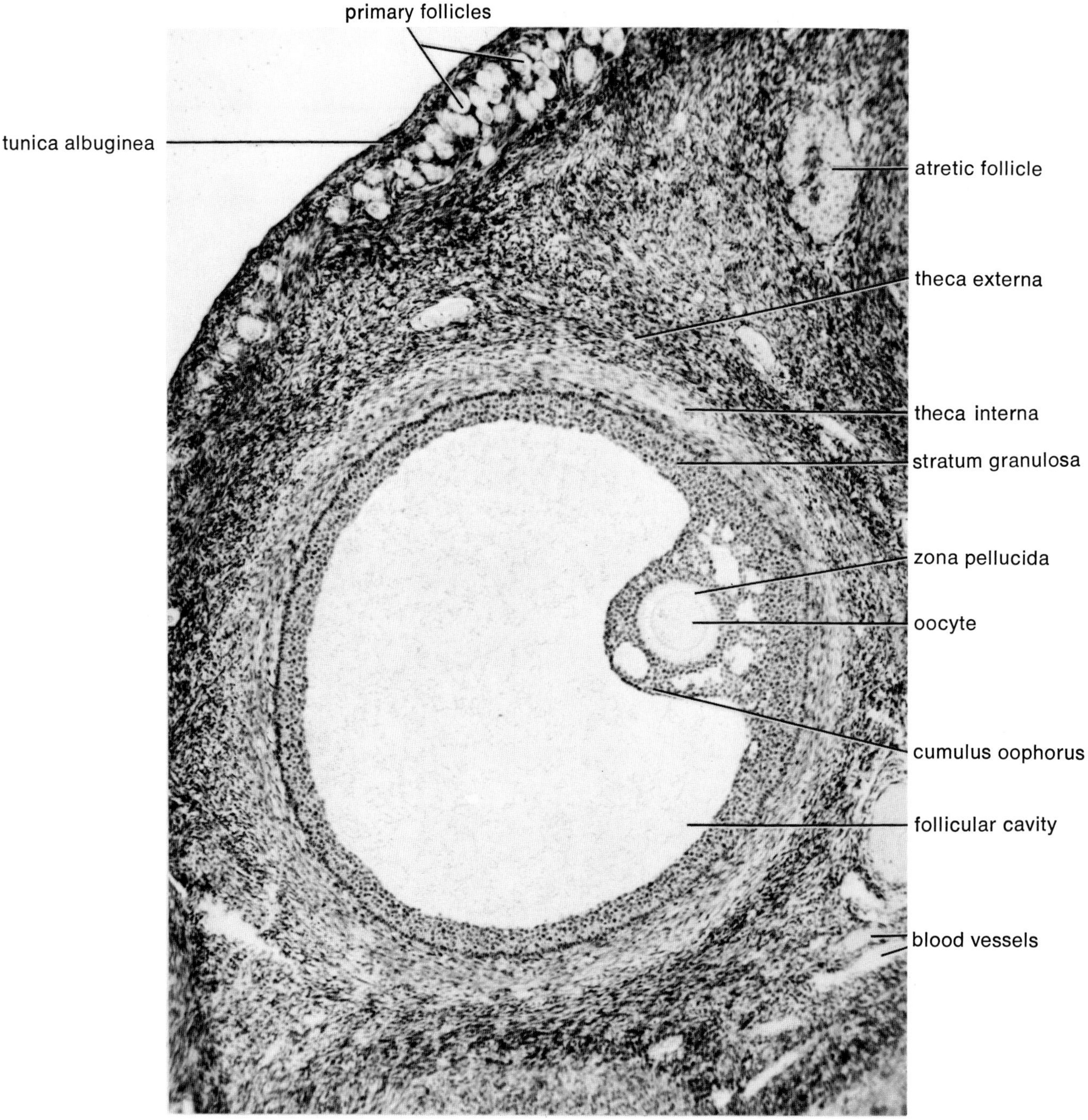

Figure 5 Mature cat ovary, section through large Graafian follicle (mag. 90X)

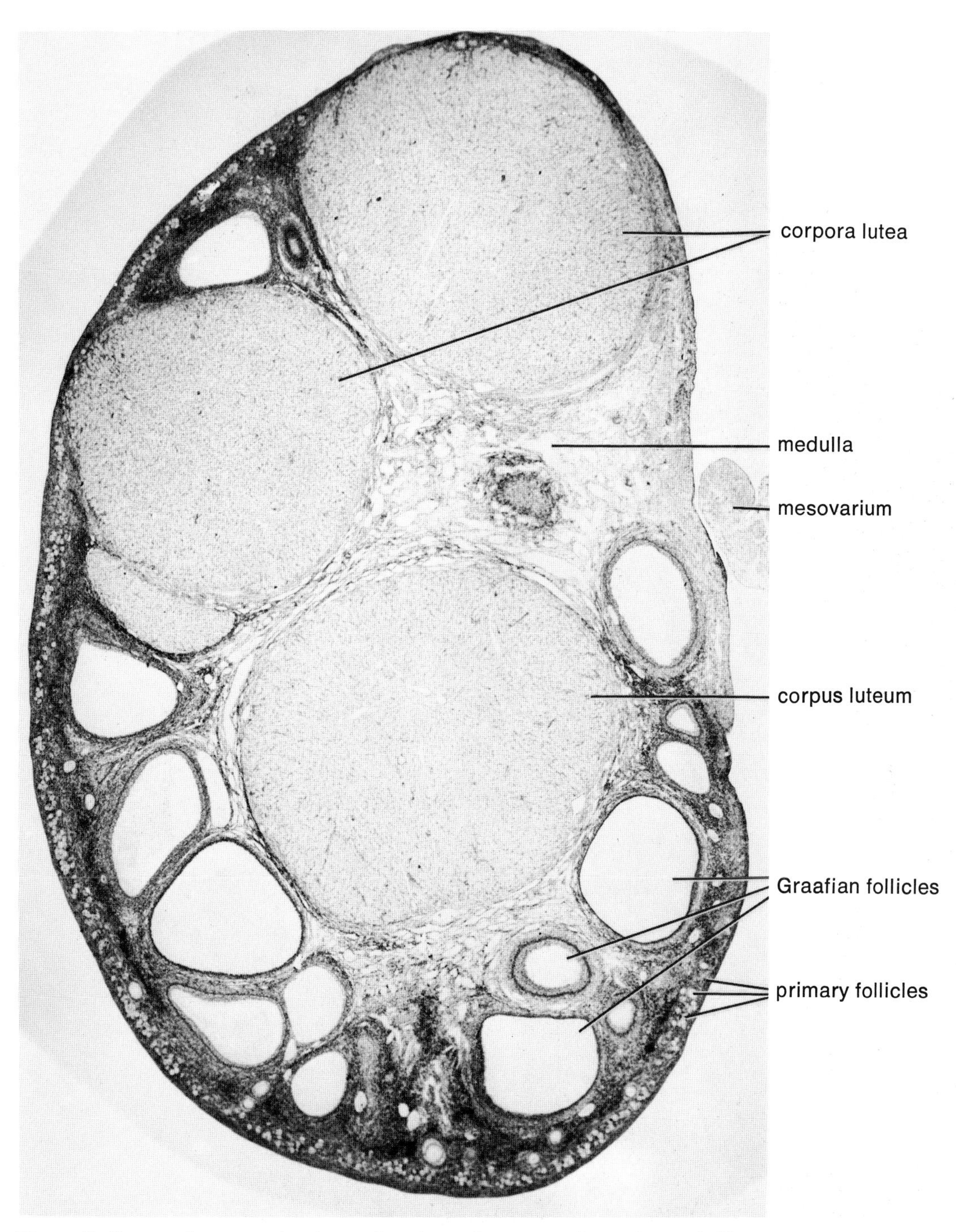

Figure 6 Ovary of pregnant cat, section through corpora lutea (mag. 25X)

8

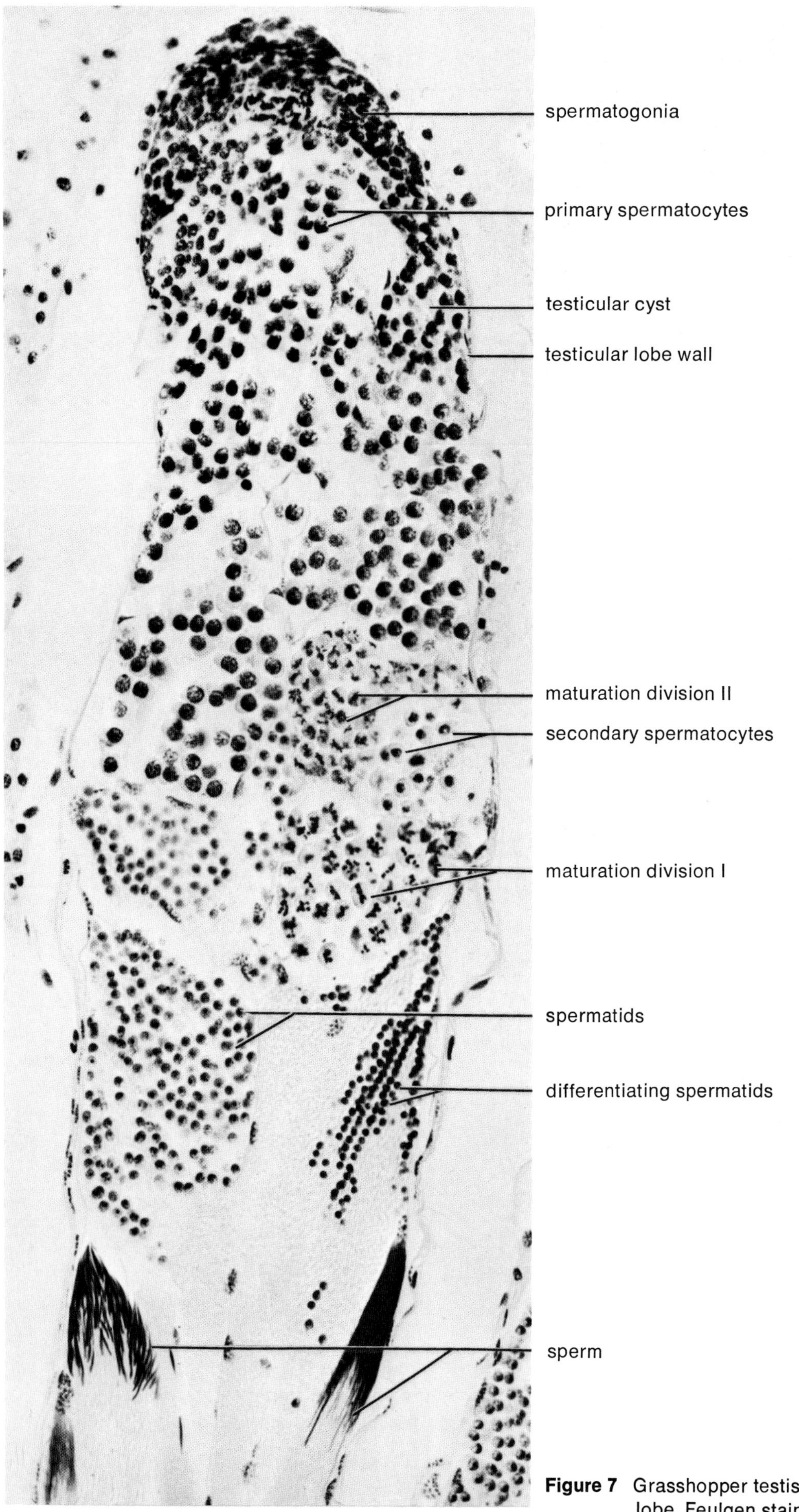

Figure 7 Grasshopper testis, longitudinal section of testicular lobe, Feulgen stain for DNA (mag. 190 X)

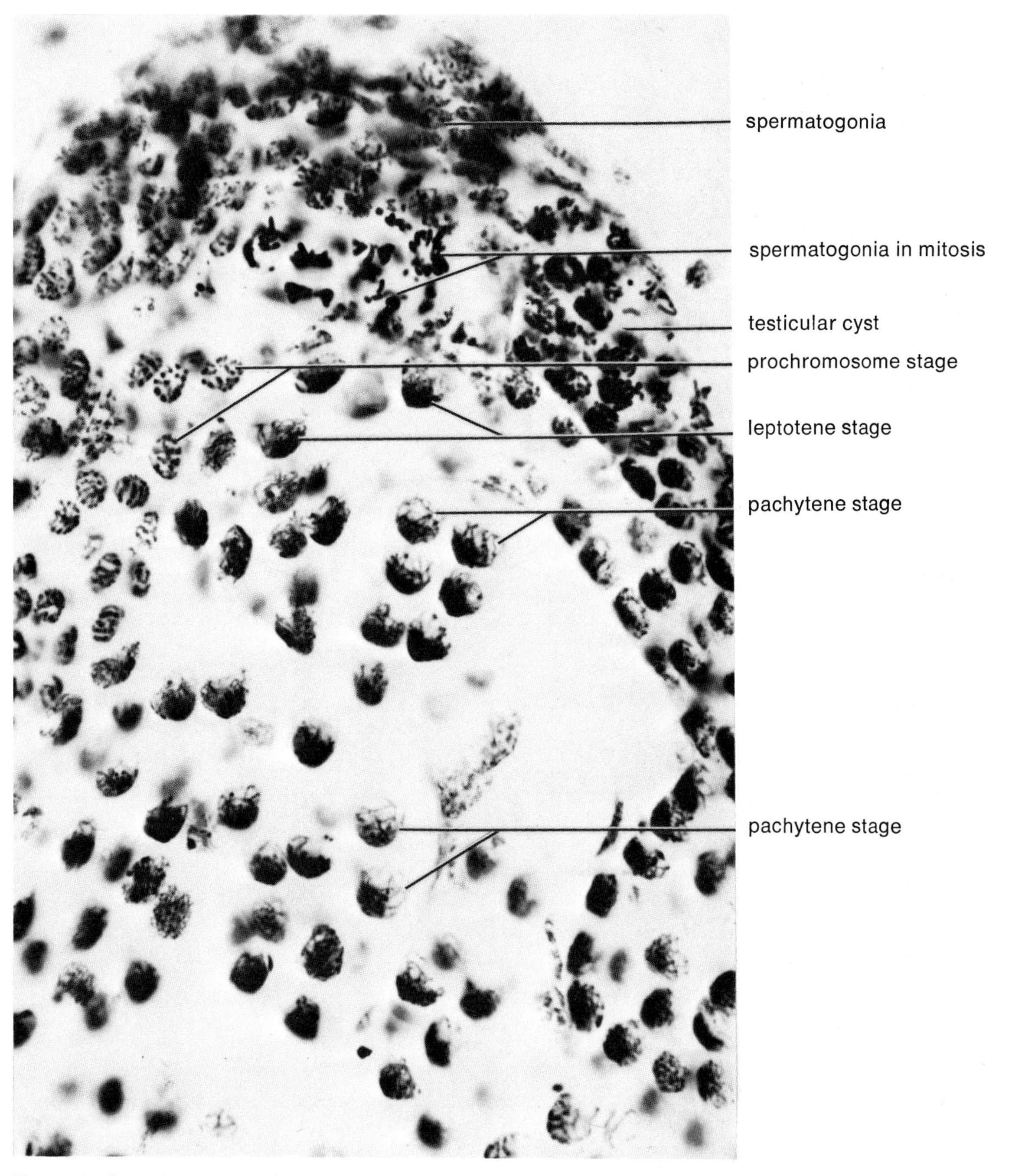

Figure 8 Grasshopper testis, longitudinal section showing area from fig. 7, Feulgen stain for DNA (mag. 550X)

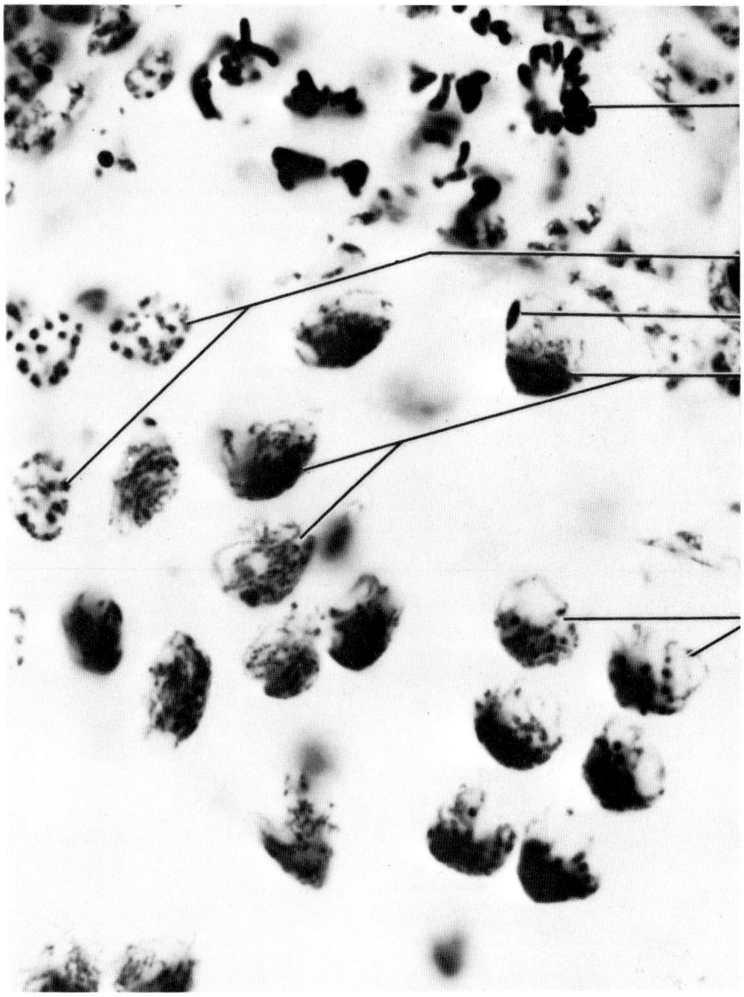

spermatogonia in mitosis

prochromosome stage

X-chromosome

leptotene stage

pachytene stage

Figure 9 Grasshopper testis, longitudinal section showing area from fig. 8, Feulgen stain for DNA (mag. 880 X)

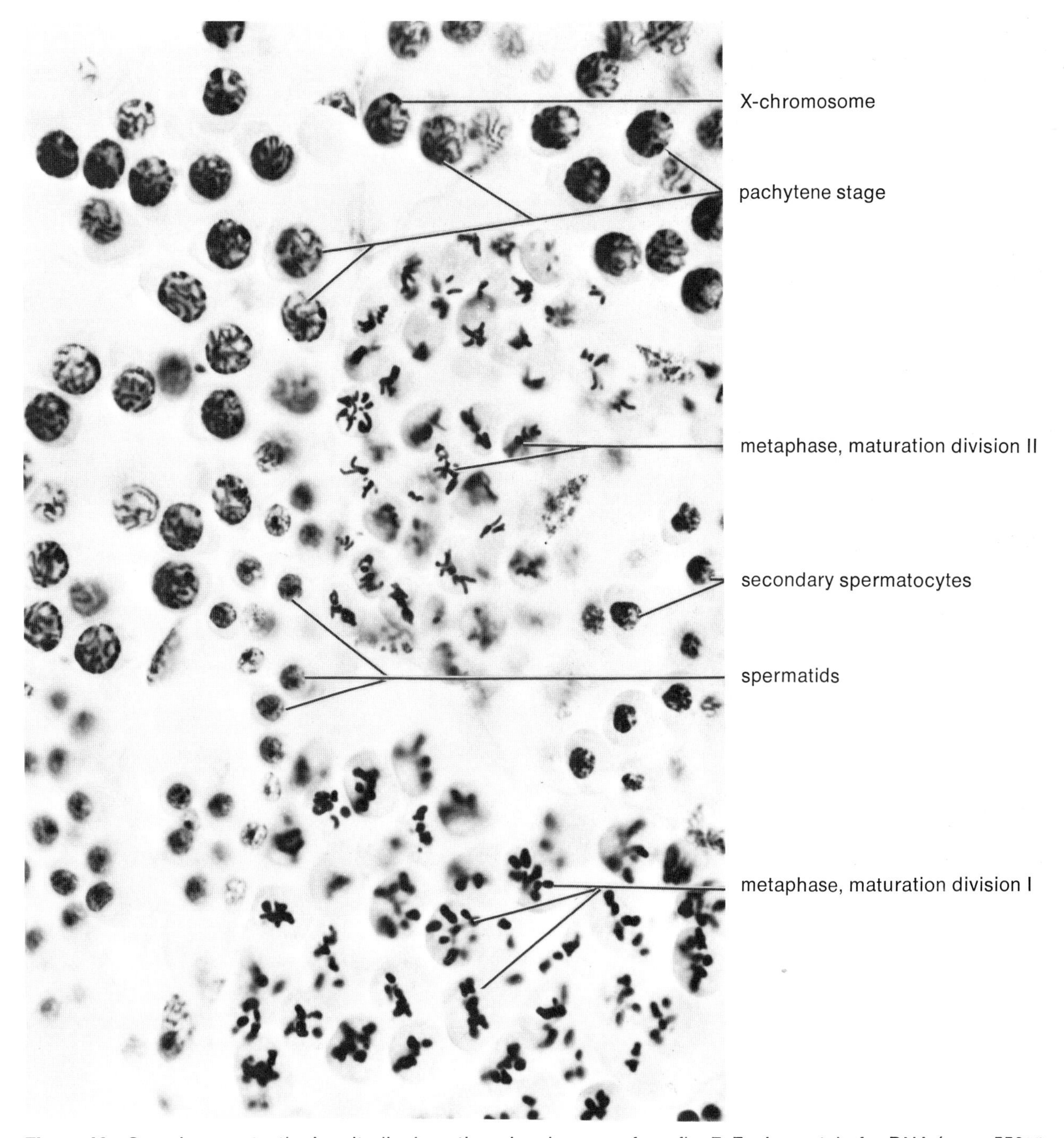

Figure 10 Grasshopper testis, longitudinal section showing area from fig. 7, Feulgen stain for DNA (mag. 550 X)

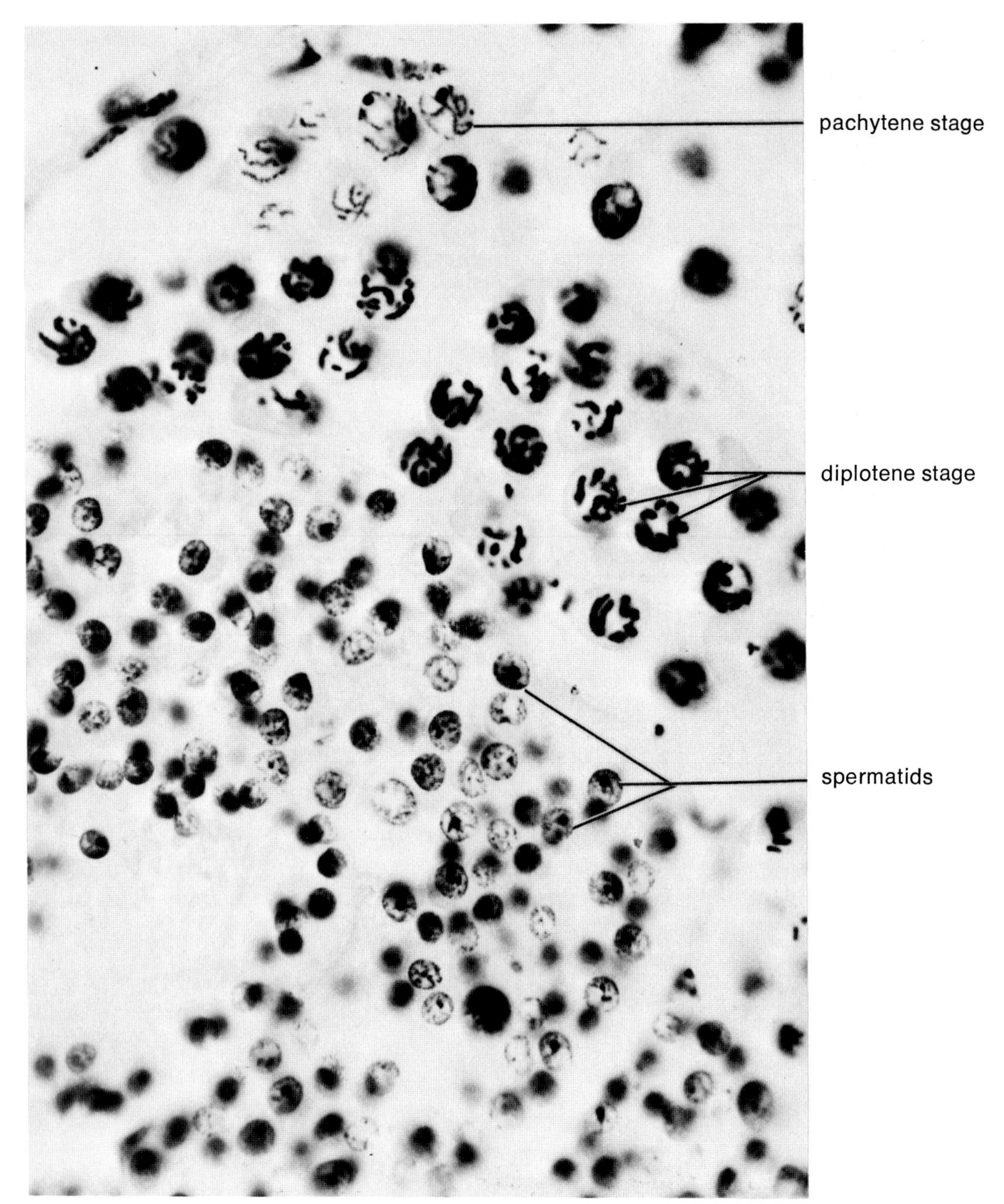

Figure 11 Grasshopper testis, longitudinal section of testicular lobe, Feulgen stain for DNA (mag. 550X)

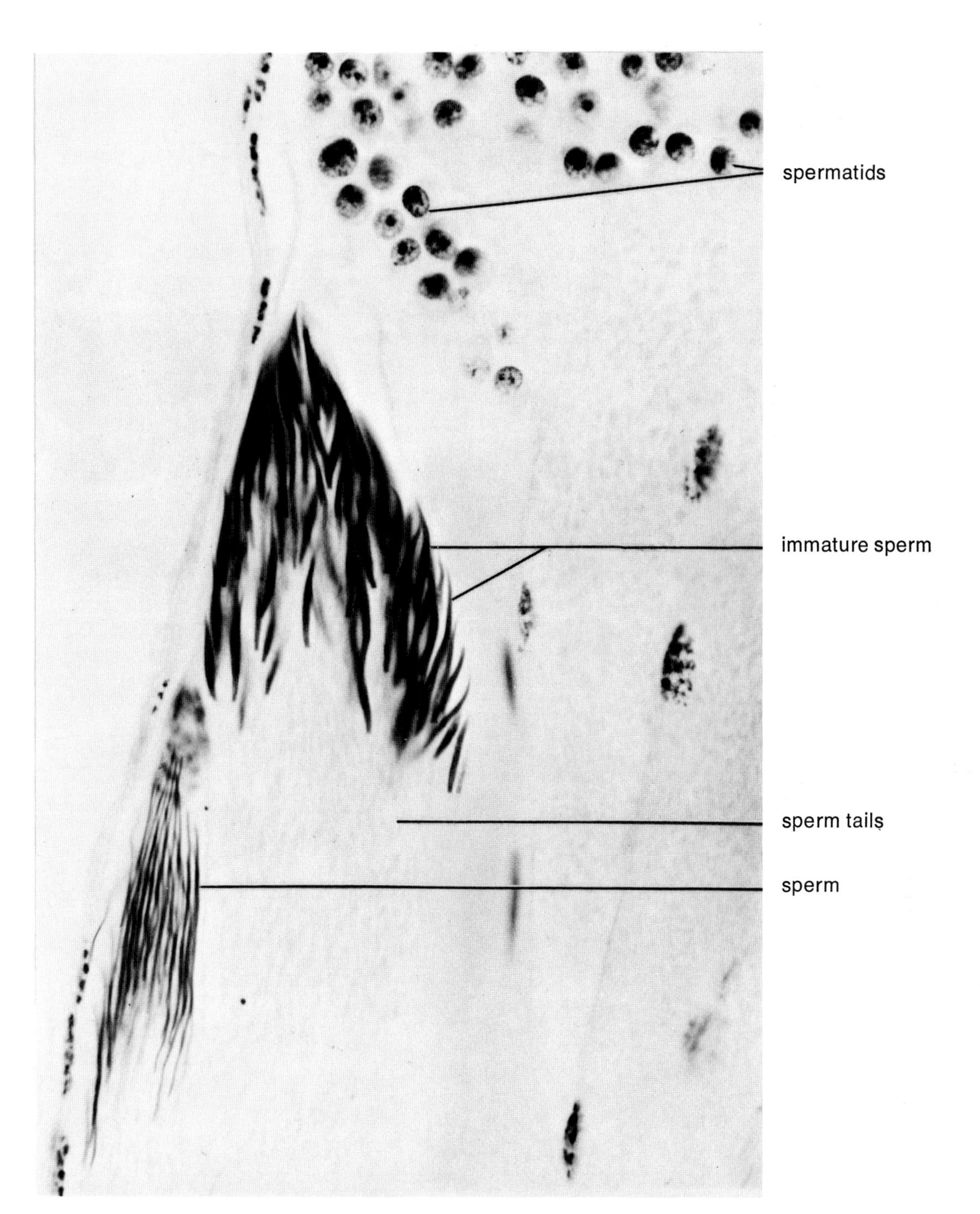

Figure 12 Grasshopper testis, longitudinal section showing area from fig. 7, Feulgen stain for DNA (mag. 550 X)

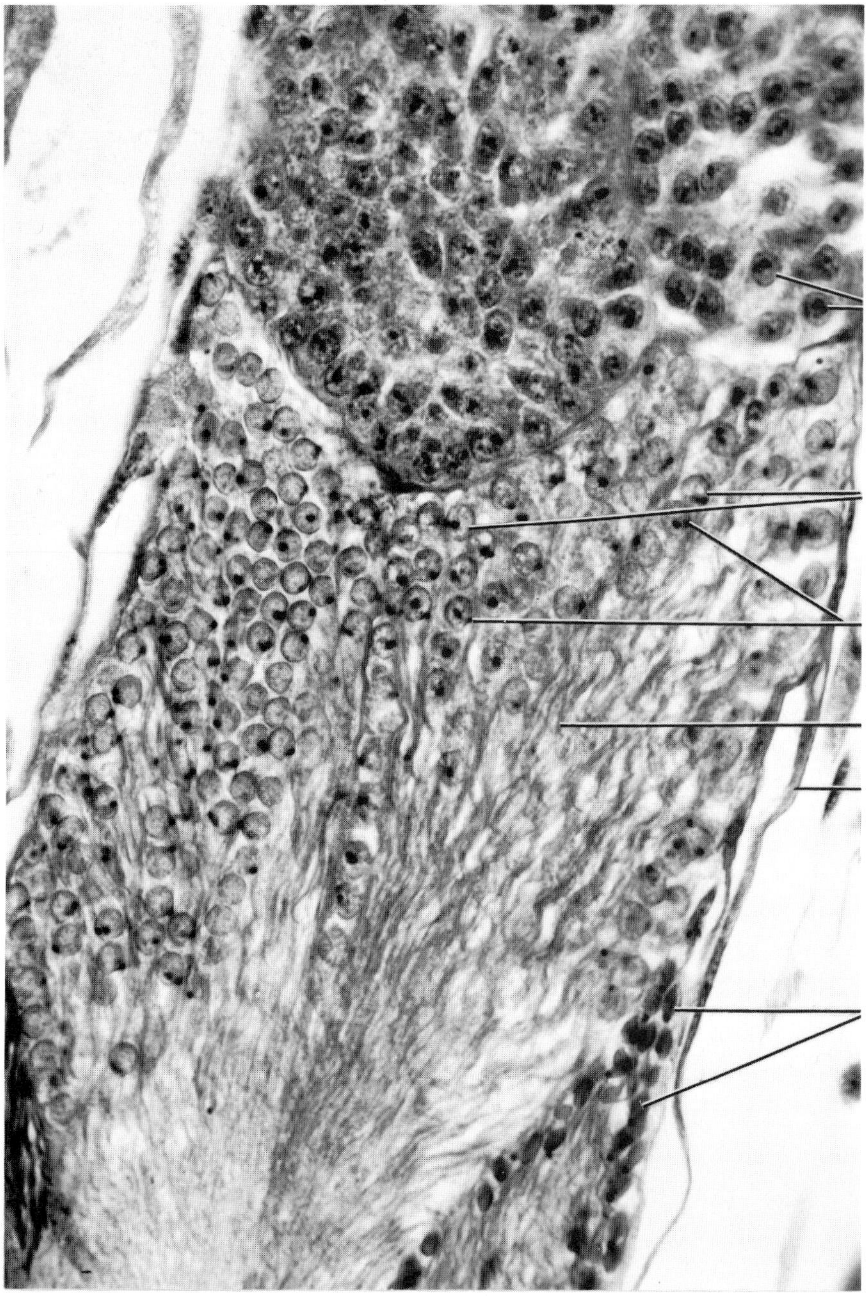

spermatids
differentiating spermatids
centrioles
axial filament
testicular lobe wall
immature sperm

Figure 13 Grasshopper testis, longitudinal section of testicular lobe, iron hematoxylin stain (mag. 550X)

2. Maturation and Fertilization in *Ascaris*

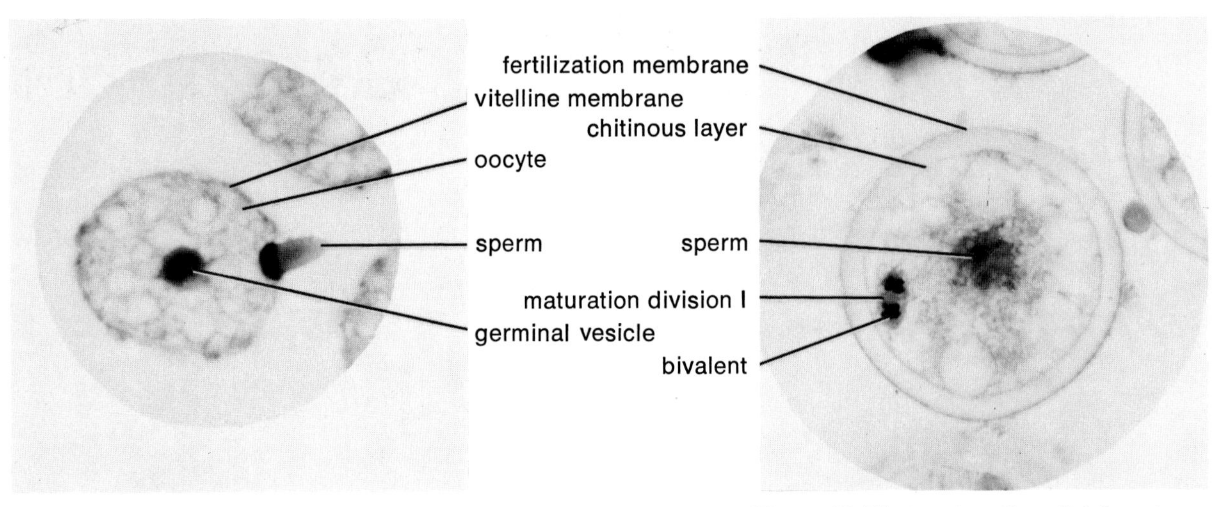

Figure 14 Sperm penetration stage (mag. 500X)

Figure 15 First maturation division stage (mag. 500X)

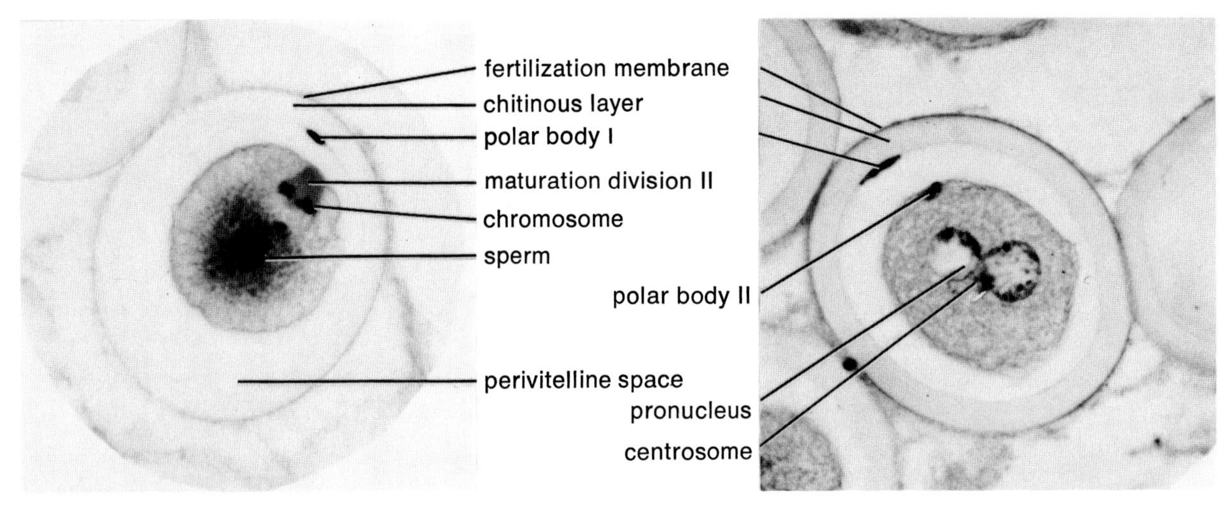

Figure 16 Second maturation division stage (mag. 500X)

Figure 17 Pronuclear stage (mag. 500X)

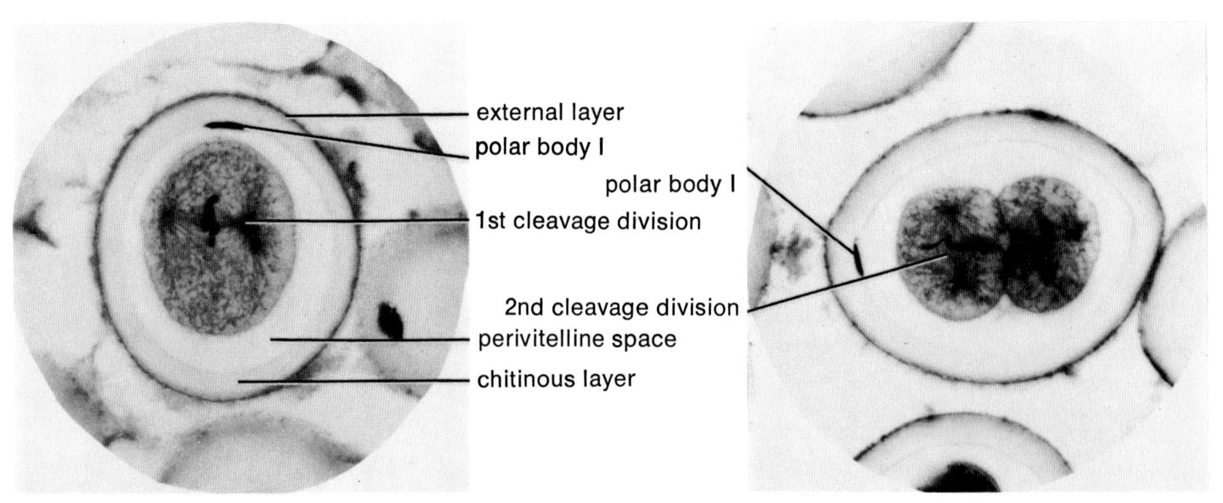

Figure 18 First cleavage division stage (mag. 500X)

Figure 19 Second cleavage division stage (mag. 500X)

3. Sea Urchin Development

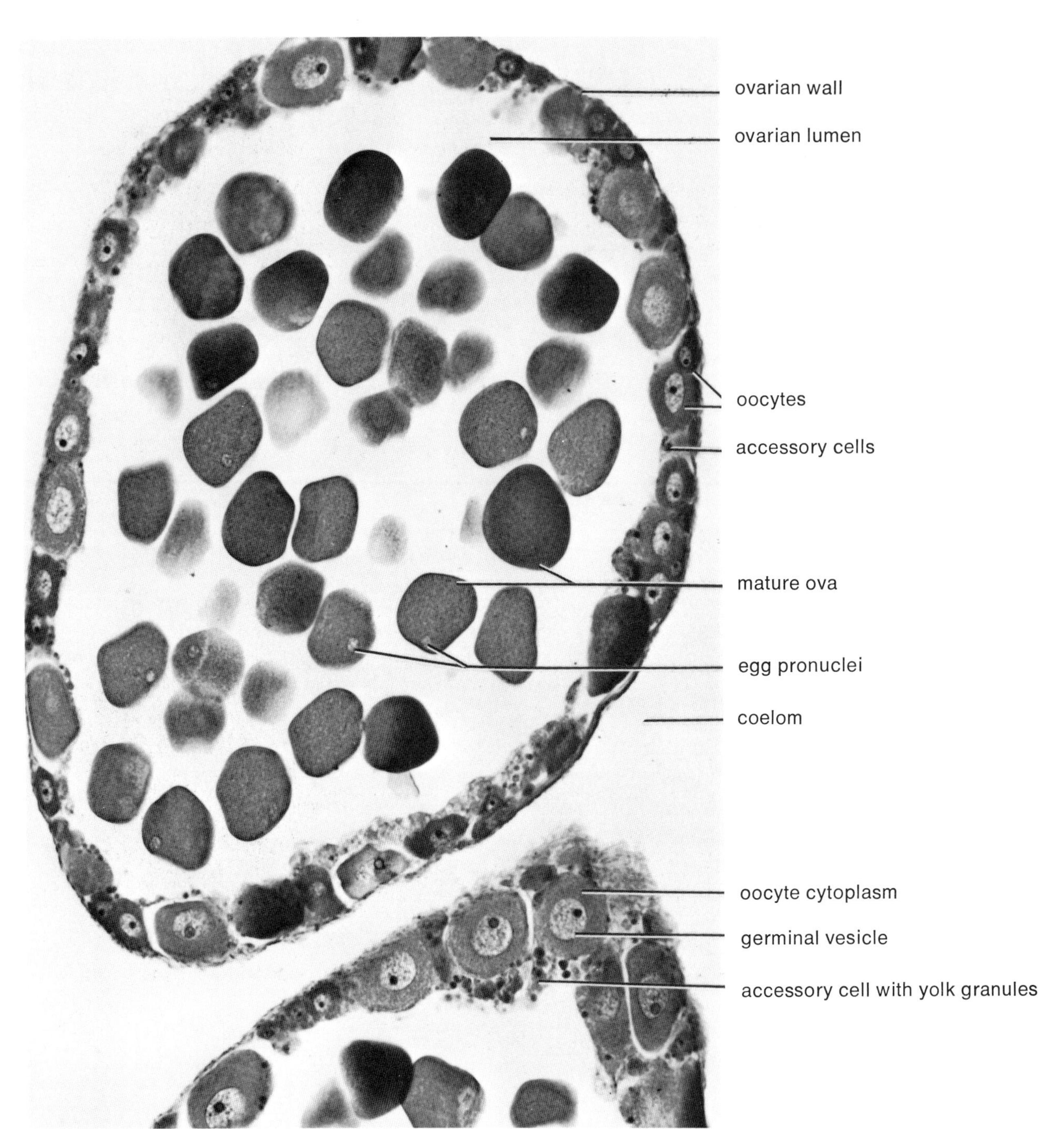

Figure 20 Sea urchin ovary, *Arbacia* (mag. 225 X)

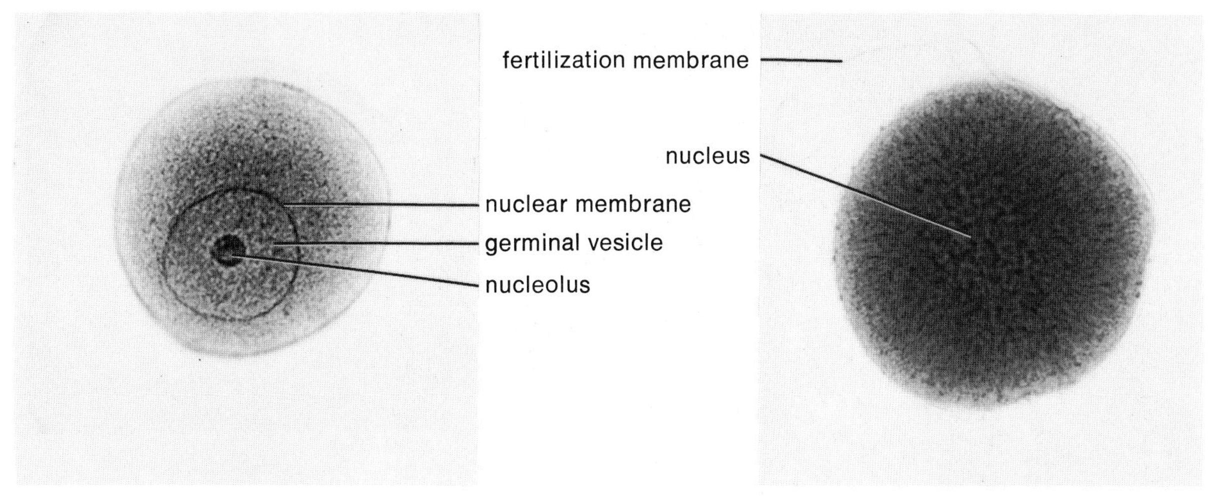

Figure 21 Primary oocyte (mag. 300X)

Figure 22 Fertilized egg (mag. 300X)

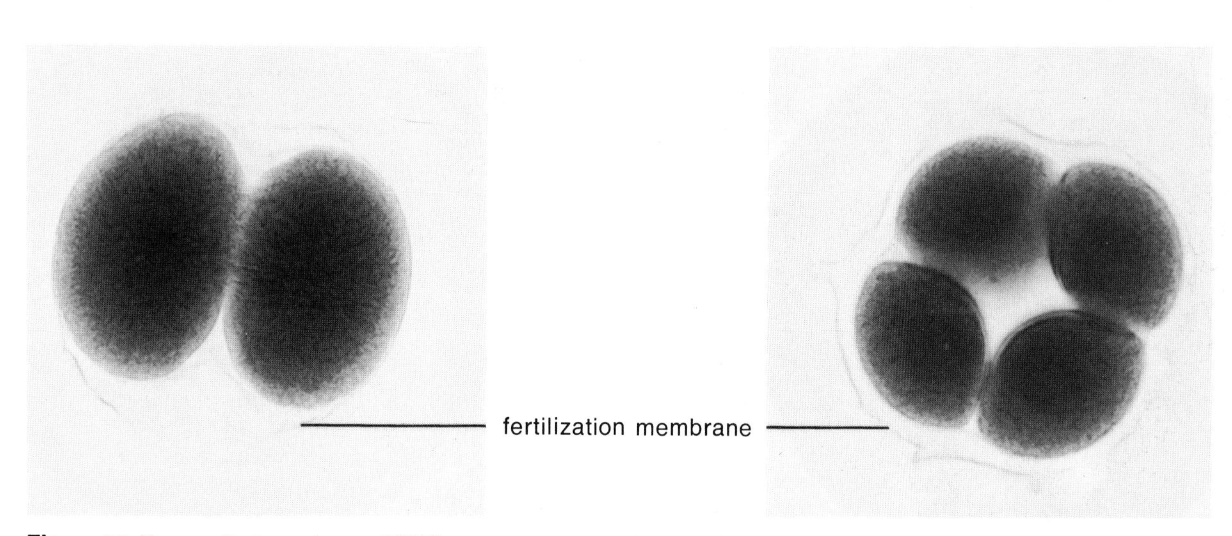

Figure 23 Two-cell stage (mag. 300X)

Figure 24 Four-cell stage (mag. 300X)

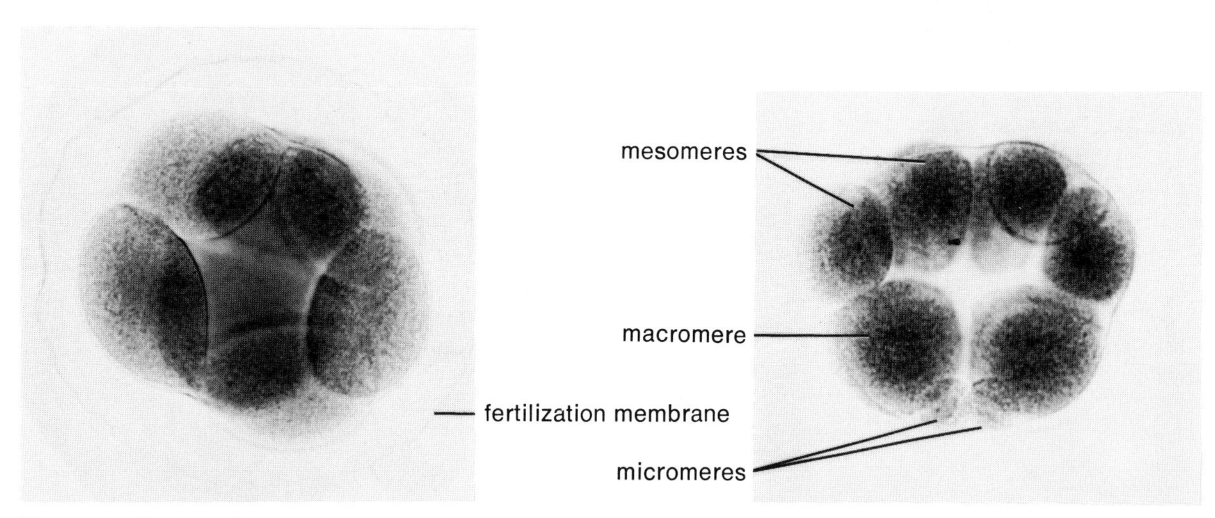

Figure 25 Eight-cell stage (mag. 300X)

Figure 26 Sixteen-cell (mag. 300X)

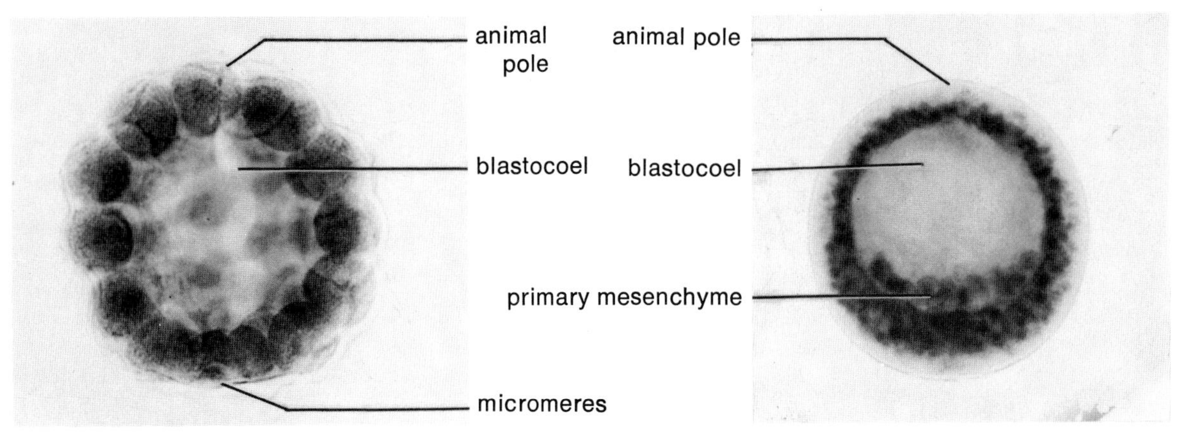

Figure 27 Early blastula stage (mag. 300X)

Figure 28 Late blastula stage (mag. 300X)

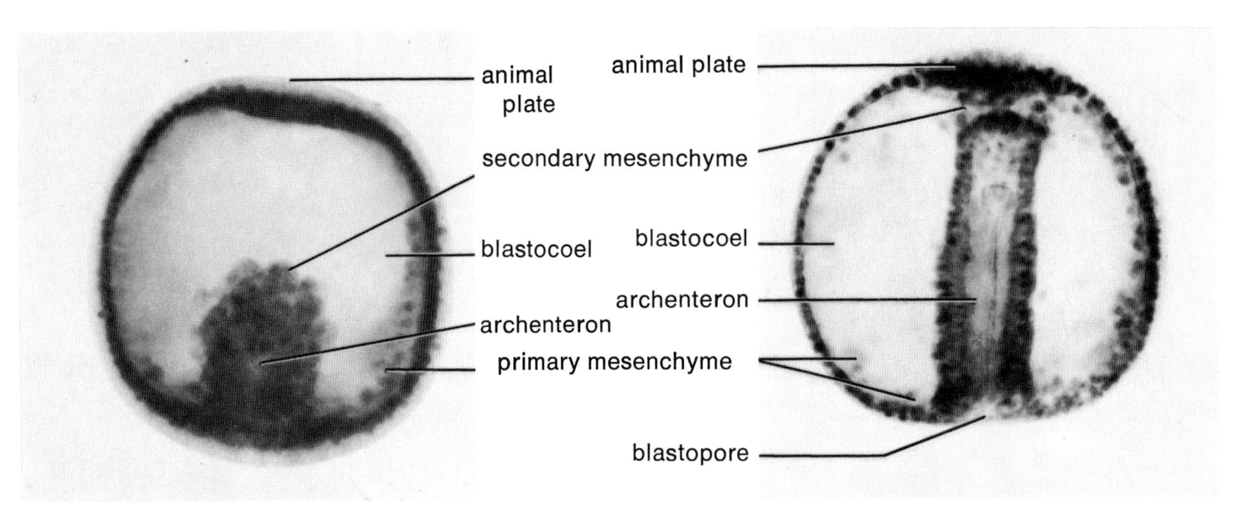

Figure 29 Early gastrula stage (mag. 300X)

Figure 30 Gastrula stage (mag. 300X)

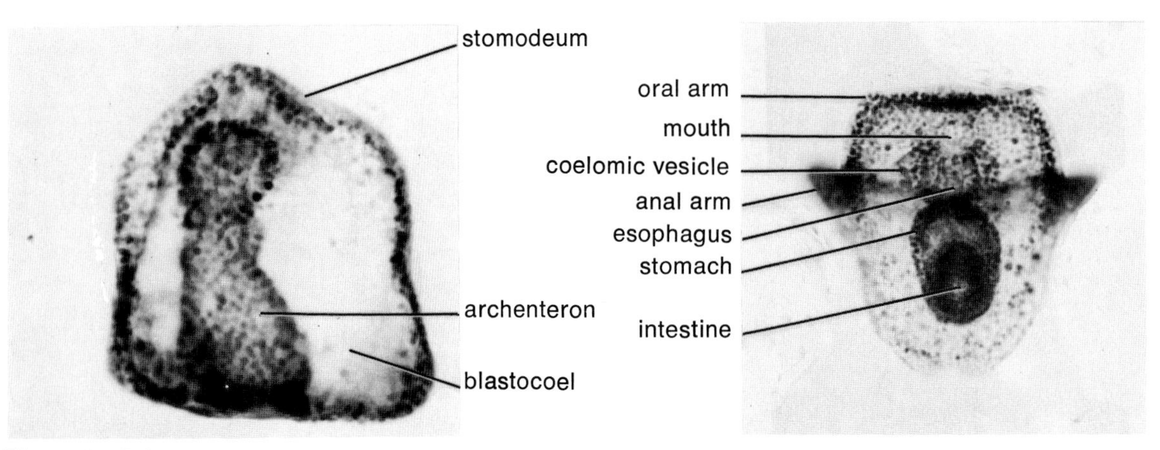

Figure 31 Prism stage (mag. 300X)

Figure 32 Early pluteus larval stage (mag. 200X)

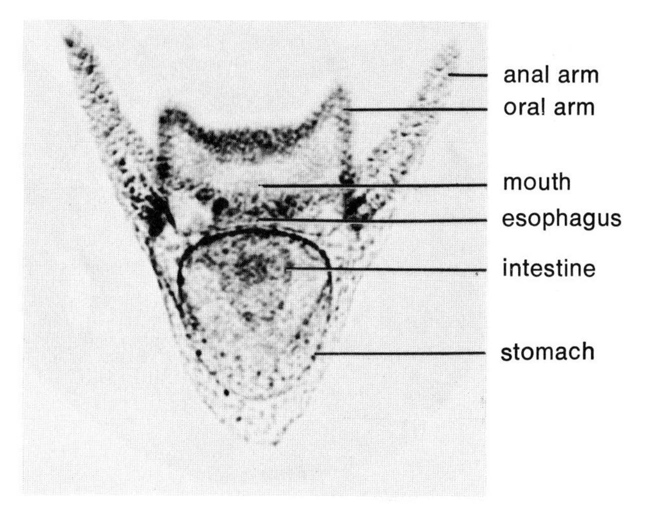

Figure 33 Late pluteus larval stage, ventral view (mag. 200X)

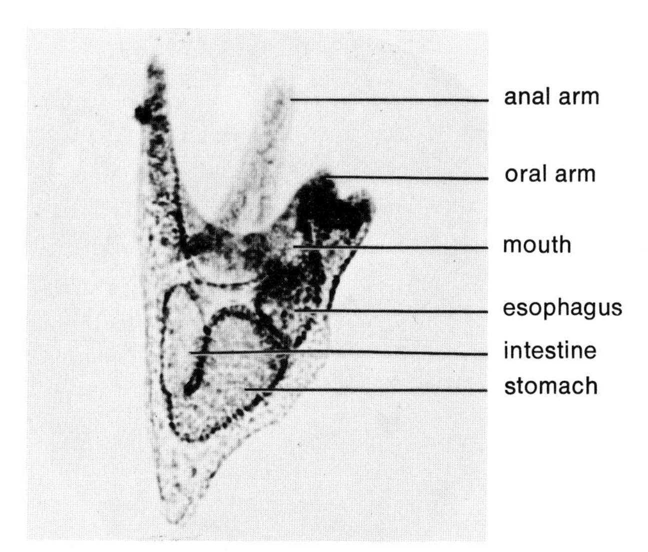

Figure 34 Late pluteus larval stage, lateral view (mag. 200X)

4. Early Development of the Frog

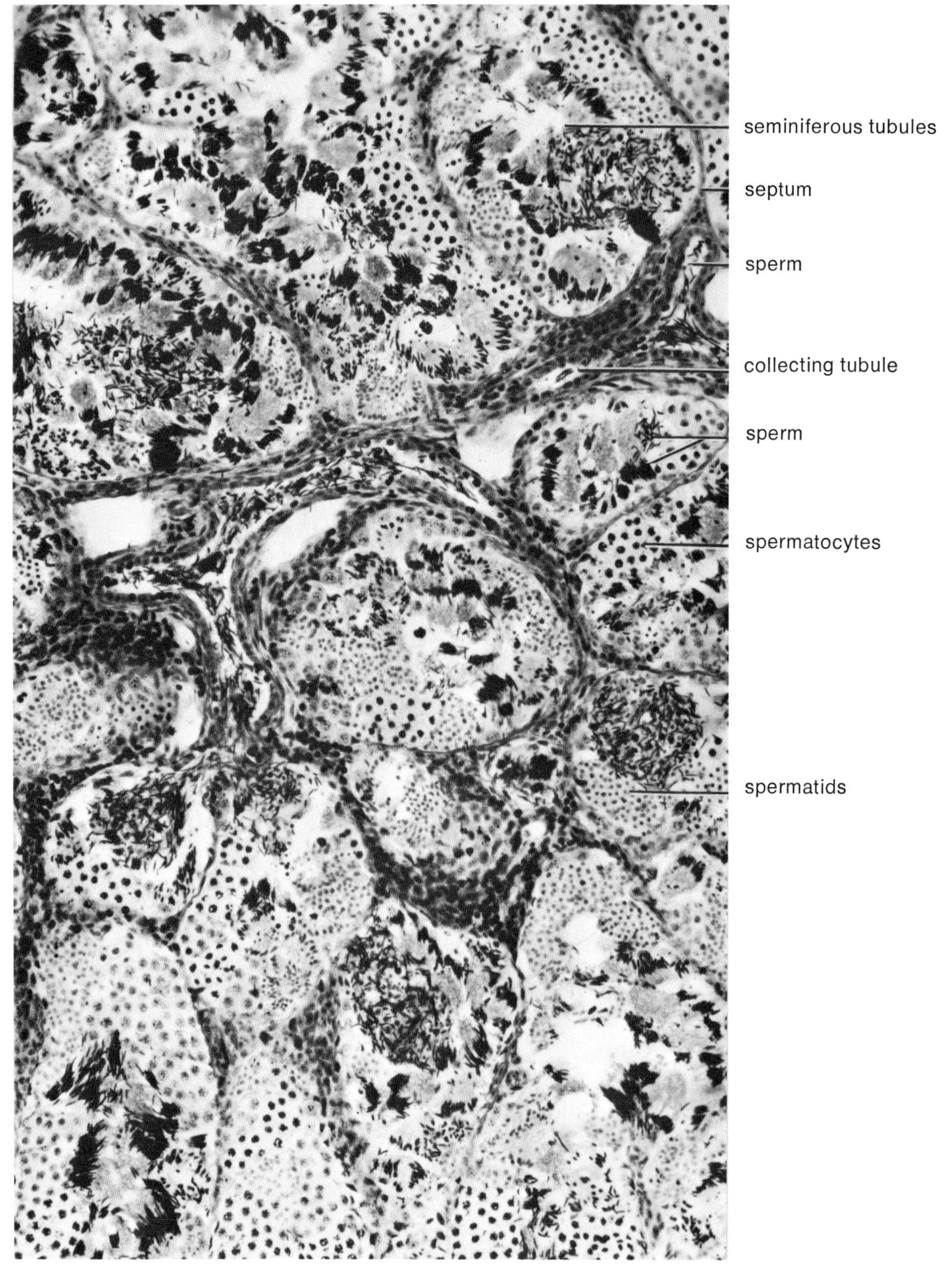

Figure 35 Frog testis, section (mag. 180 X)

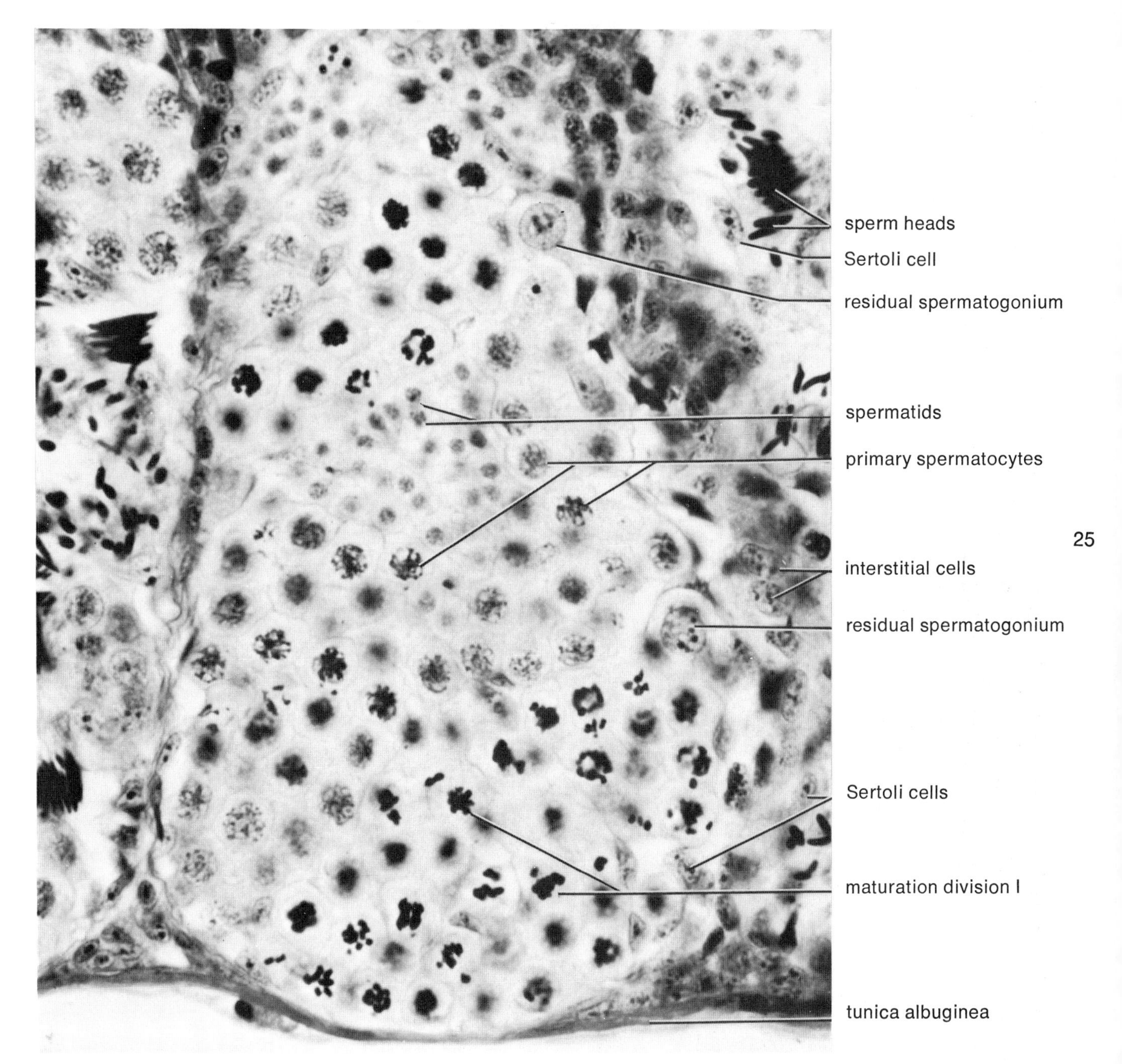

Figure 36 Frog testis, section (mag. 725 X)

Figure 37 Frog ovary, section showing growing oocytes (mag. 135X)

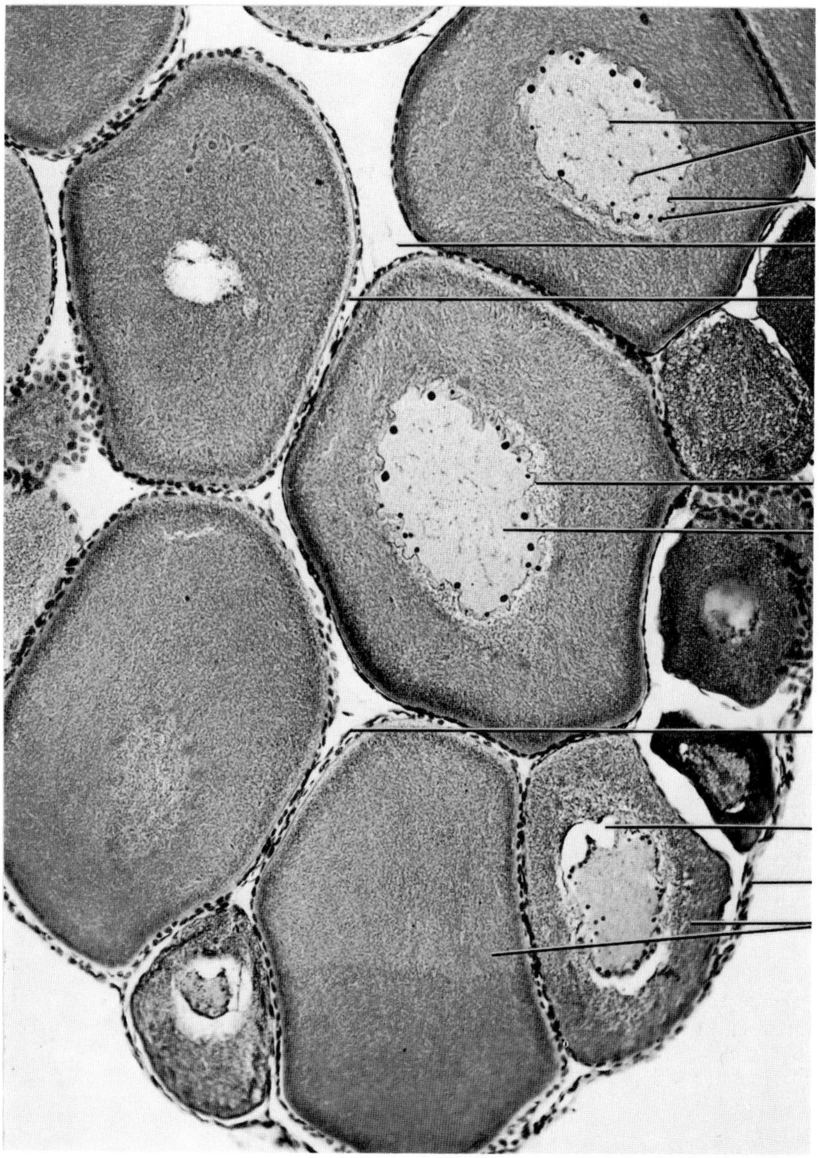

chromosomes
nucleoli
lymph sinus
follicle cell
nuclear membrane
germinal vesicle
theca interna
shrinkage artifact
theca externa
growing oocytes

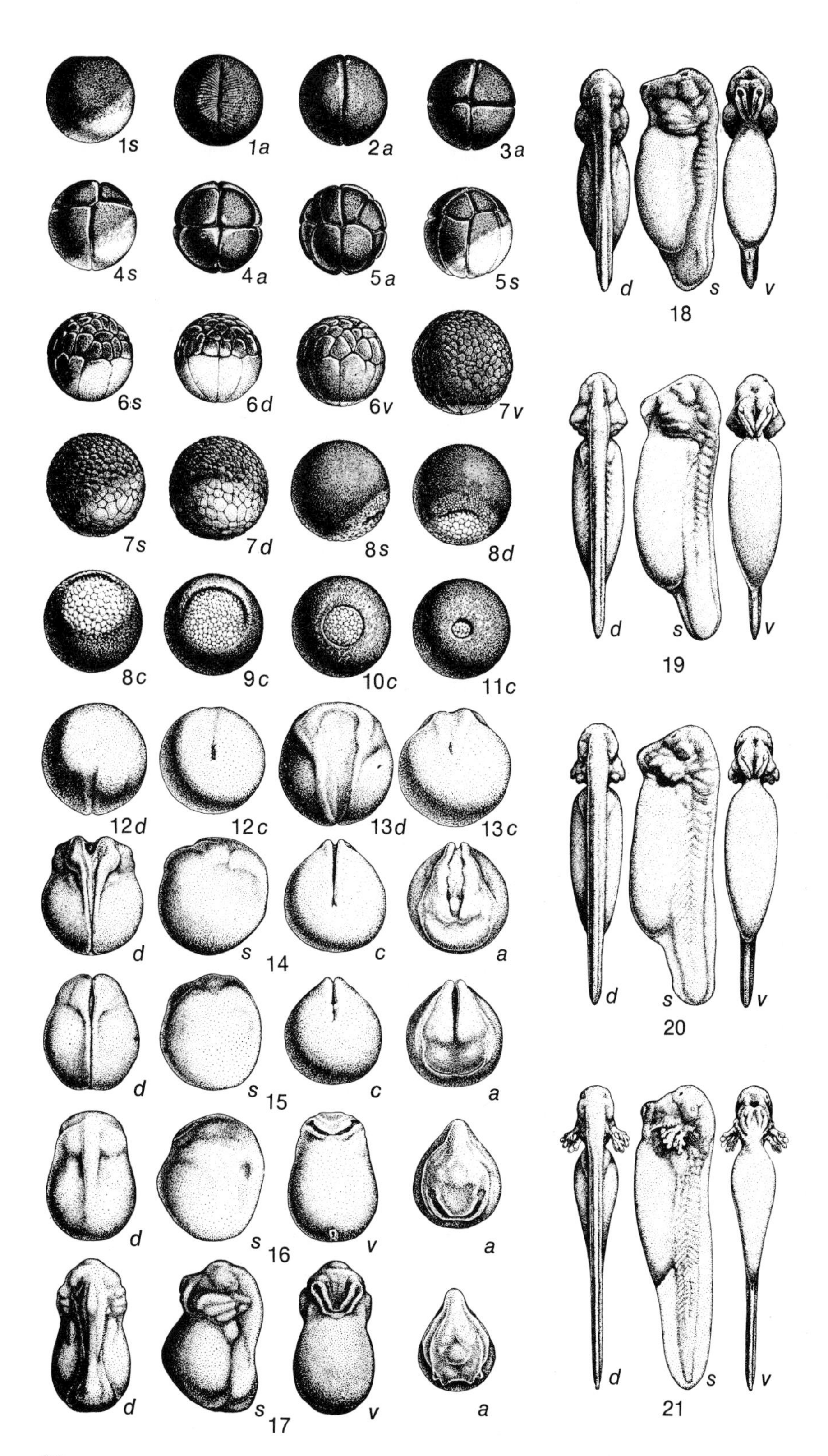

Figure 38

Stages of frog development *(Rana pipiens).* Numbers 1-33 designate developmental stages (see Table I). *a,* View from animal pole (frontal view); *c,* caudal (blastoporal) view; *d,* dorsal view; *s,* left lateral view; *v,* ventral view. Stages 1-25 mag. 6.5×; stages 26-28 mag. 3×; stages 29-33 mag. 1.2×. (From *Development of the Vertebrates* by Emil Witschi. Copyright 1956 by W. B. Saunders Co. Used with permission of the W. B. Saunders Co.)

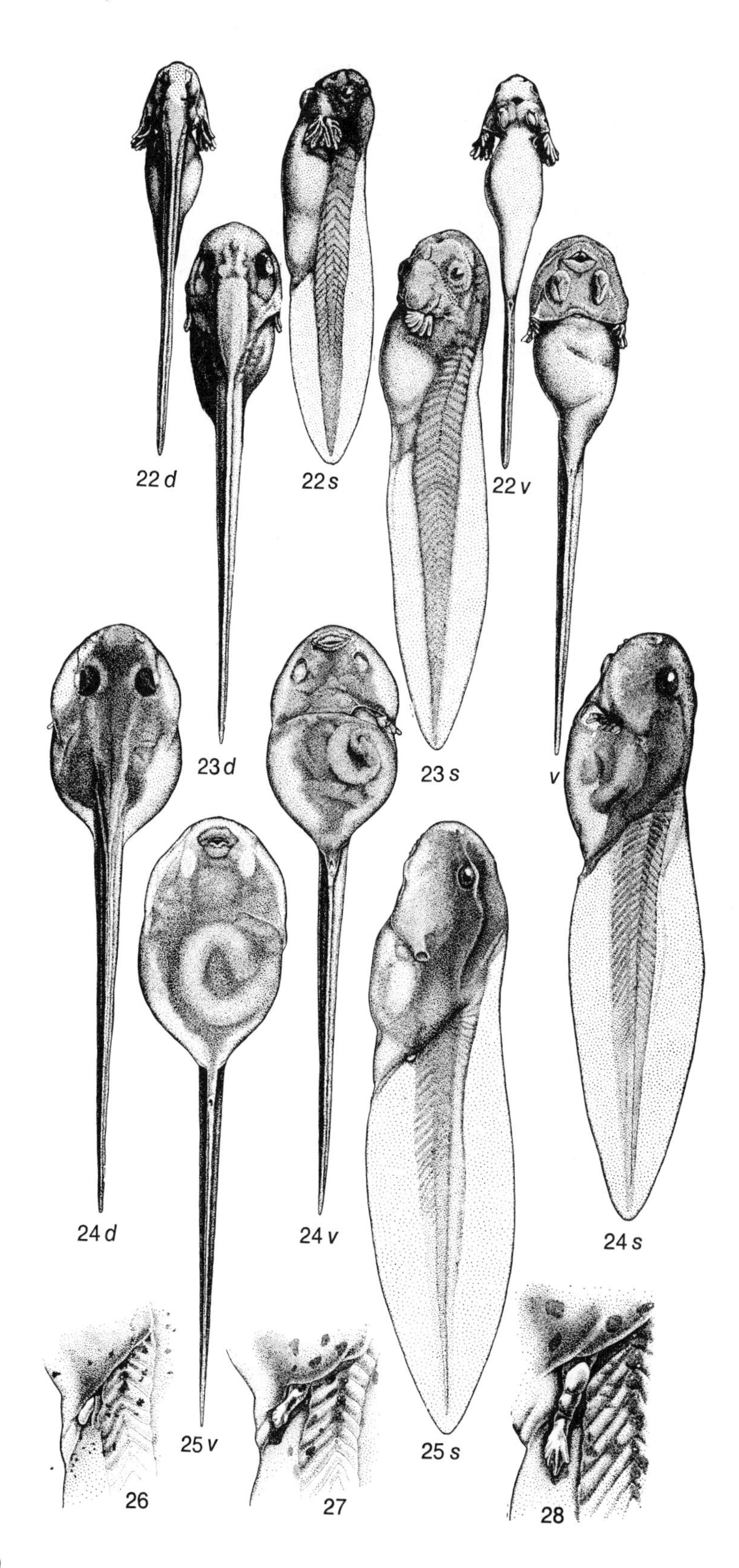

Figure 39

Stages of frog development *(Rana pipiens).* Numbers 1-33 designate developmental stages (see Table I). *a,* View from animal pole (frontal view); *c,* caudal (blastoporal) view; *d,* dorsal view; *s,* left lateral view; *v,* ventral view. Stages 1-25 mag. 6.5×; stages 26-28 mag. 3×; stages 29-33 mag. 1.2×. (From *Development of the Vertebrates* by Emil Witschi. Copyright 1956 by W. B. Saunders Co. Used with permission of the W. B. Saunders Co.)

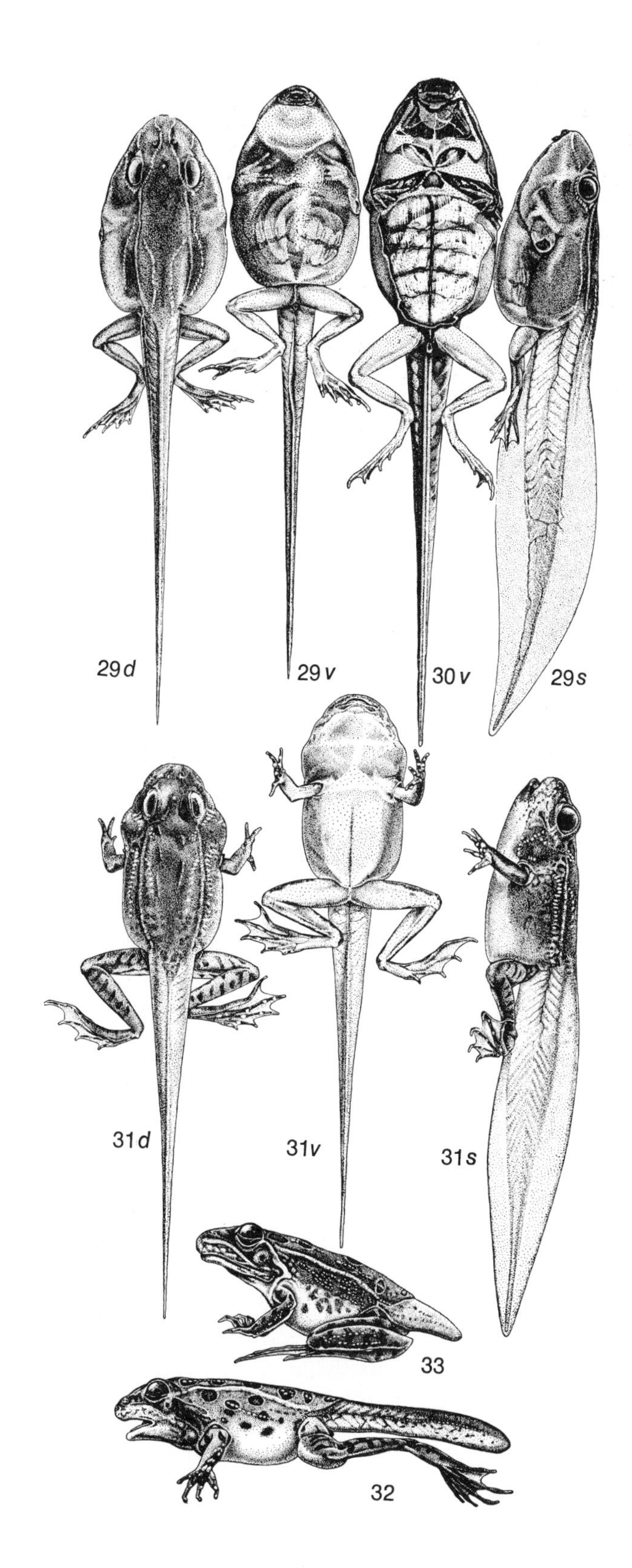

Figure 40

Stages of frog development *(Rana pipiens)*. Numbers 1-33 designate developmental stages (see Table I). *a,* View from animal pole (frontal view); *c,* caudal (blastoporal) view; *d,* dorsal view; *s,* left lateral view; *v,* ventral view. Stages 1-25 mag. 6.5×; stages 26-28 mag. 3×; stages 29-33 mag. 1.2×. (From *Development of the Vertebrates* by Emil Witschi. Copyright 1956 by W. B. Saunders Co. Used with permission of the W. B. Saunders Co.)

Table I
Frog Development Stages, *Rana pipiens* (see figs. 38–40)

Witschi Stage No.*	Approximate Lengths in mm†	Description of Stages†
1	1.7	fertilized egg
2		two-cell stage
3		four-cell stage
4		eight-cell stage
5		sixteen-cell stage
6		early blastula
7		late blastula
8		early gastrula
9		middle gastrula
10		yolk plug stage
11		late gastrula
12		neural plate stage
13		neural fold stage
14		early neural groove stage
15		late neural groove stage
16	2.5–2.7	early neural tube stage
17	2.8–3.0	early tail bud stage
18	4	tail bud stage
19	5	gill buds
20	6	hatching, gill circulation
21	7	mouth open
22	8	tail fin circulation
23	9	opercular fold
24	10	right operculum closed
25	11	operculum complete
26-33		metamorphosis

*Similar to but not identical with Shumway stage numbers.
†Based on Shumway, W., Stages in the normal development of *Rana pipiens*. I. External form. *Anatomical Record* **78**:139-147, 1940.

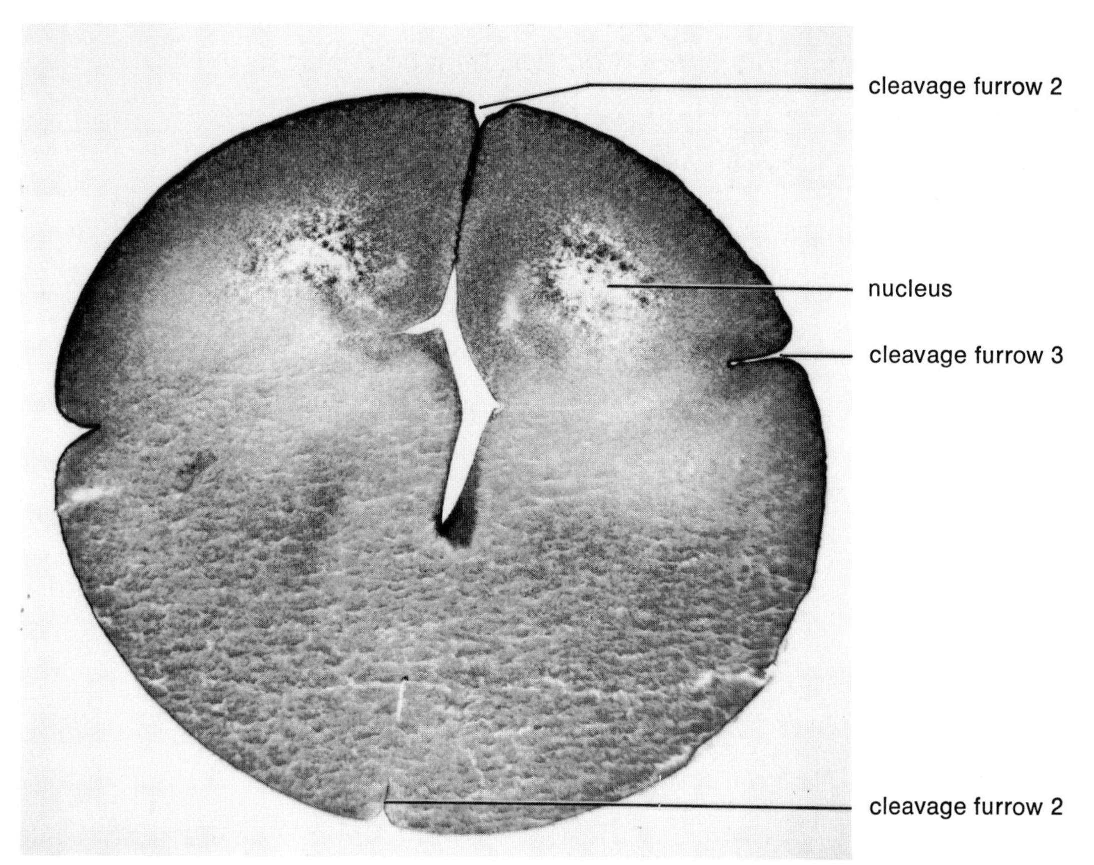

Figure 41 Frog embryo, early cleavage, 8-cell stage (stage 5*), median section (mag. 65X)
*Based on Shumway, W., "Stages in the normal development of *Rana pipiens,*" *Anatomical Record* **78**(2) 1940.

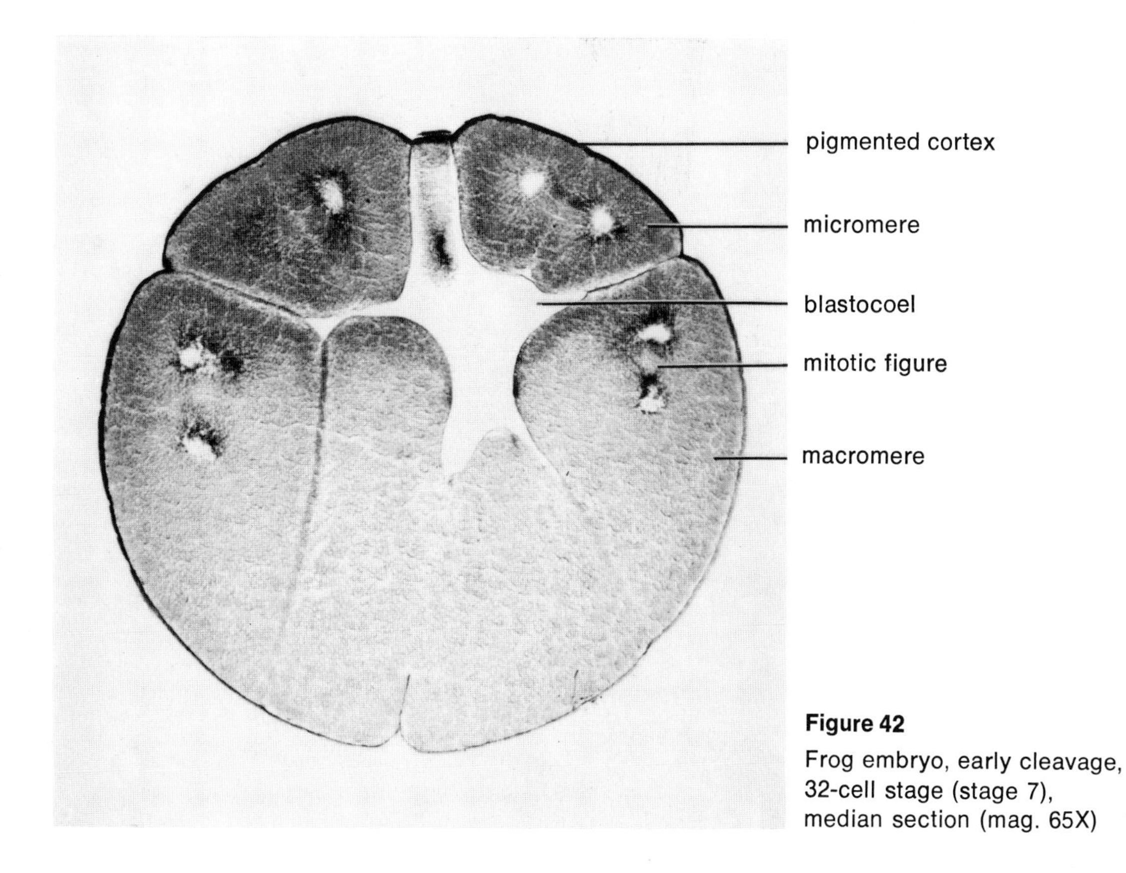

Figure 42
Frog embryo, early cleavage, 32-cell stage (stage 7), median section (mag. 65X)

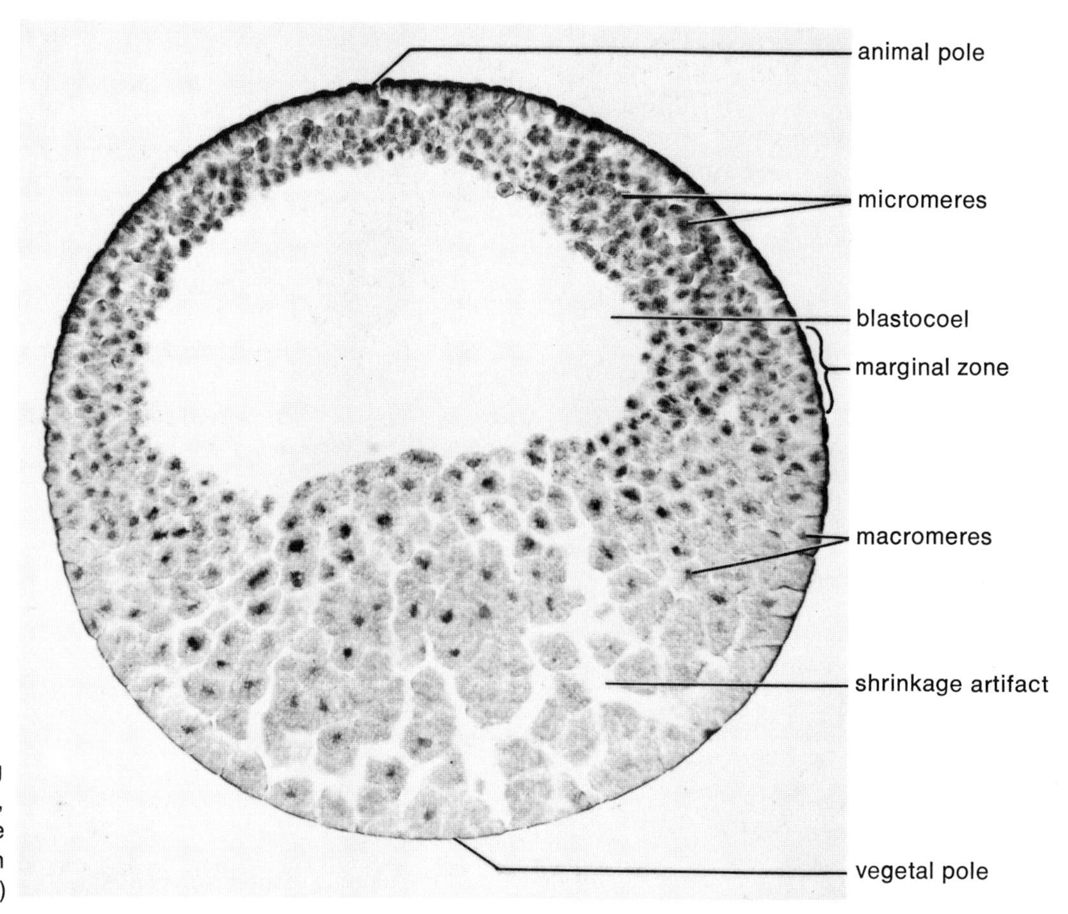

Figure 43 Frog embryo, late cleavage, blastula stage (stage 8), median section (mag. 65X)

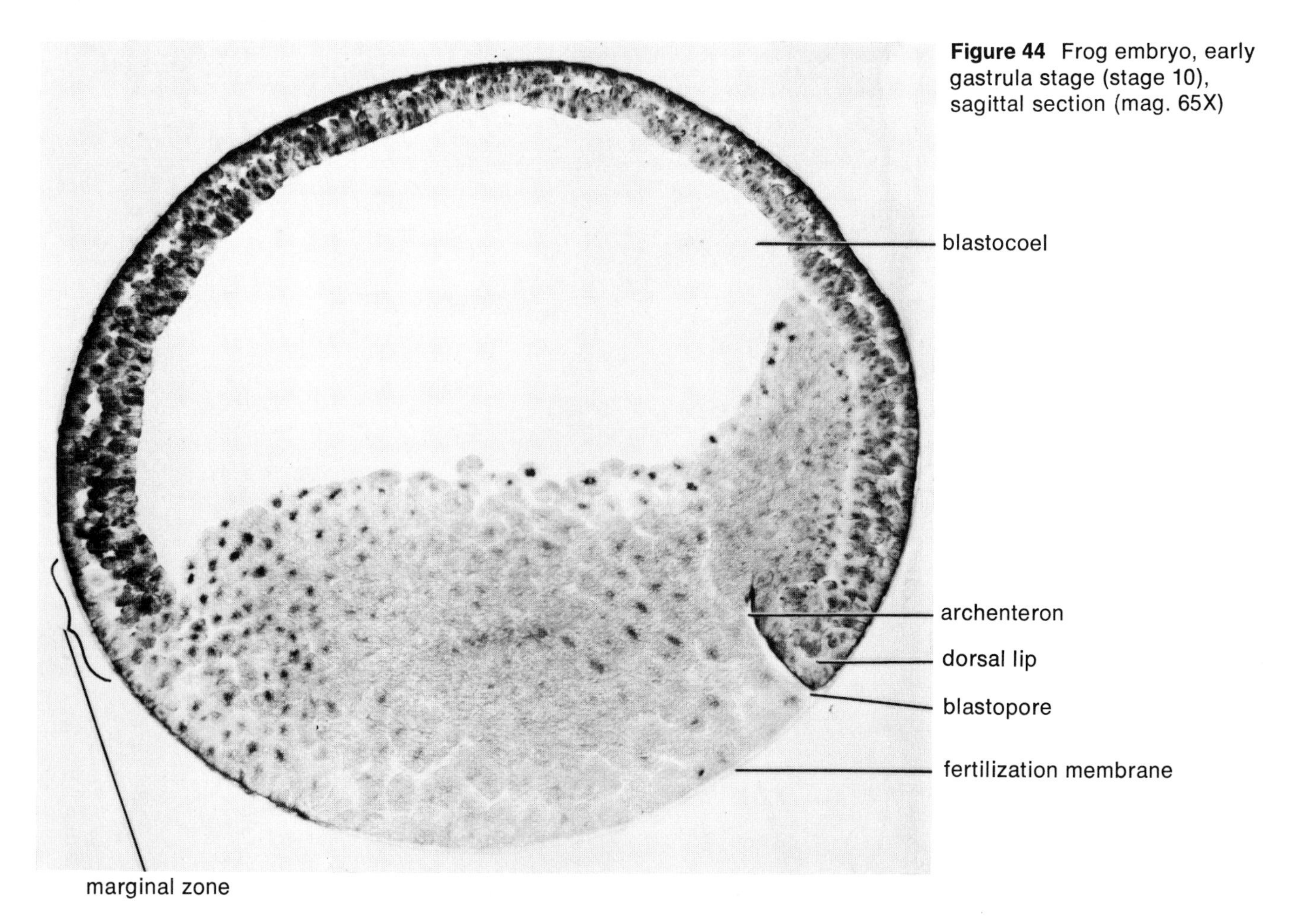

Figure 44 Frog embryo, early gastrula stage (stage 10), sagittal section (mag. 65X)

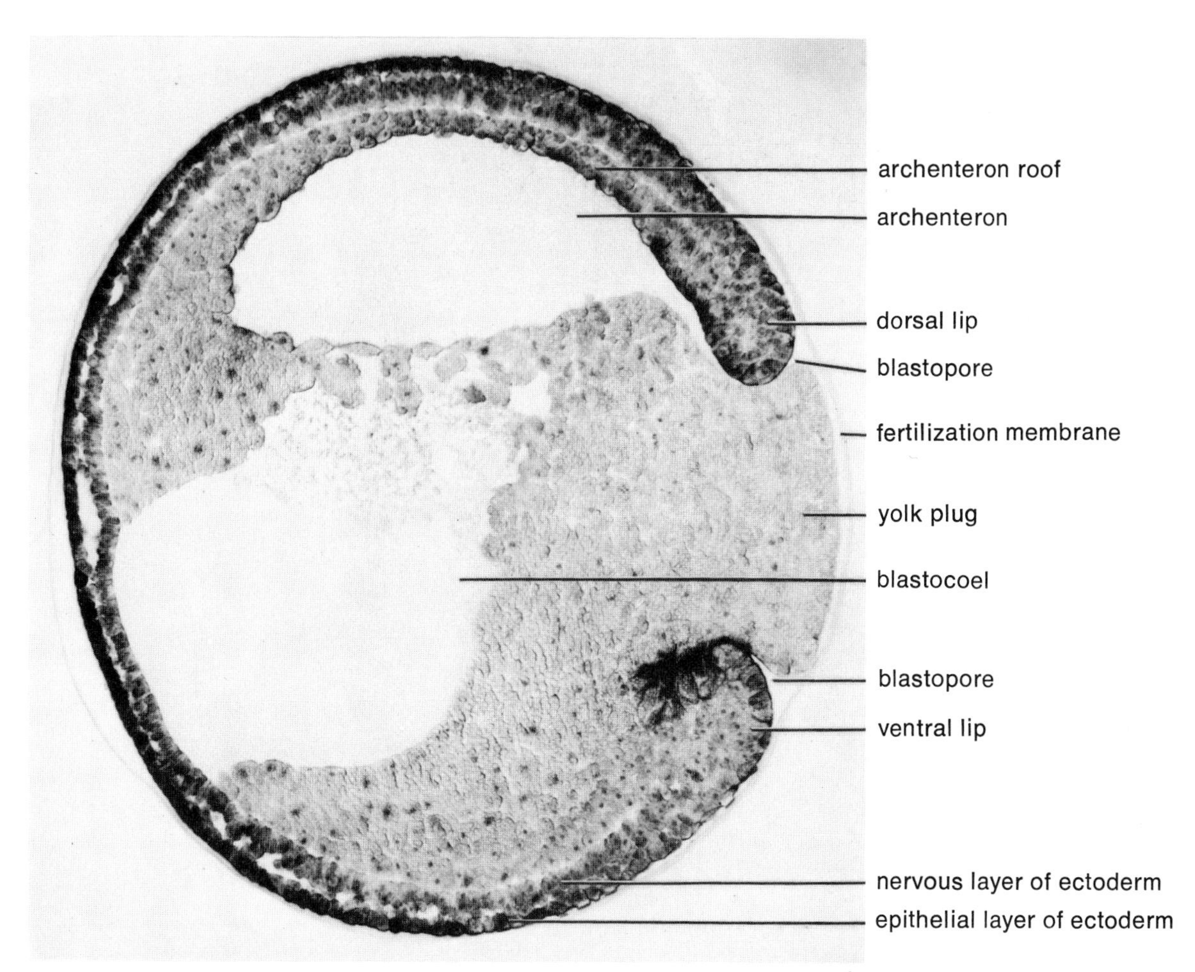

Figure 45 Frog embryo, late gastrula stage (stage 12), sagittal section (mag. 65X)

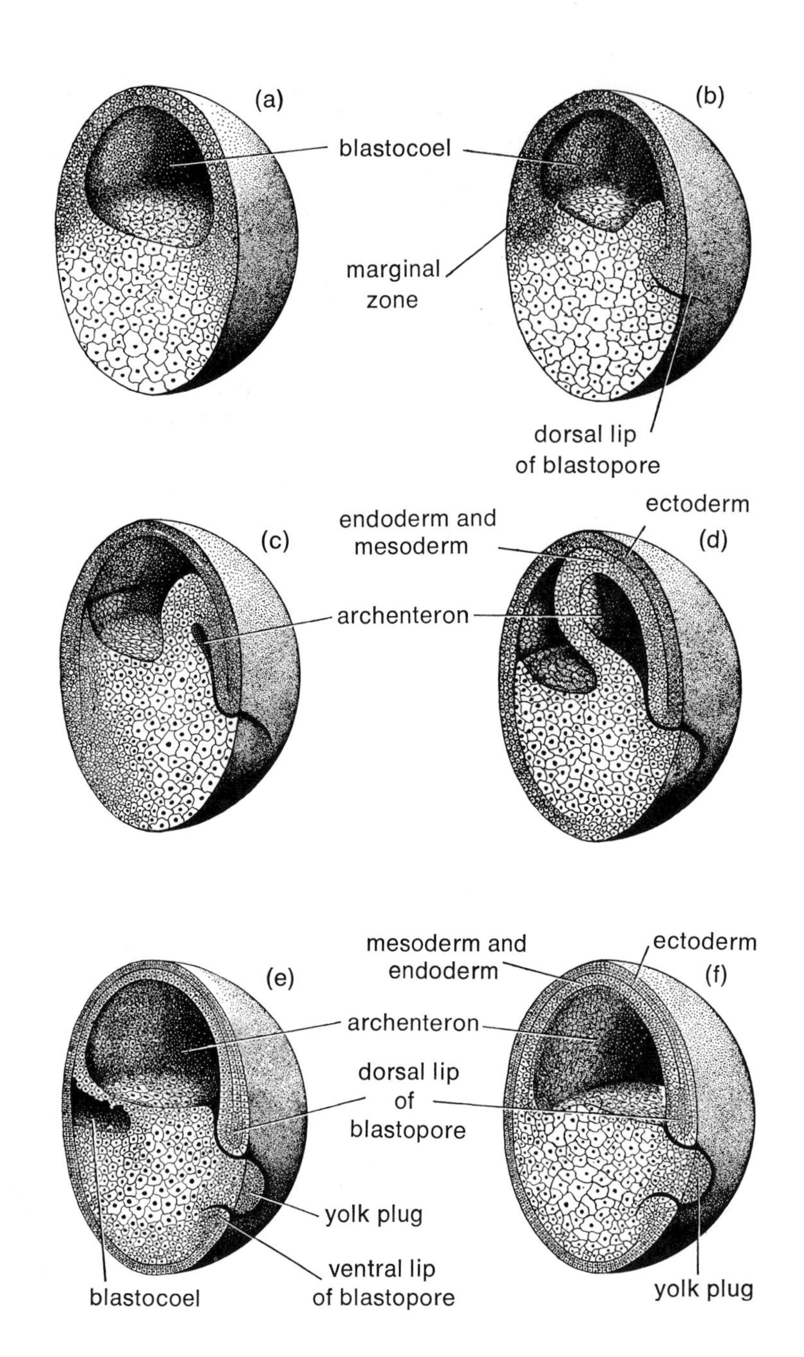

Figure 46

The blastula (*a*) of the frog and its transformation into the gastrula (*b-f*). *b*, beginning of gastrulation. *c-e*, elimination of the blastocoel or segmentation cavity by the gastrocoel or archenteron. *f*, completed gastrula with mesoderm and endoderm beneath the ectoderm. (From *Fundamentals of Comparative Embryology of the Vertebrates* by Alfred F. Heuttner. Copyright 1941 by Macmillan Publishing Co., Inc., New York. Used with permission of Macmillan Publishing Co.)

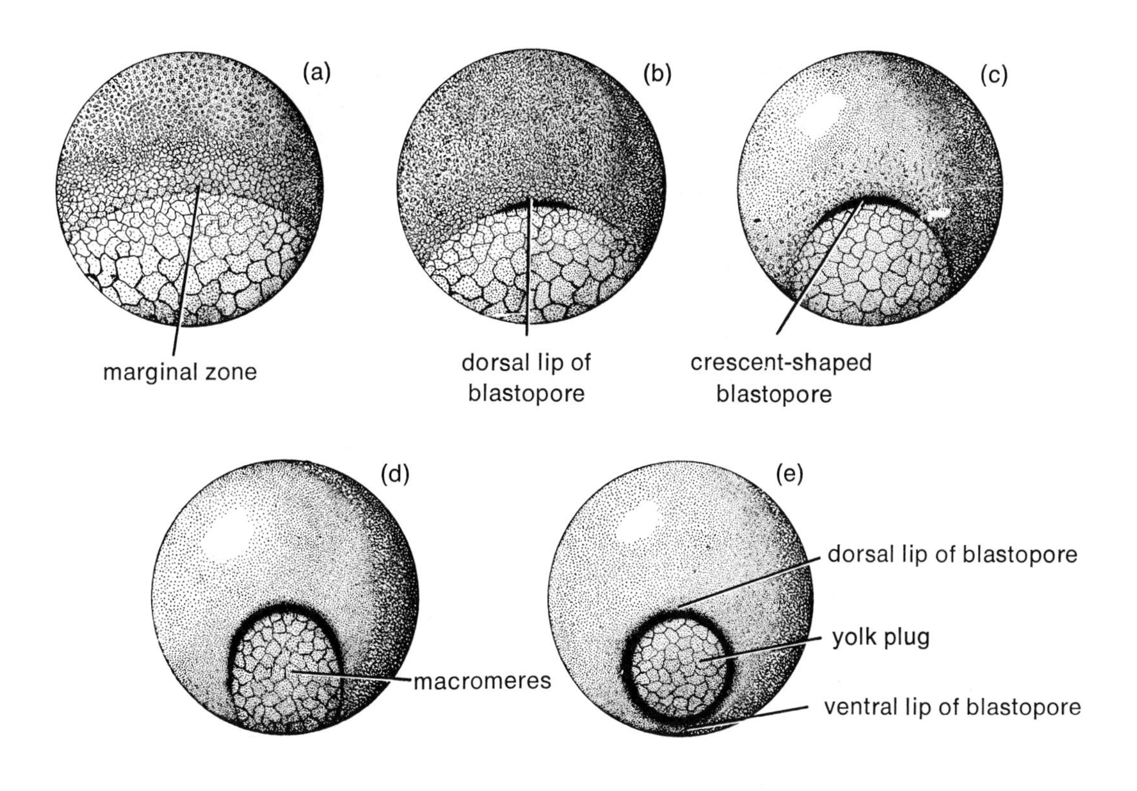

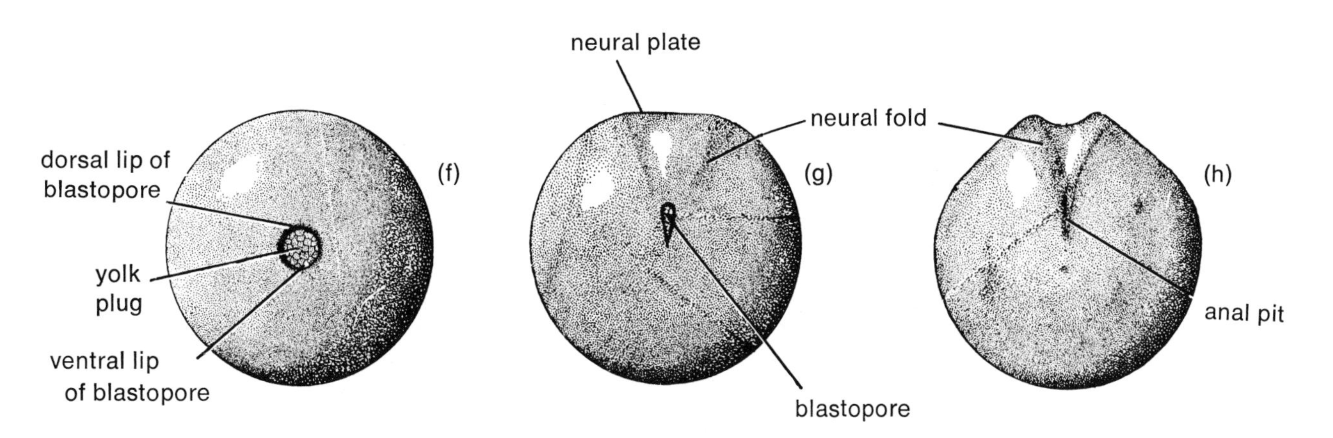

Figure 47

The process of gastrulation in the frog embryo seen from the posterior or blastoporal point of view. Stages *a-e* are equivalent to those contained in fig. 46. Stages *f* and *g* are almost the same as those of *d* and *e* of fig. 48. (From *Fundamentals of Comparative Embryology of the Vertebrates* by Alfred F. Huettner. Copyright 1941 by Macmillan Publishing Co., Inc., New York. Used with permission of Macmillan Publishing Co.)

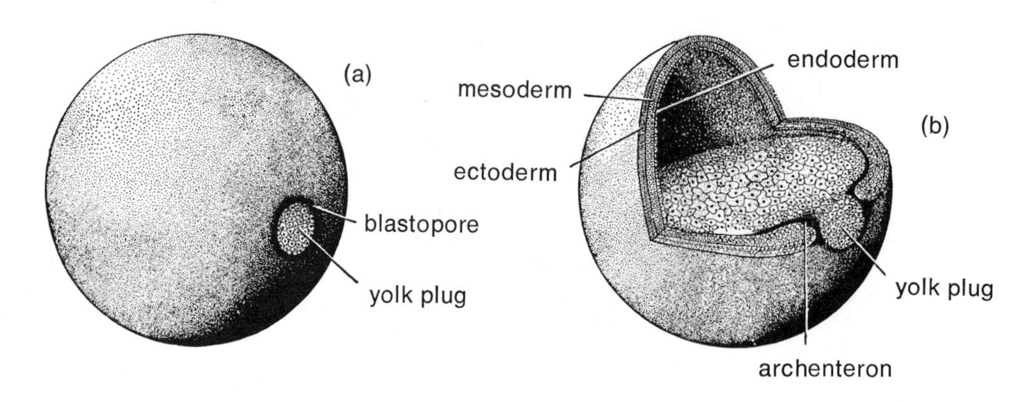

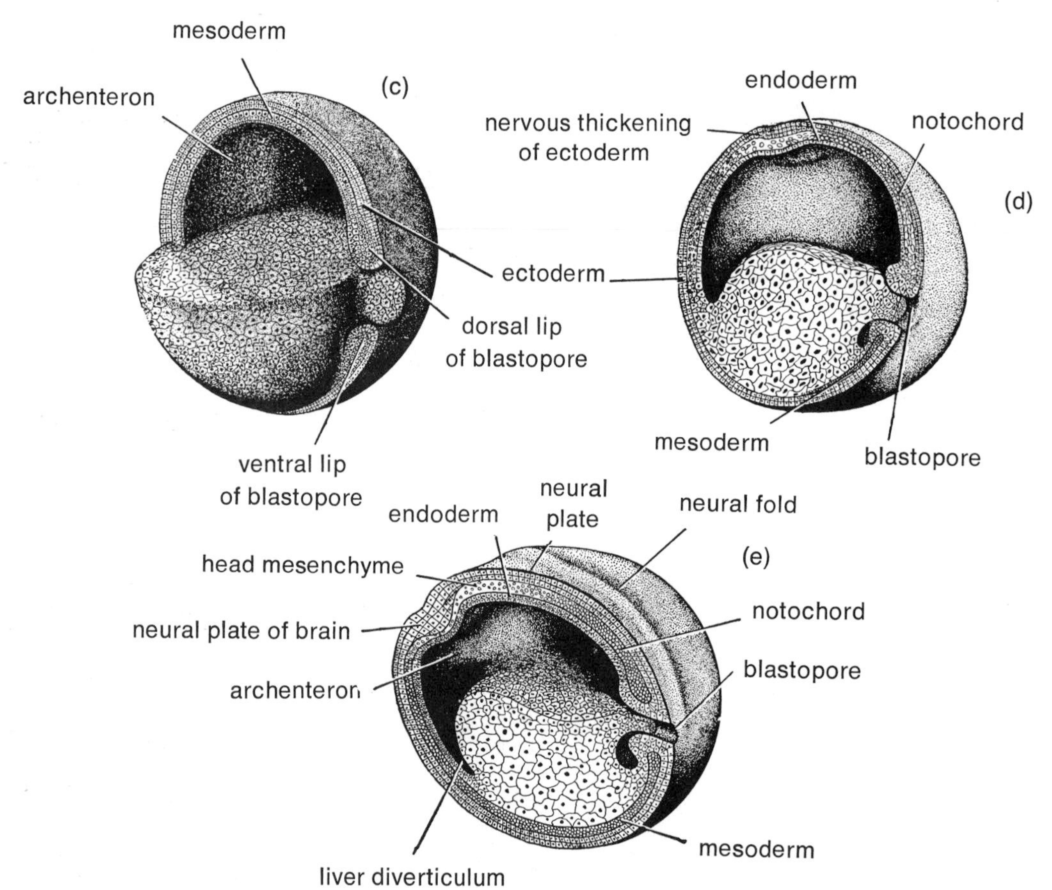

Figure 48

This figure is continuous with fig. 46. *a,* external appearance of gastrula. *b, c,* different partial sections of the gastrula. The mesoderm is in the process of migrating up from the yolk cells. *d, e,* late gastrula stages. The blastopore becomes smaller, the yolk plug is withdrawn, and the embryo is elongating in the antero-posterior axis. (From *Fundamentals of Comparative Embryology of the Vertebrates* by Alfred F. Huettner. Copyright 1941 by Macmillan Publishing Co., Inc., New York. Used with permission of Macmillan Publishing Co.)

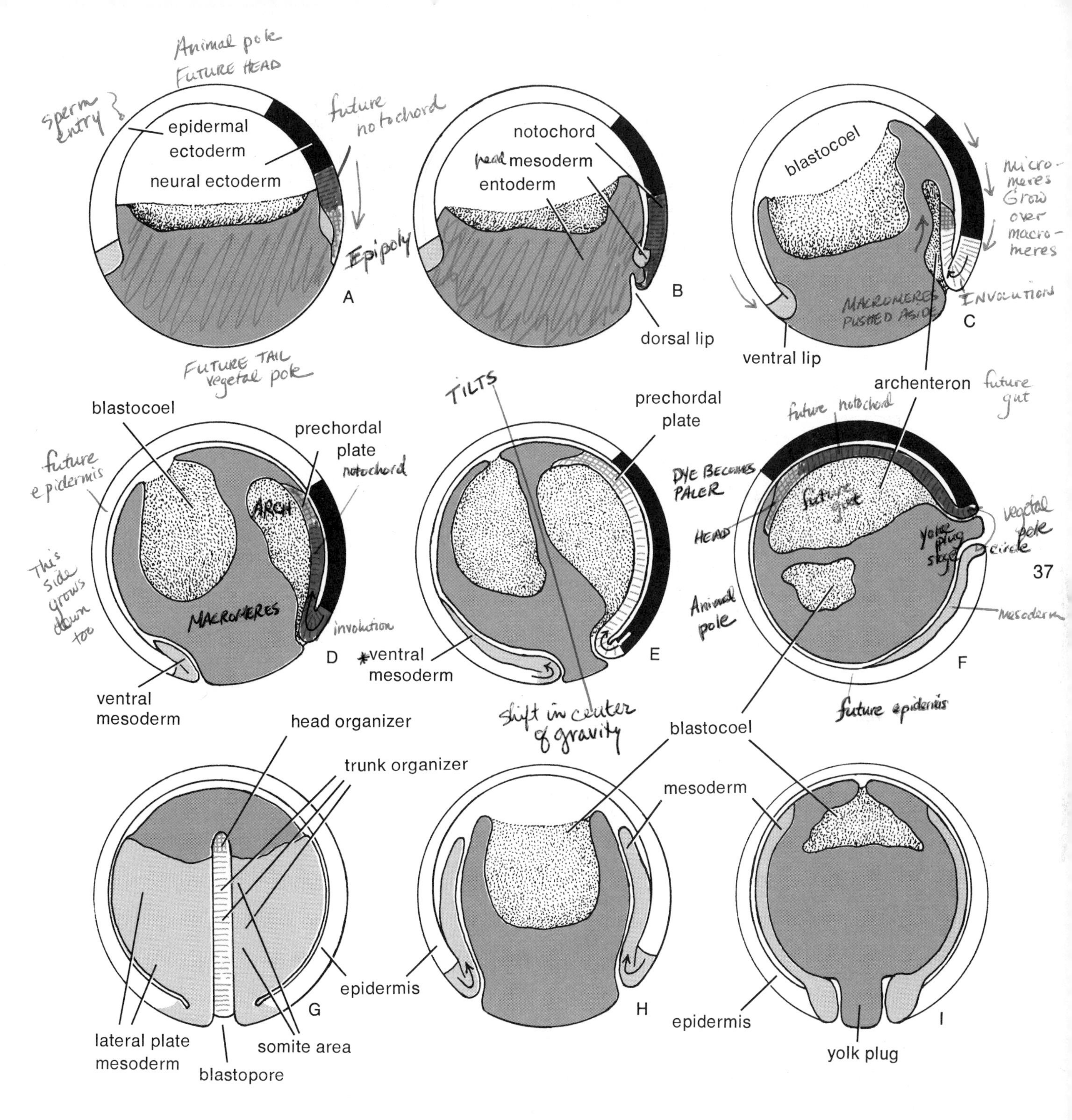

Figure 49

Migration of the presumptive organ-forming areas of the blastula during gastrulation in the amphibia (with reference particularly to the frog). *A,* late blastula, sagittal section through midplane of future embryo. *B-F,* processes of epiboly and emboly. In epiboly, the black (neural) and white (epidermal) areas become extended and gradually envelop the inward moving notochord, endoderm, and mesoderm. The processes concerned with emboly bring about the inward migration of the latter presumptive areas. *G,* late gastrular condition, with neural area and upper portion of the epidermal area removed to show relationships of the middle germ layer of chordamesoderm. *H,* horizontal section of middle gastrular condition, showing involution of mesoderm between endoderm and ectoderm. *L,* late gastrula, horizontal section, showing yolk plug, mesoderm, and final engulfment of blastocoelic space by endoderm. (From *Comparative Embryology of the Vertebrates* by Olin E. Nelsen. Copyright 1953 by The Blakiston Co., Inc. Used with permission of the McGraw-Hill Book Co.)

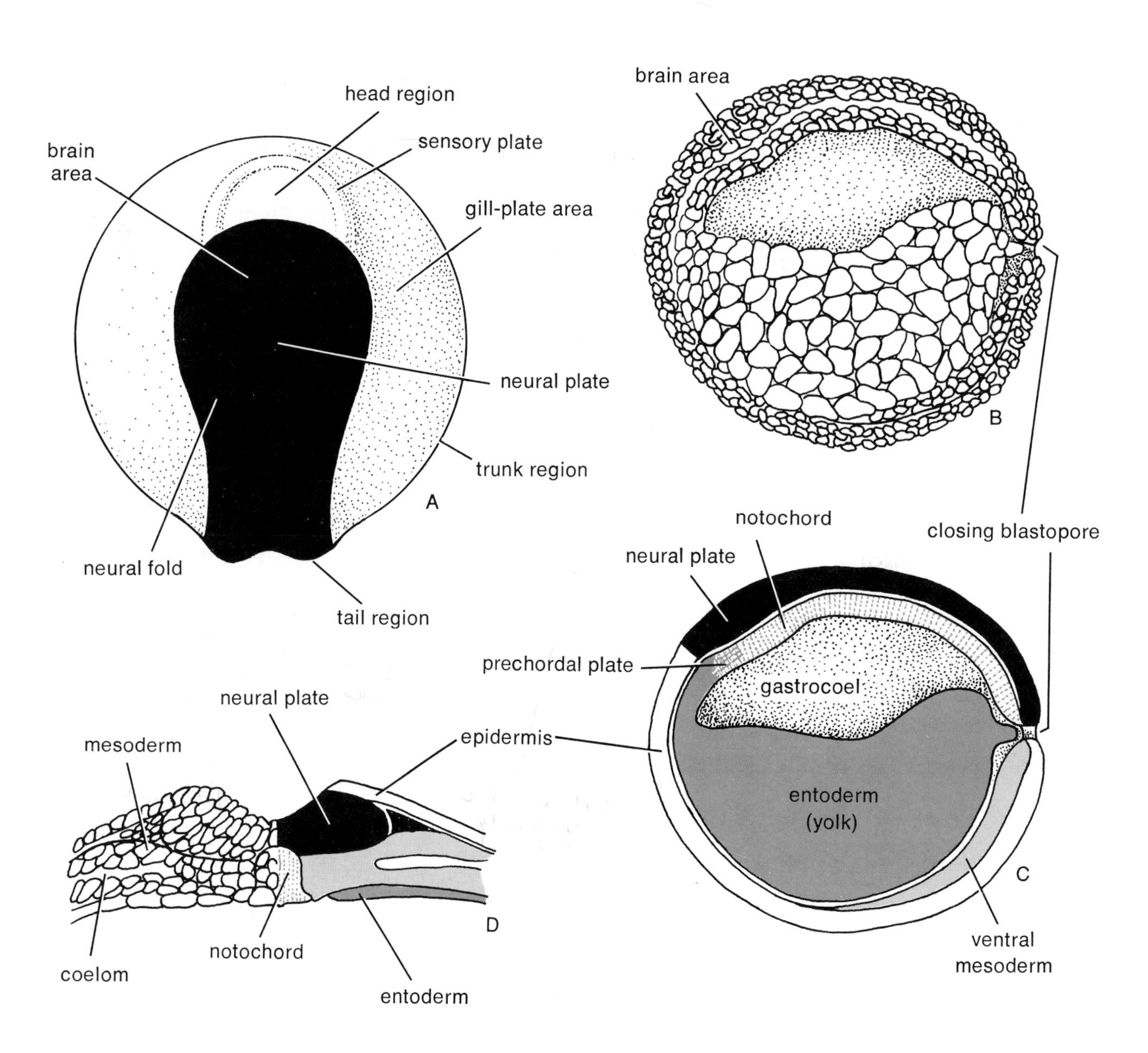

Figure 50

Relationships of the major presumptive organ-forming areas at the end of gastrulation in the anuran amphibia. *A,* external view of gastrula, showing the ectodermal layer composed of presumptive epidermis (white) and presumptive neural plate (black), as viewed from the dorsal aspect. *B,* diagrammatic median sagittal section of condition shown in *A. C,* same as *B,* showing major organ-forming areas. *D,* section through middorsal area of conditions *B* and *C,* a short distance caudal to foregut and prechordal plate region. Observe that the notochord occupies the middorsal area of the gut roof. (From *Comparative Embryology of the Vertebrates* by Olin E. Nelsen. Copyright 1953 by The Blakiston Co., Inc. Used with permission of the McGraw-Hill Book Co.)

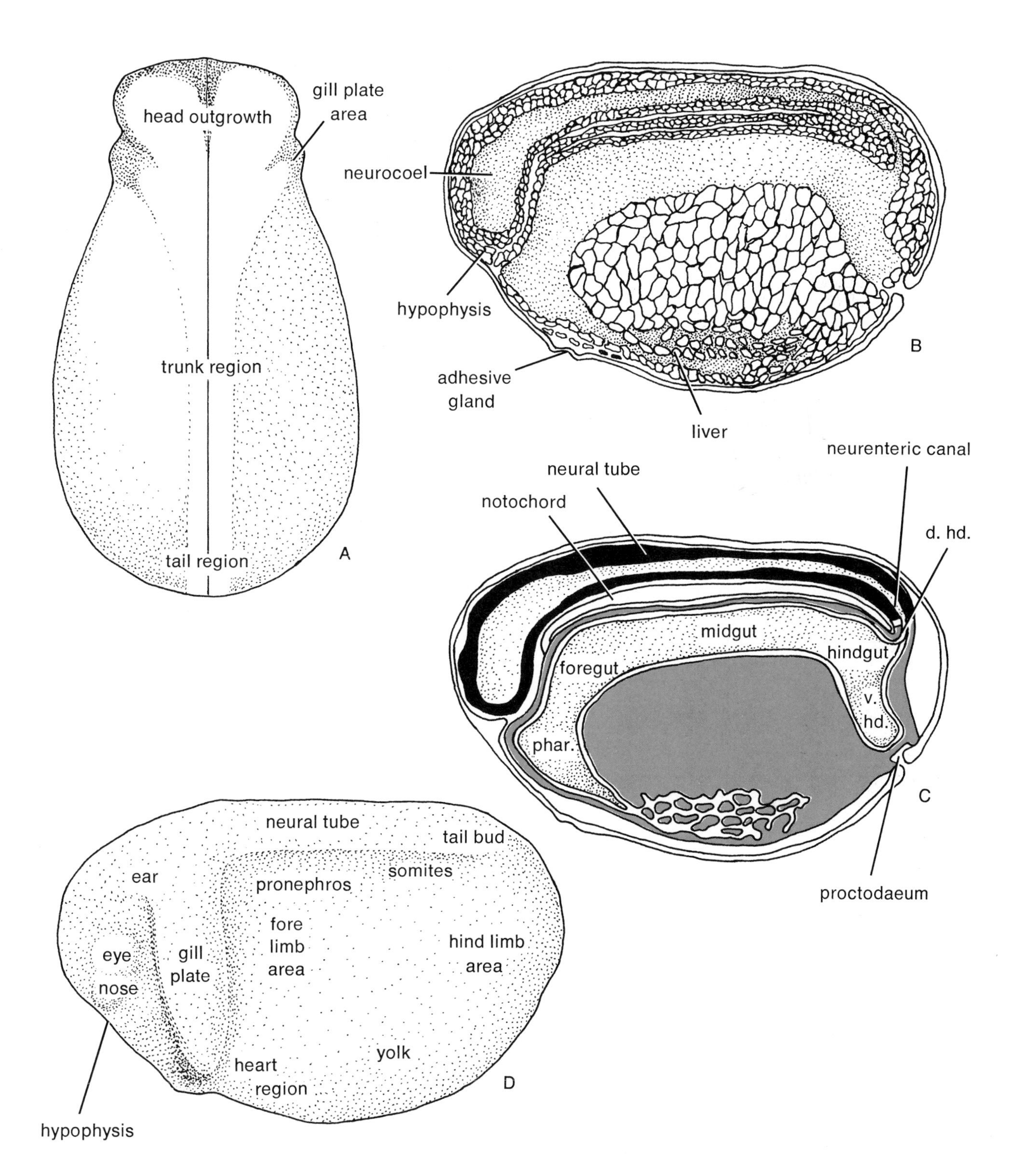

Figure 51

Early neural tube stage of the frog, *Rana pipiens,* 2½ to 3 mm in length. *A,* dorsal view. *B,* midsagittal section of embryo similar to *A. C,* same as *B,* showing organ-forming areas. Abbreviations: v. hd. = ventral hindgut diverticulum; d. hd. = dorsal hindgut diverticulum; phar. = pharynx. *D,* Lateral view of *A.* (From *Comparative Embryology of the Vertebrates* by Olin E. Nelsen. Copyright 1953 by The Blakiston Co., Inc. Used with permission of the McGraw-Hill Book Co.)

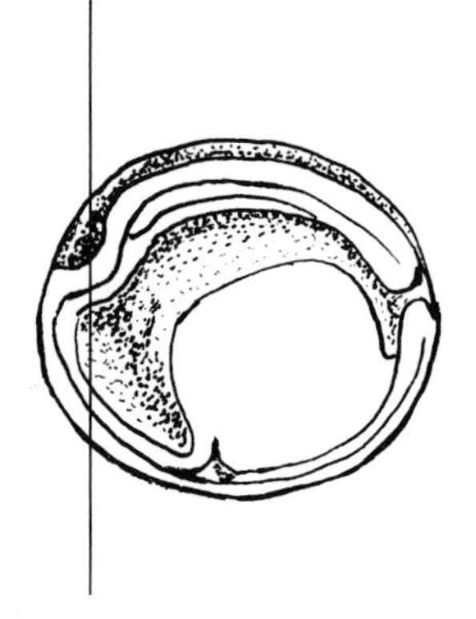

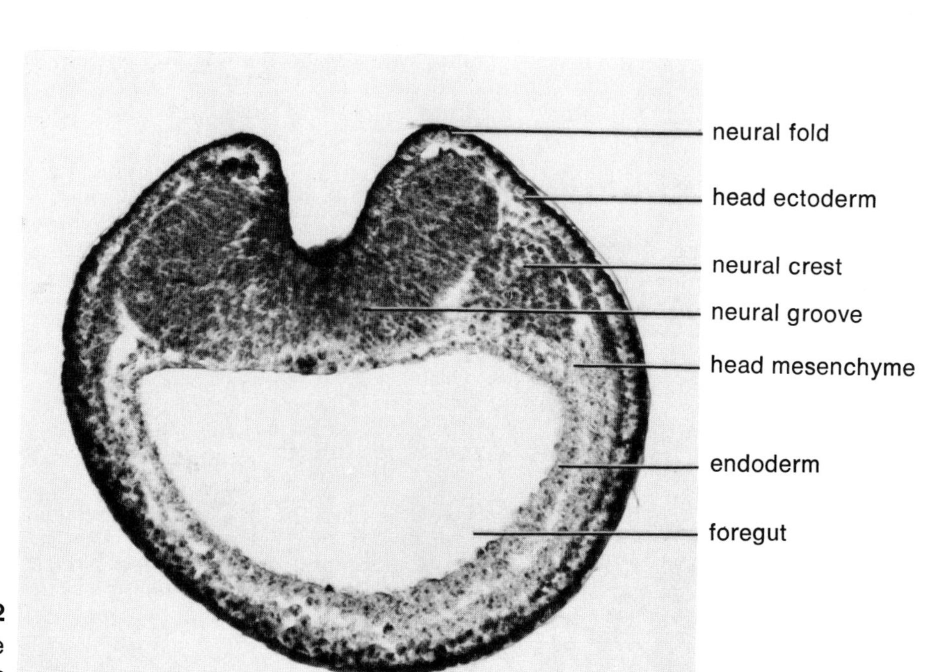

Figure 52
Frog embryo, neural fold stage (stage 14), transverse section through head region (mag. 65X)

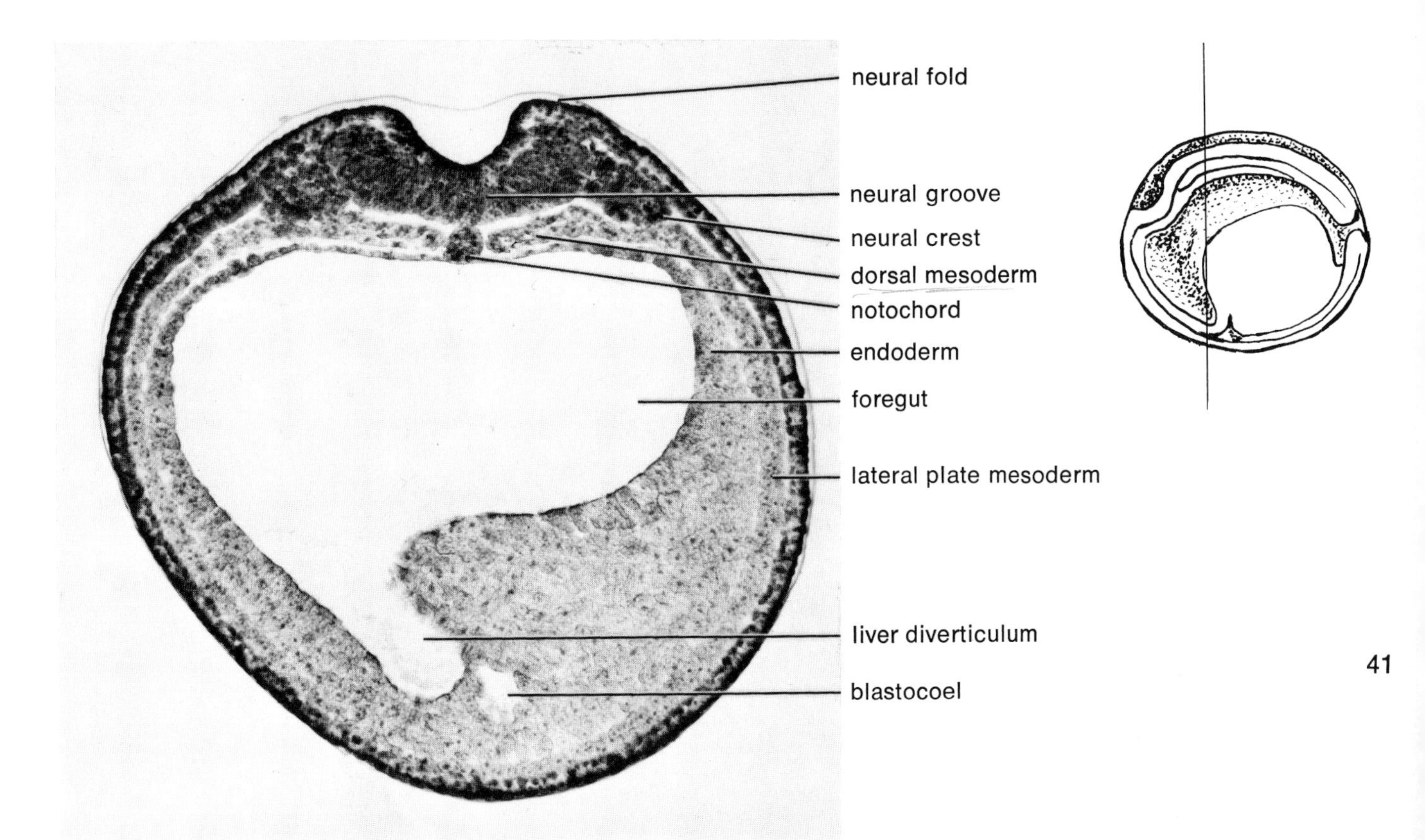

Figure 53 Frog embryo, neural fold stage (stage 14), transverse section through foregut region (mag. 65X)

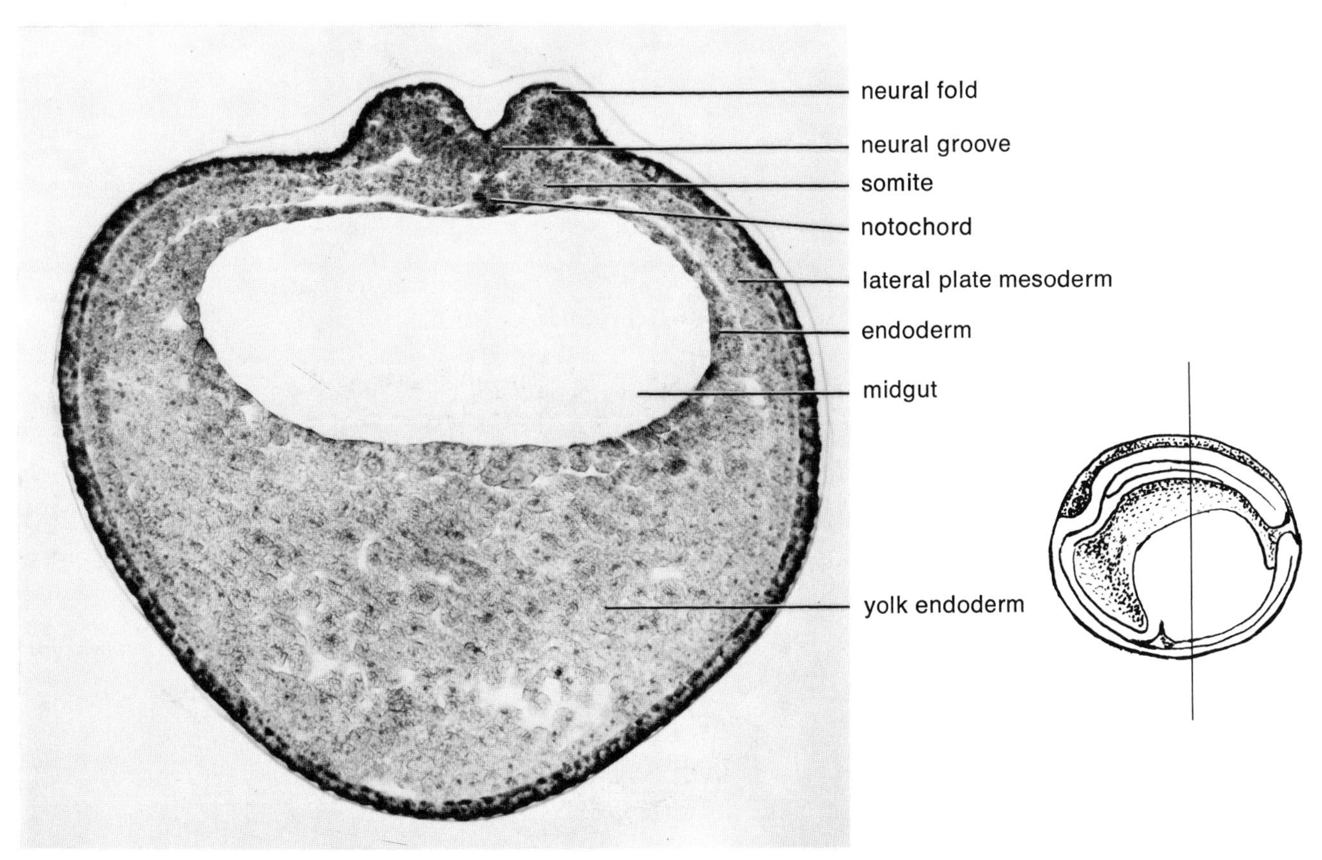

Figure 54 Frog embryo, neural fold stage (stage 14), transverse section through midgut region (mag. 65X)

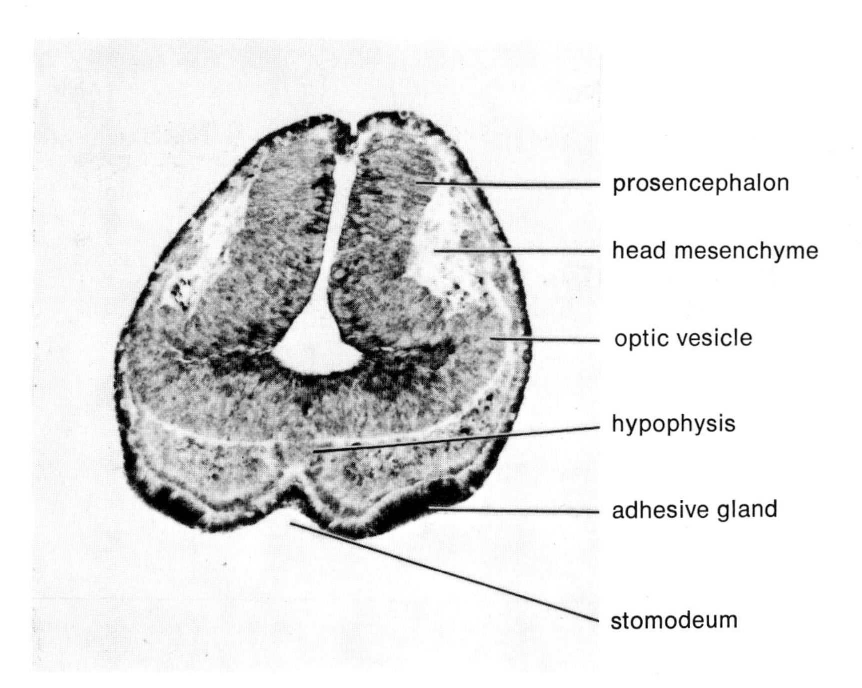

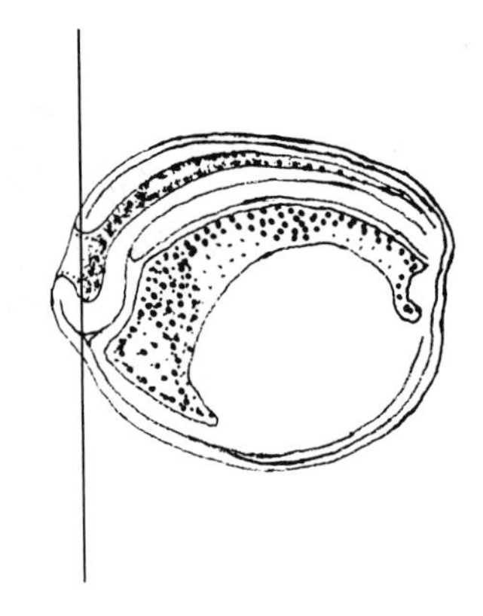

Figure 55 Frog embryo, neural tube stage (stage 16), transverse section through optic vesicles (mag. 65X)

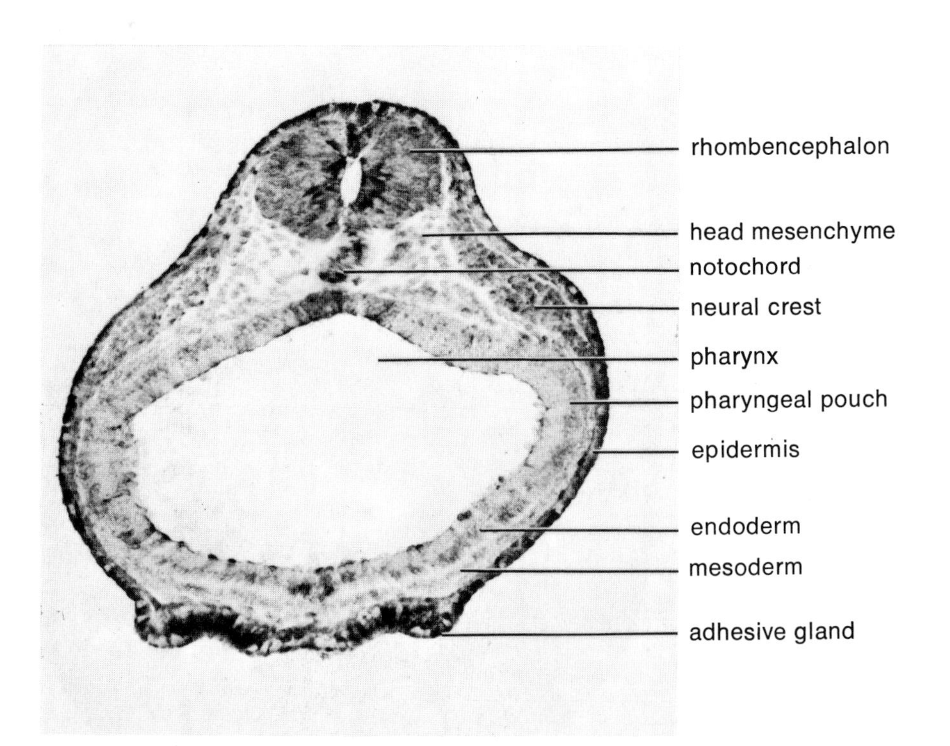

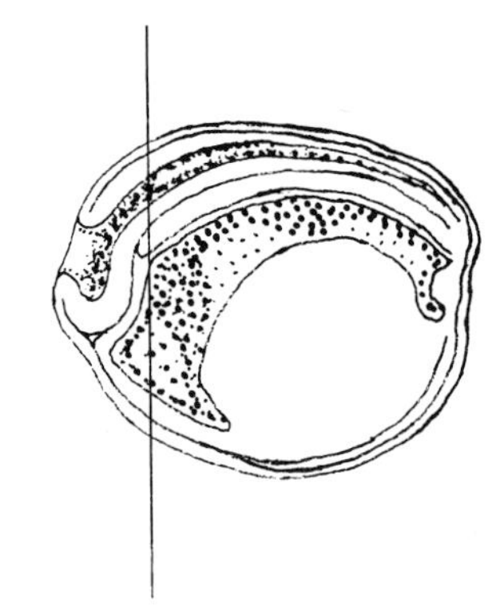

Figure 56 Frog embryo, neural tube stage (stage 16), transverse section through pharynx (mag. 65X)

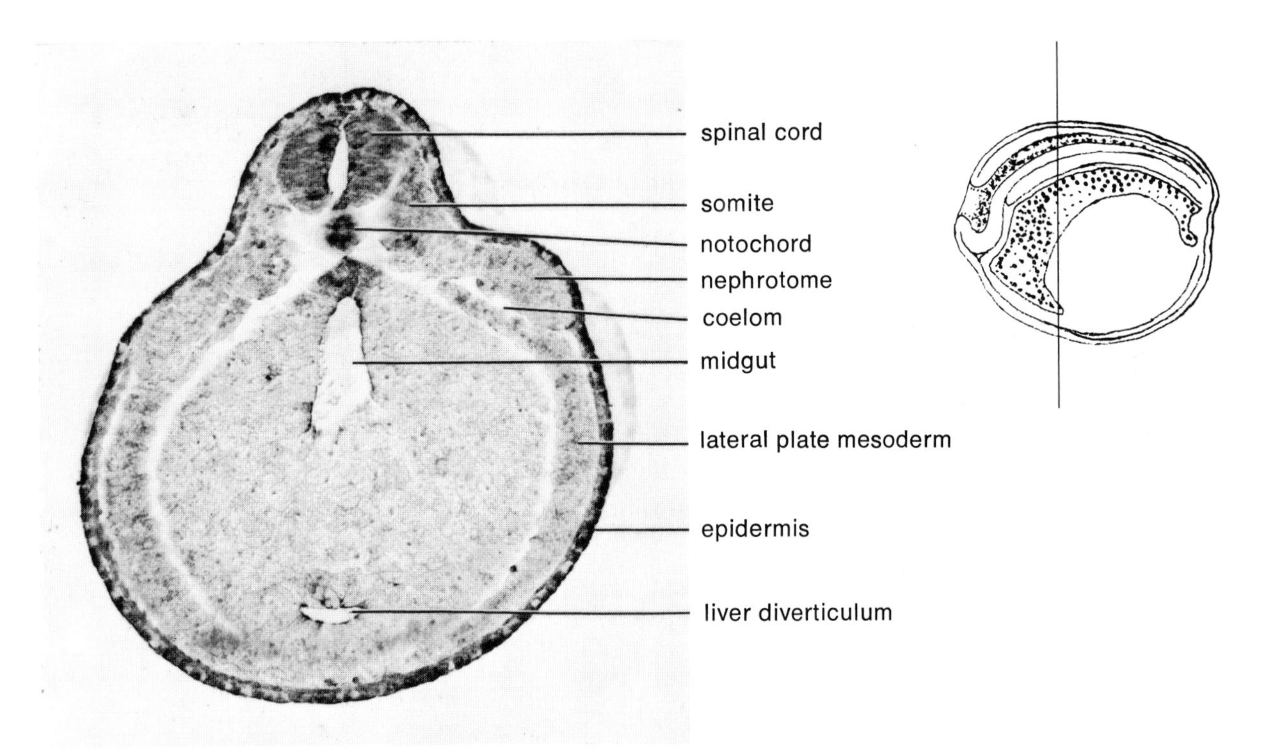

Figure 57 Frog embryo, neural tube stage (stage 16), transverse section through nephrotome (mag. 65X)

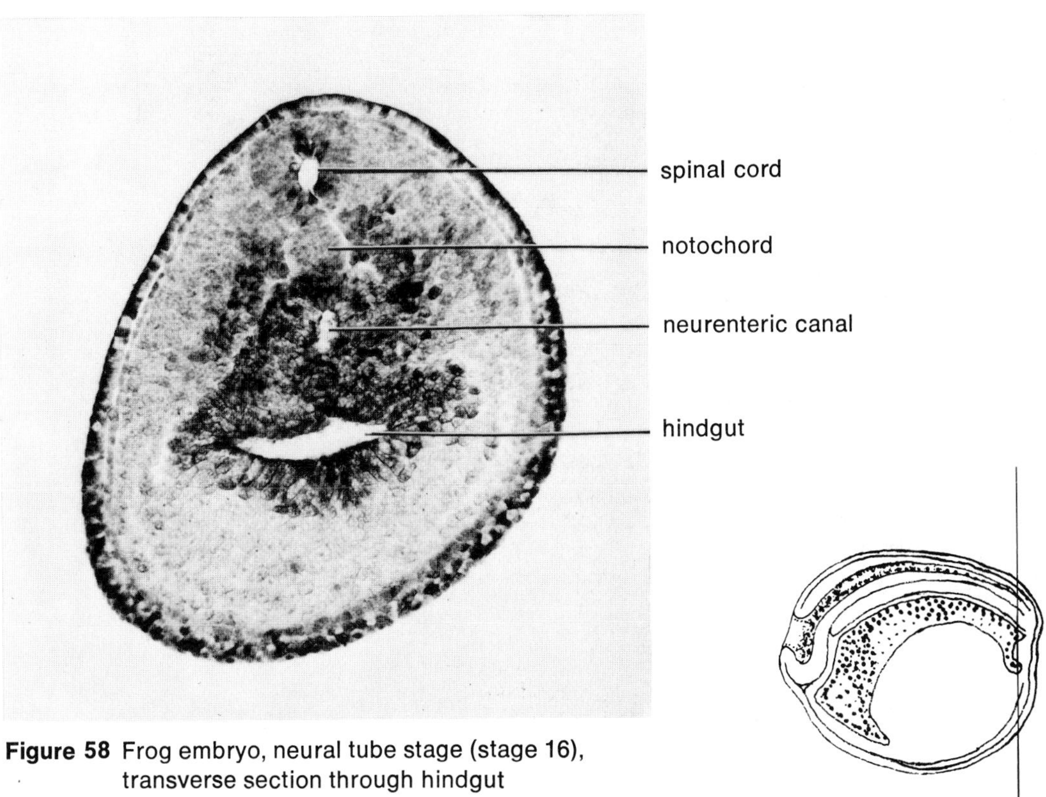

Figure 58 Frog embryo, neural tube stage (stage 16), transverse section through hindgut (mag. 65X)

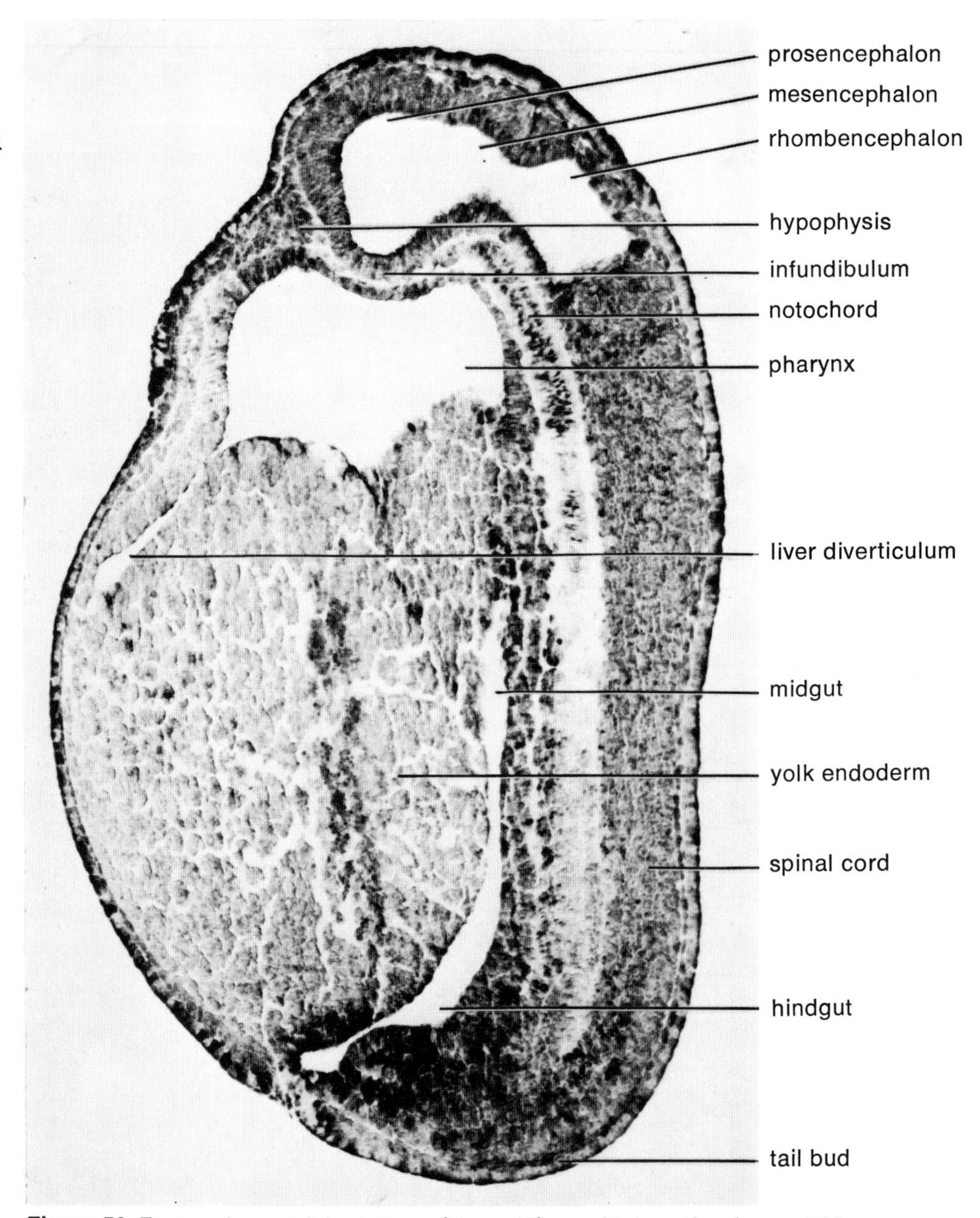

Figure 59 Frog embryo, tail bud stage (stage 17), sagittal section (mag. 65X)

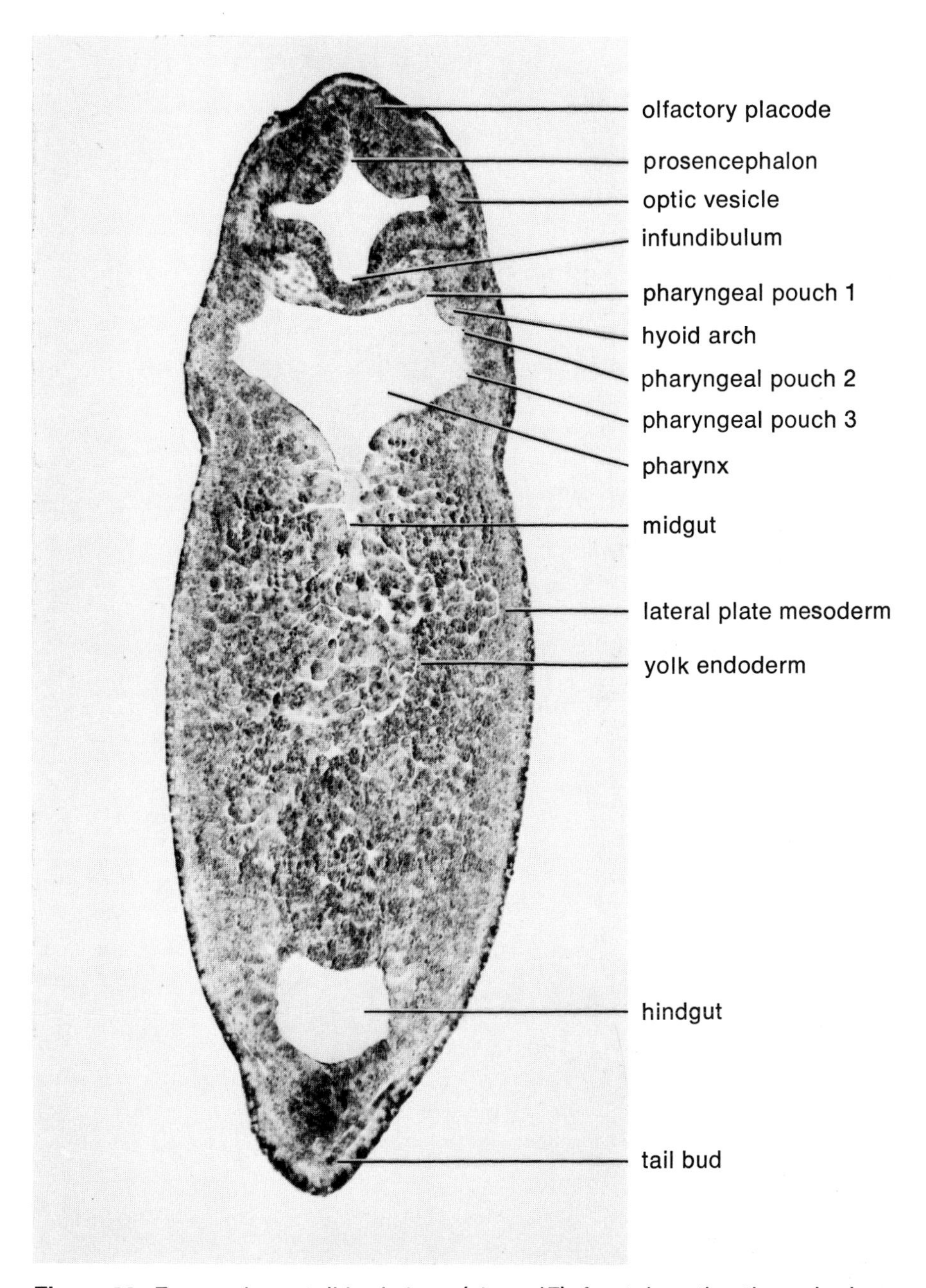

Figure 60 Frog embryo, tail bud stage (stage 17), frontal section through pharynx (mag. 50X)

5. The 4-mm Frog Embryo (Stage 18)

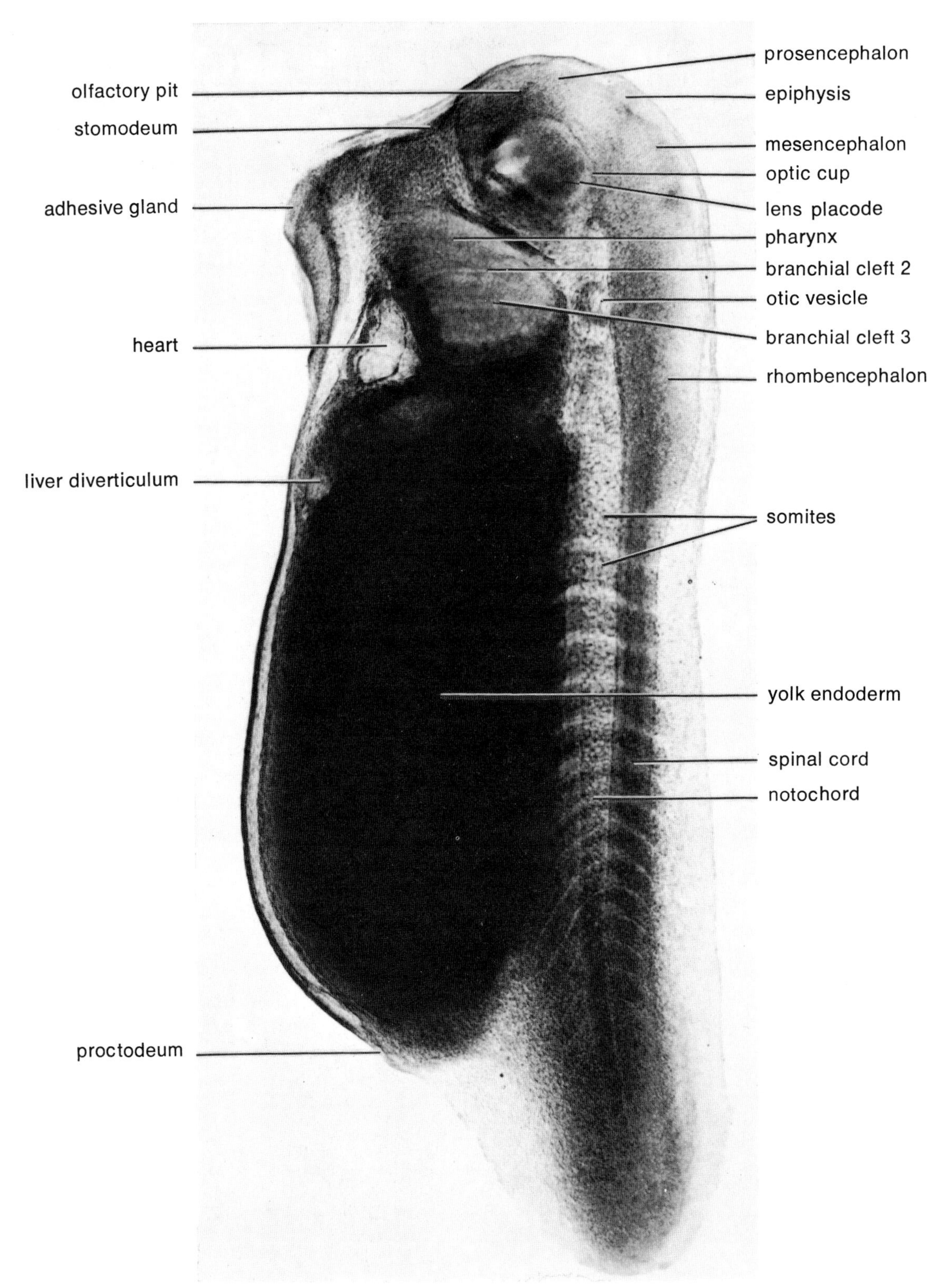

Figure 61 Whole mount (mag. 50X)

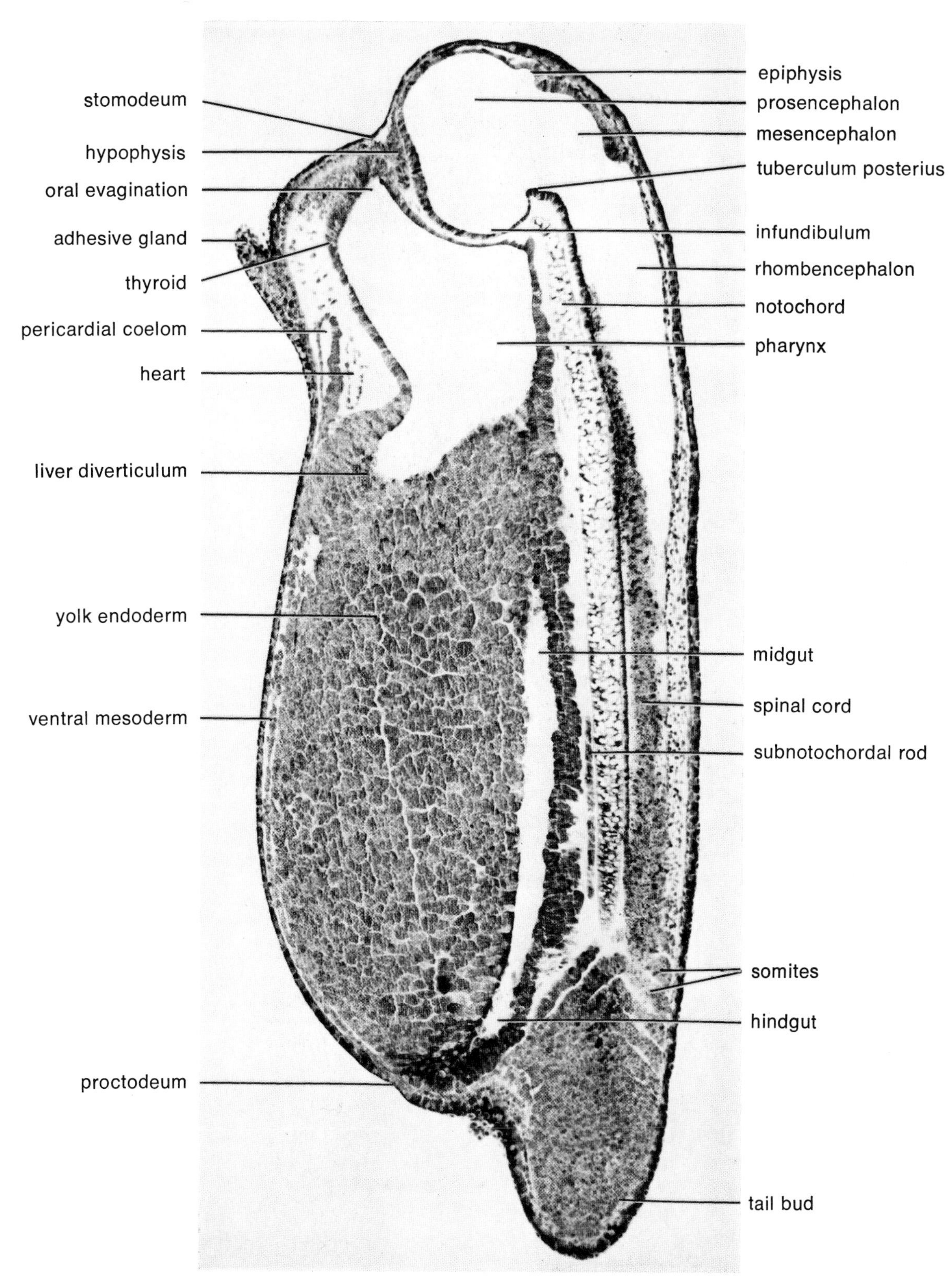

Figure 62 4-mm frog embryo, sagittal section (mag. 50X)

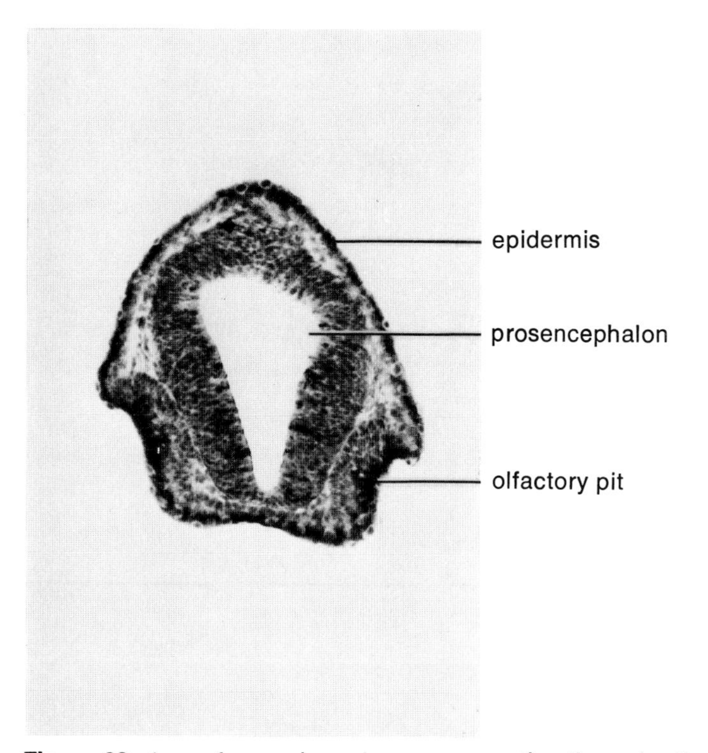

Figure 63 4-mm frog embryo, transverse section through olfactory pits (mag. 65X)

Figure 64 4-mm frog embryo, transverse section through optic cups (mag. 65X)

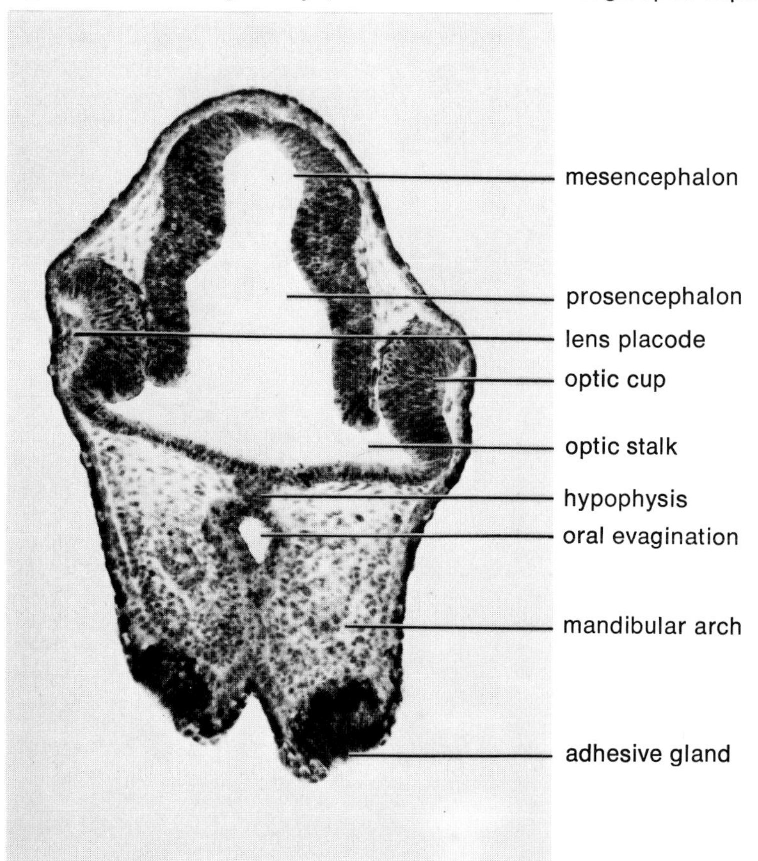

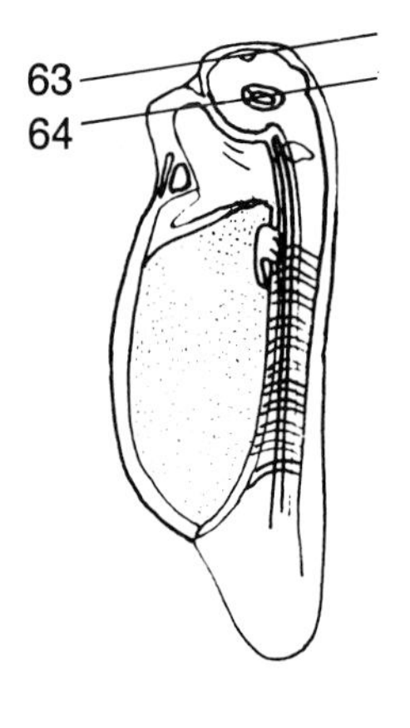

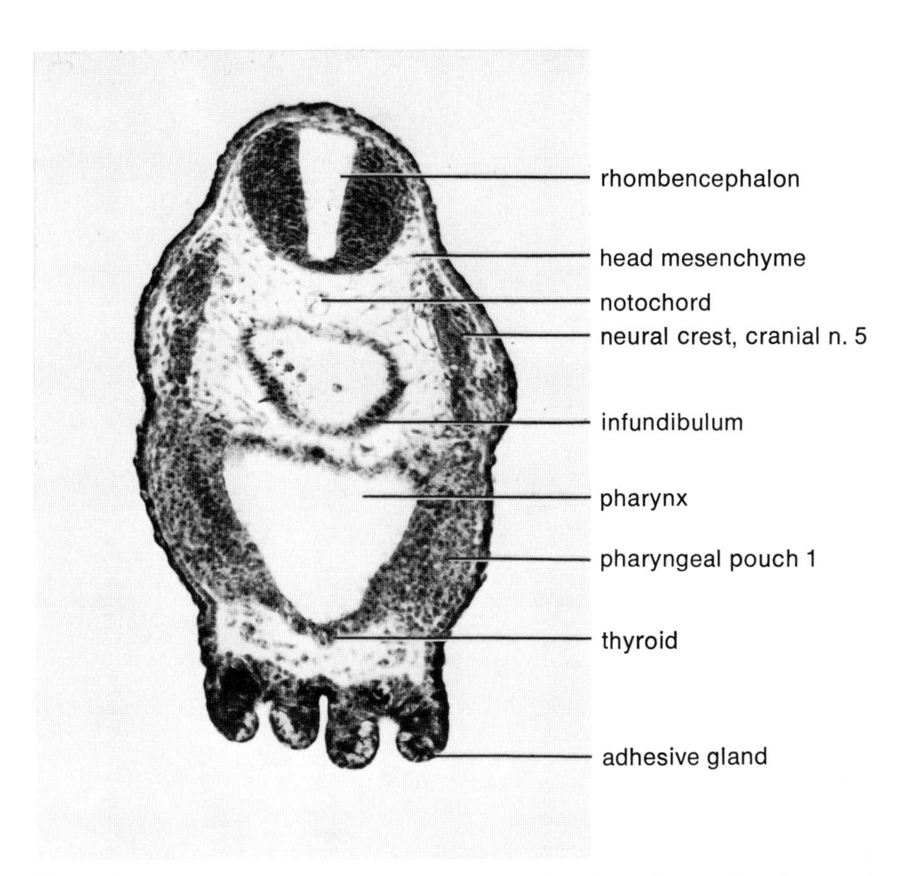

Figure 65 4-mm frog embryo, transverse section through anterior pharynx (mag. 65X)

Figure 66 4-mm frog embryo, transverse section through otic vesicles (mag. 65X)

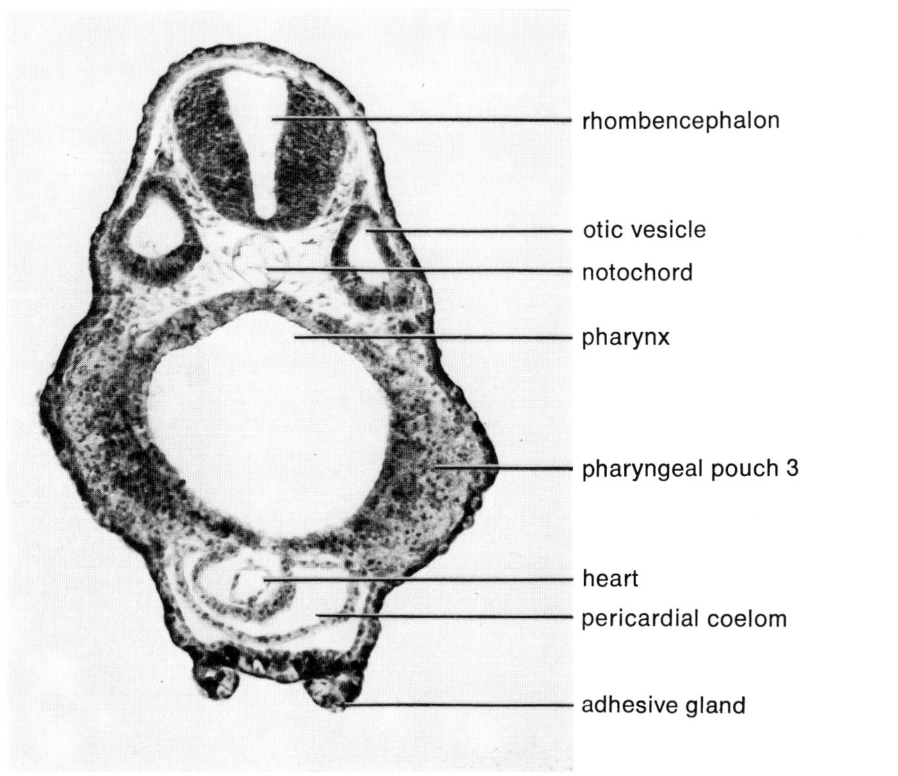

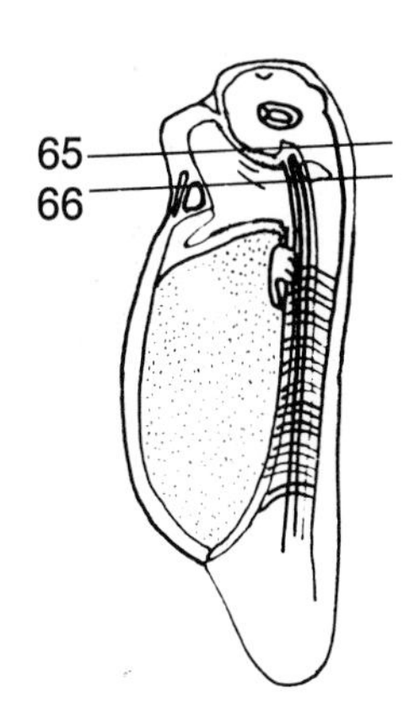

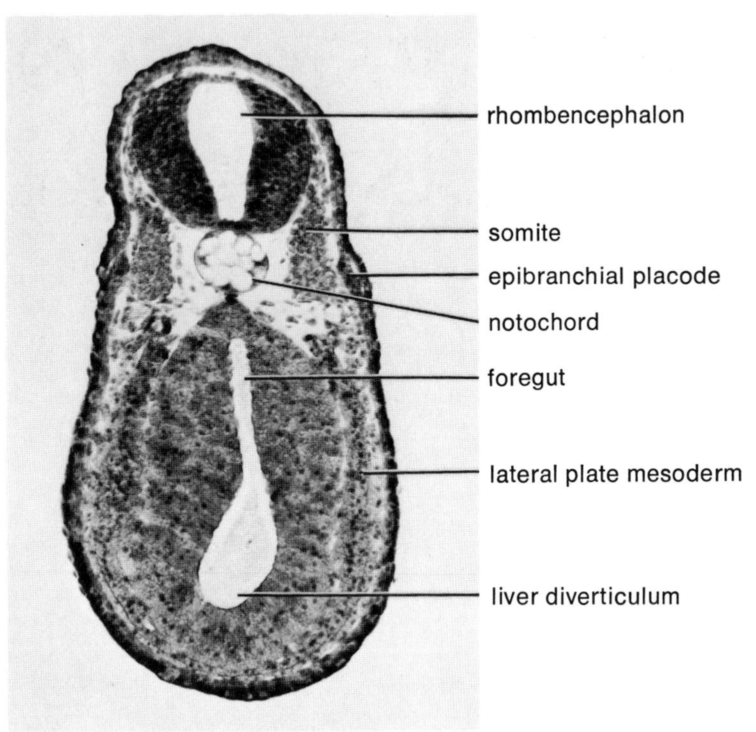

Figure 67 4-mm frog embryo, transverse section through liver diverticulum (mag. 65X)

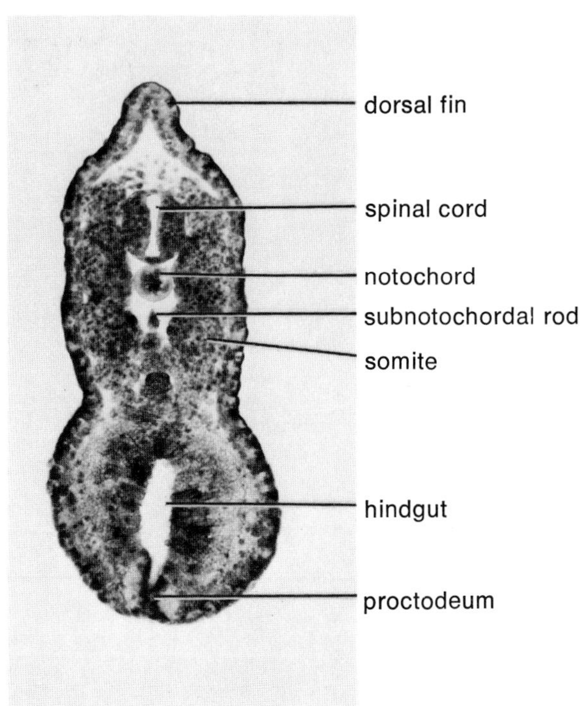

Figure 69 4-mm frog embryo, transverse section through hindgut (mag. 65X)

Figure 68 4-mm frog embryo, transverse section through pronephros (mag. 65X)

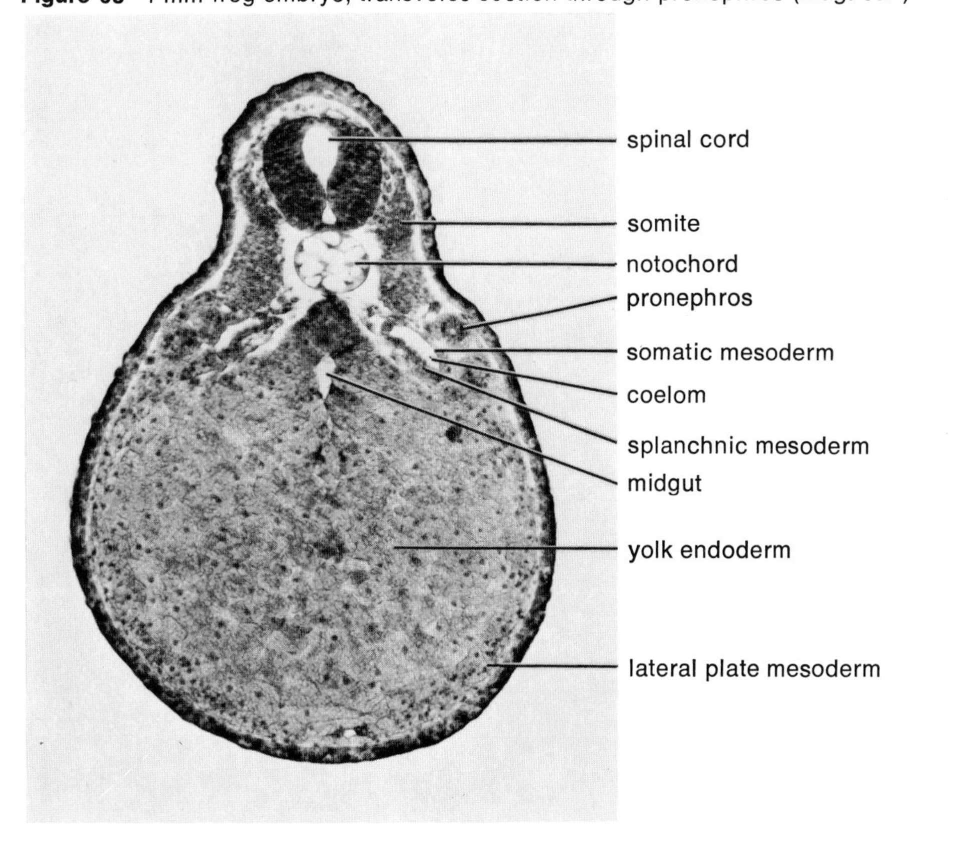

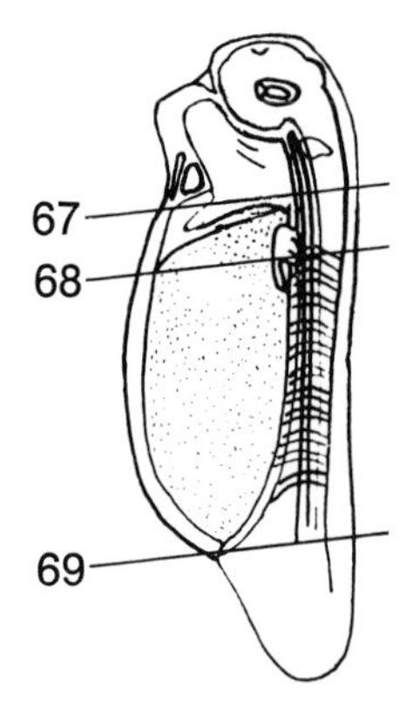

6. The 6-7-mm Frog Embryo (Stages 20-21)

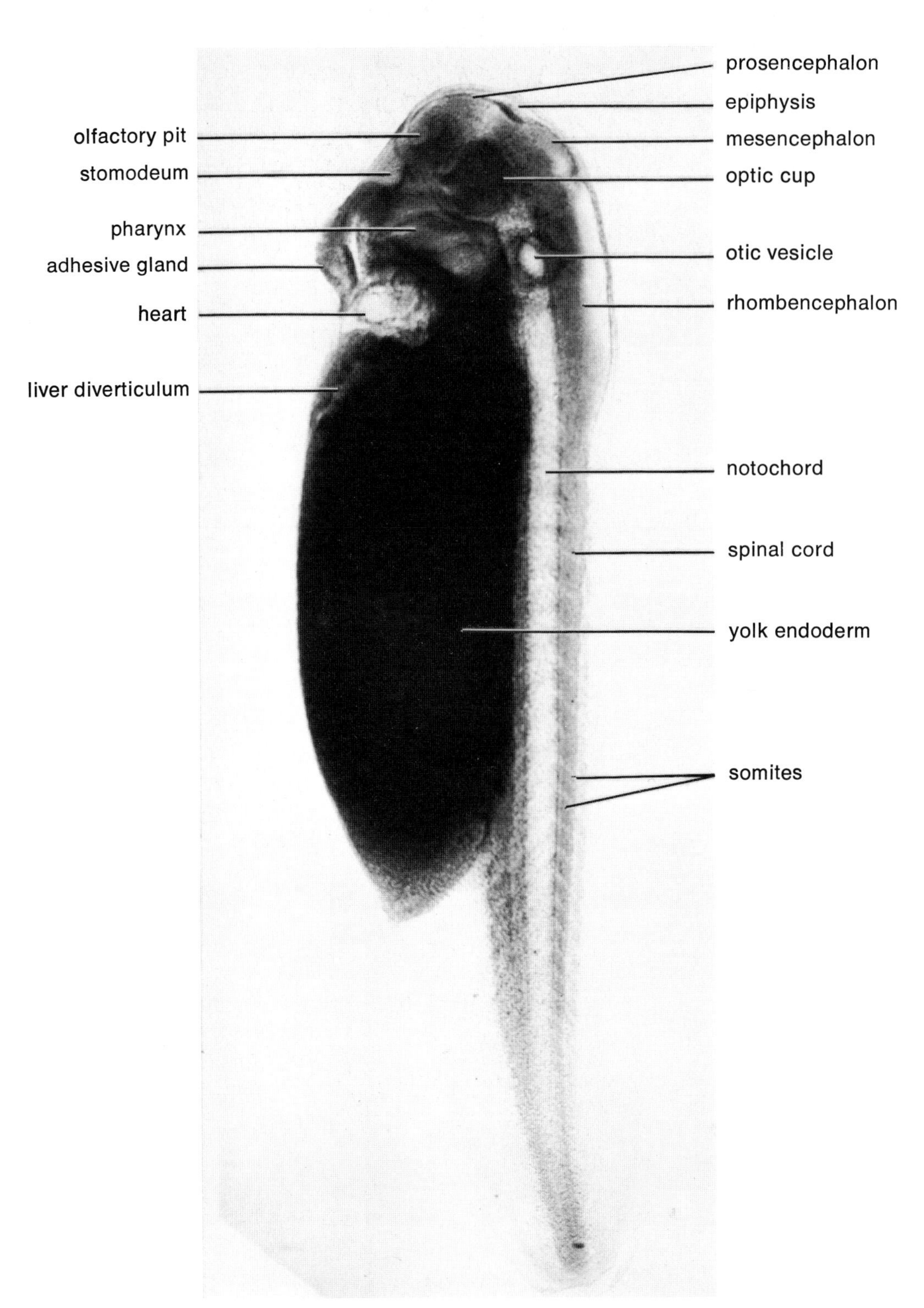

Figure 70 Whole mount, 7-mm embryo (stage 21) (mag. 35X)

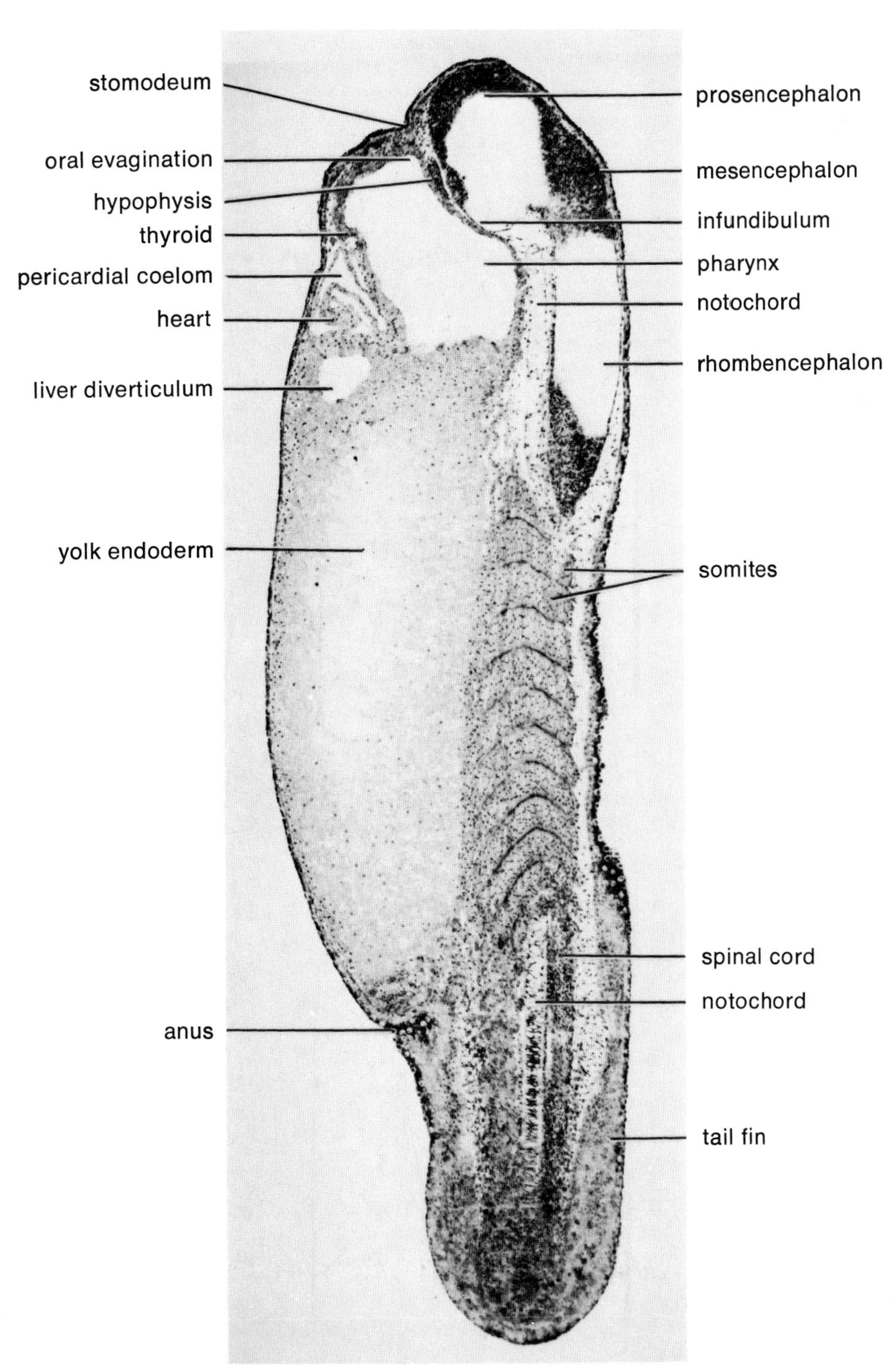

Figure 71 Sagittal section, 6-mm embryo (stage 20) (mag. 40X)

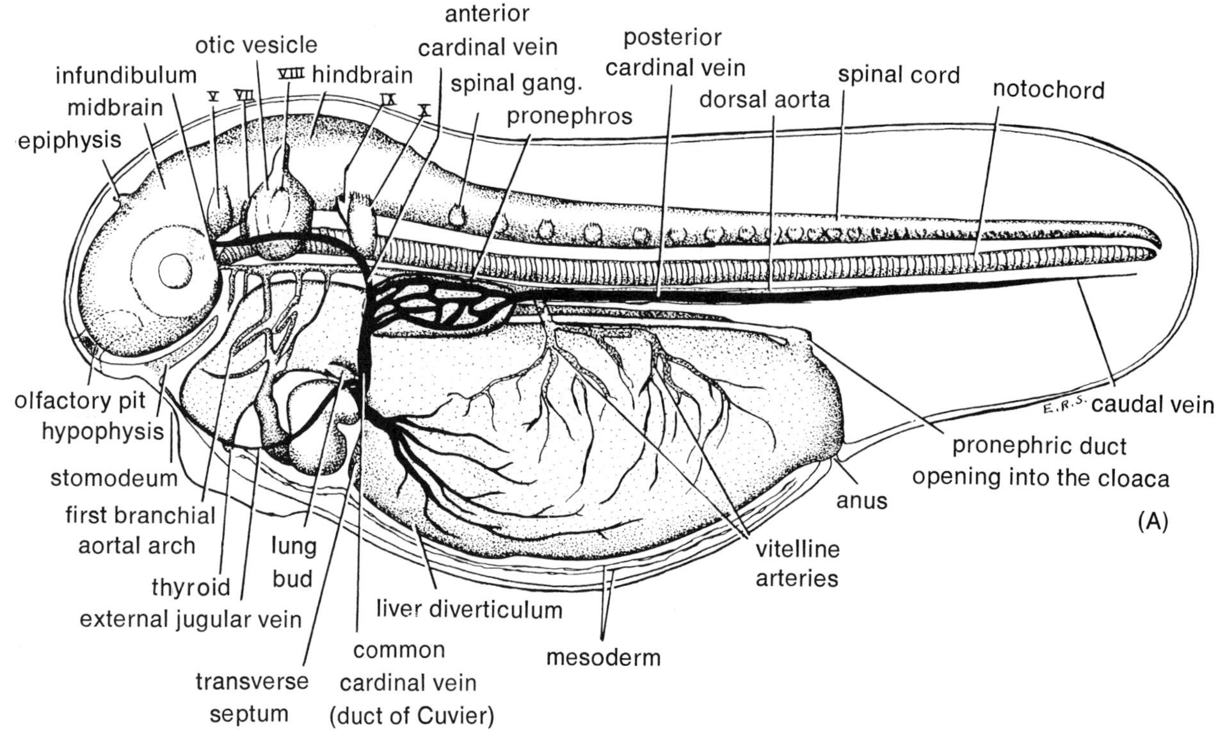

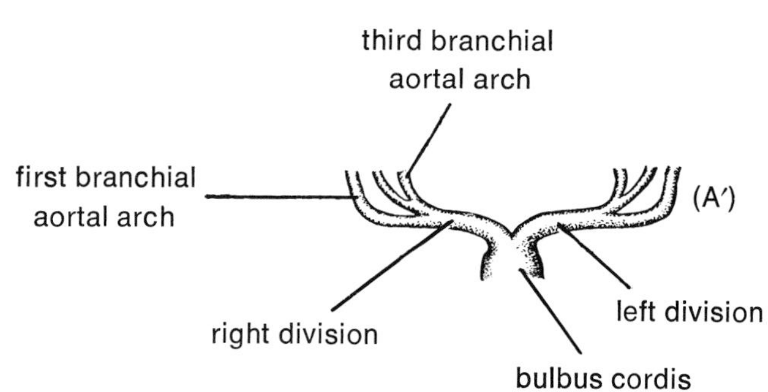

Figure 72

Drawings of early frog tadpoles showing development of early systems. *A,* frog tadpole (*R. pipiens*) of about 6–7 mm. It is difficult to determine the exact number of vitelline arteries at this stage of development and the number given in the figure is a diagrammatic representation. *A′* Shows right and left ventral aortal divisions of bulbus cordis. *B,* anatomy of frog tadpole of about 10–18 mm. (From *Comparative Embryology of the Vertebrates* by Olin E. Nelsen. Copyright 1953 by The Blakiston Co. Inc. Used with permission of the McGraw-Hill Book Co.)

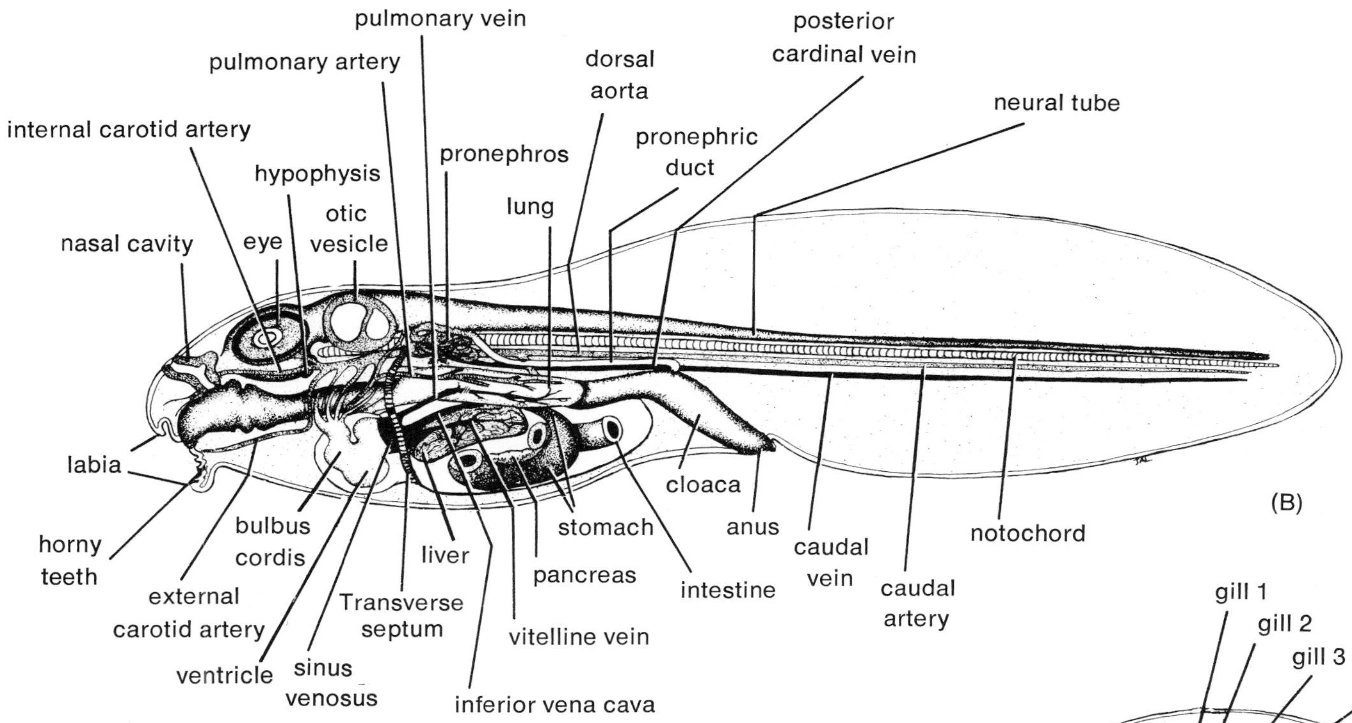

pulmonary vein
pulmonary artery
internal carotid artery
hypophysis
nasal cavity
eye
otic
vesicle
pronephros
lung
dorsal
aorta
pronephric
duct
posterior
cardinal vein
neural tube
labia
horny
teeth
bulbus
cordis
external
carotid artery
ventricle
sinus
venosus
liver
Transverse
septum
stomach
pancreas
vitelline vein
inferior vena cava
cloaca
intestine
anus
caudal
vein
caudal
artery
notochord
(B)

aortal arches 3-6 function as part of the branchial mechanism

aortal arches 1 and 2 are vestigial in the frog

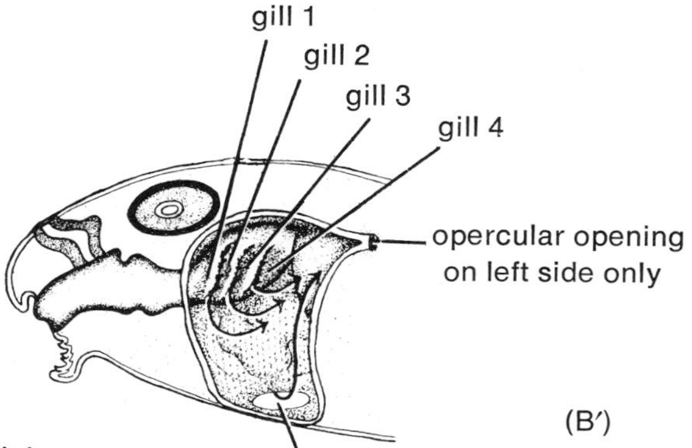

gill 1
gill 2
gill 3
gill 4
opercular opening
on left side only
(B′)
interbranchial
chamber communication
below heart

diagram showing branchial chamber on left side

arrows show water currents

7. Gametogenesis in Chickens

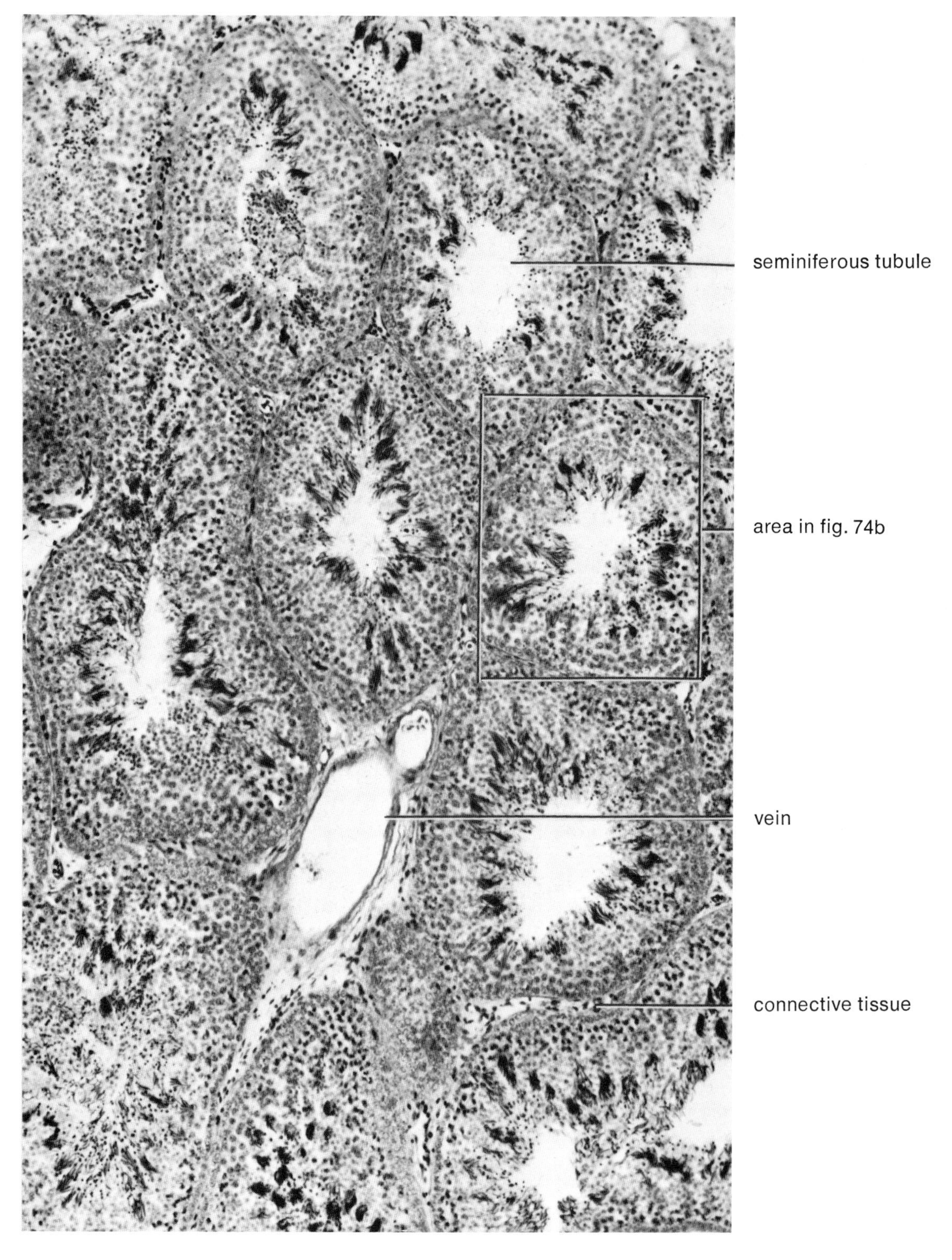

Figure 73 Chicken testis, section (mag. 180 X)

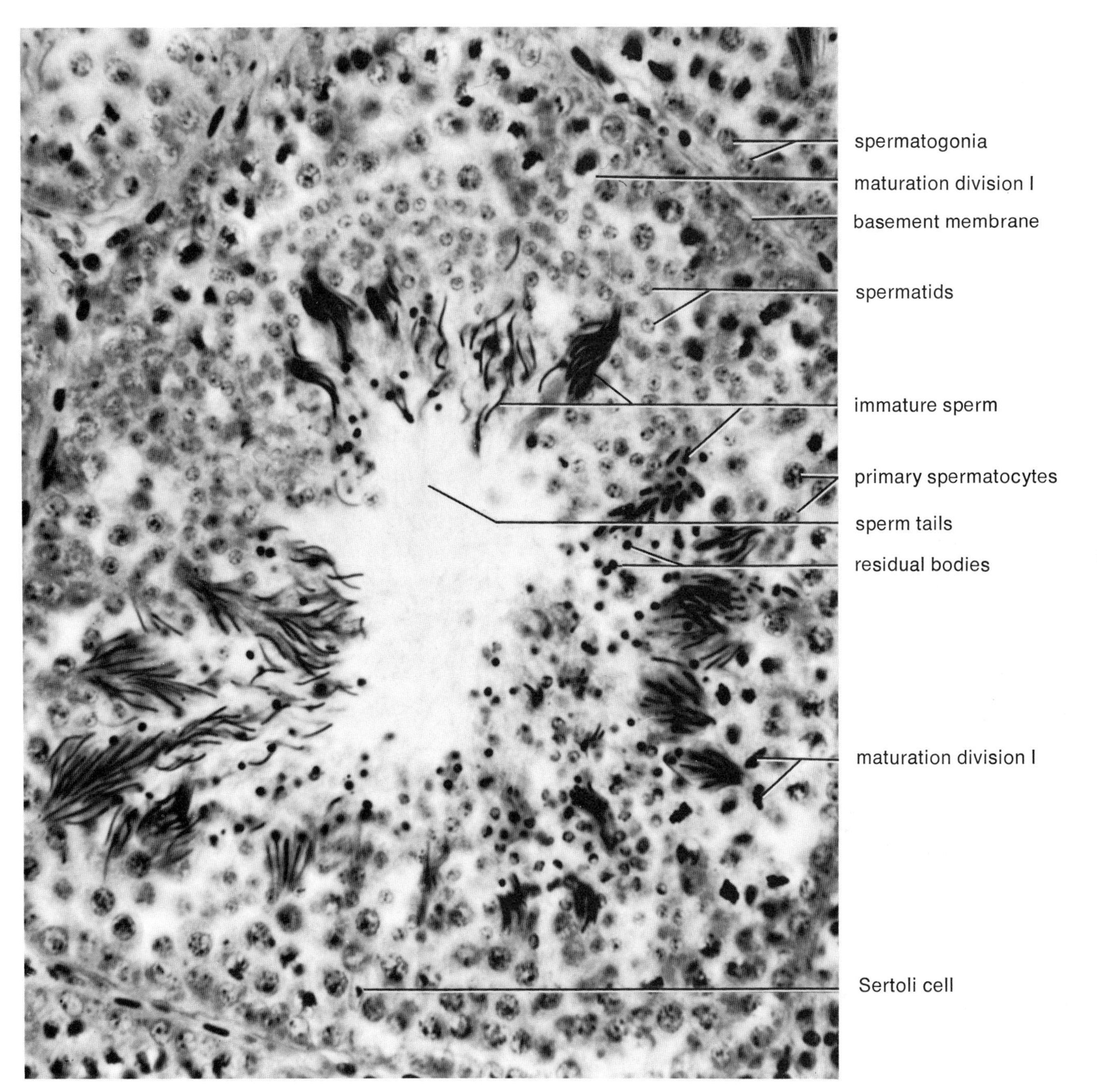

Figure 74 Chicken testis, section (mag. 680 X)

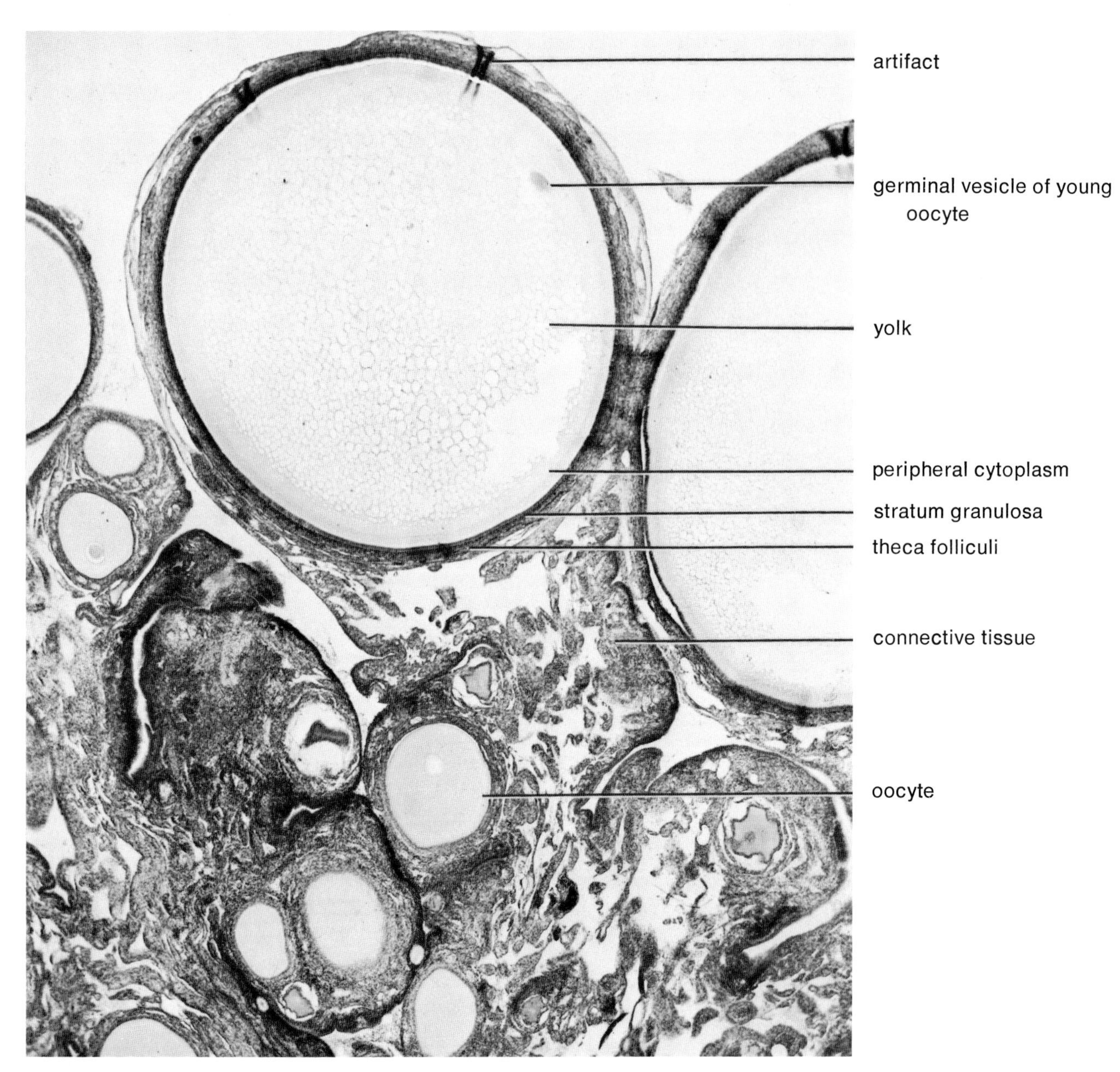

Figure 75 Chicken ovary, section (mag. 35 X)

8. The Unincubated Chick Blastoderm

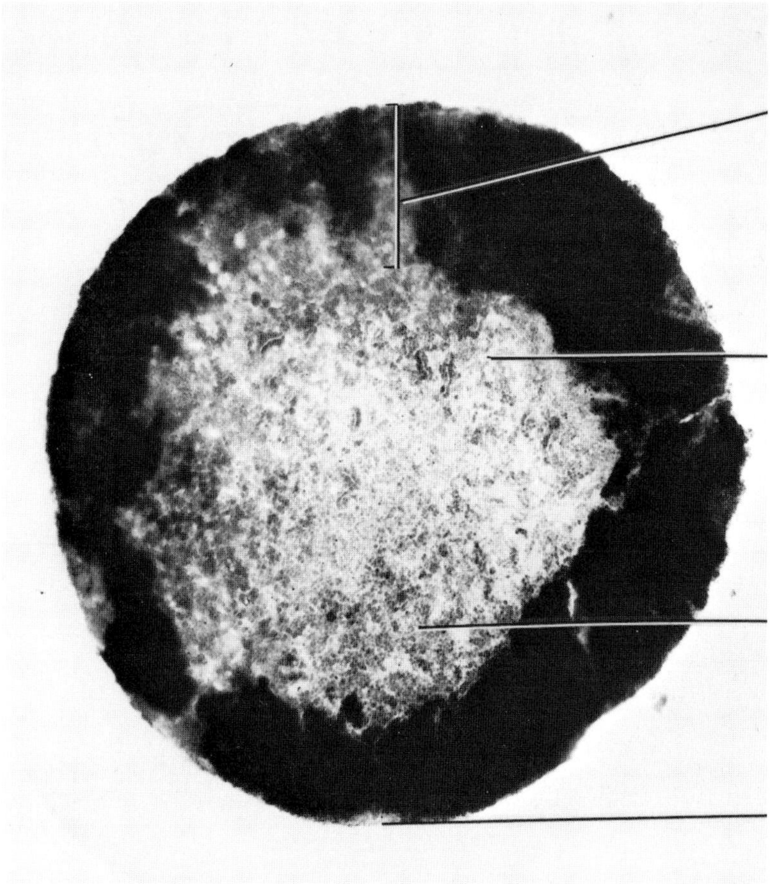

area opaca

area pellucida

embryonic shield

posterior end

Figure 76 Chick embryo, stage 1, unincubated blastoderm, whole mount (mag. 27.5 X)

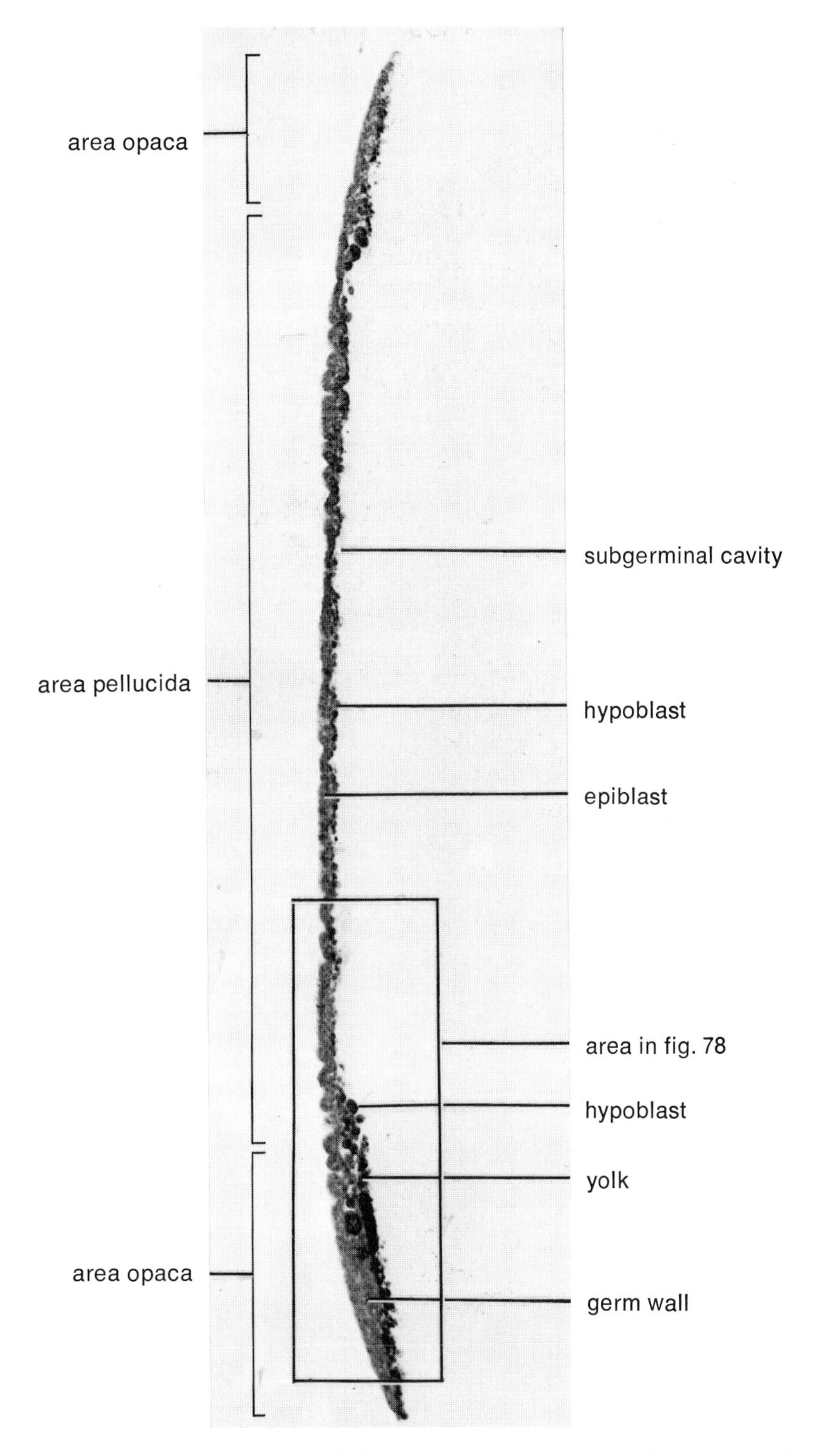

Figure 77 Chick embryo, stage 1, unincubated blastoderm, longitudinal section (mag. 50 X)

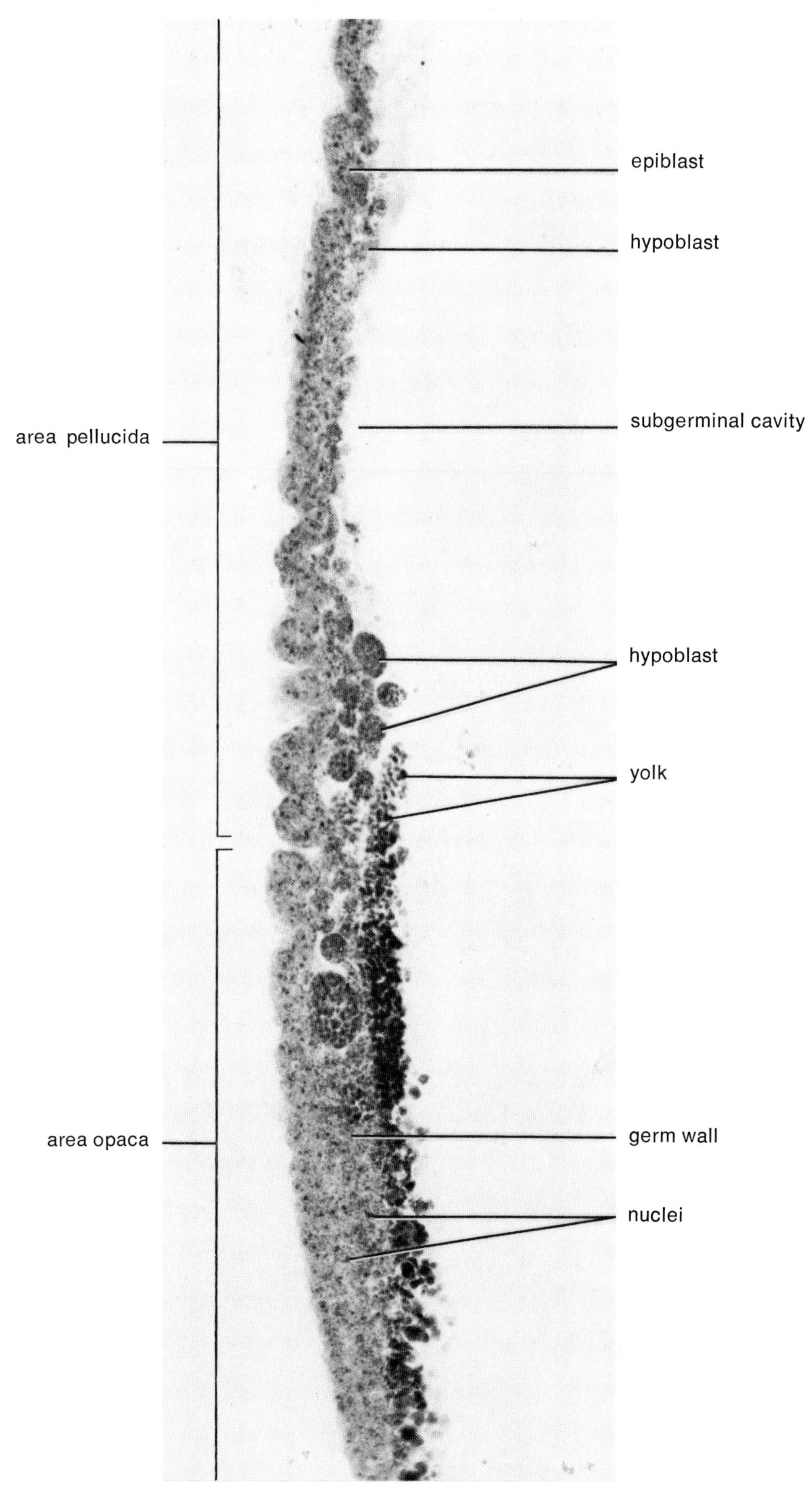

Figure 78 Chick embryo, stage 1, unincubated blastoderm, longitudinal section (mag. 150 X)

9. The Stage* 5 Chick Embryo (19-22 hours incubation)

* Stages based on Hamburger, V., and H. L. Hamilton, "A series of normal stages in the development of the chick embryo," *Journal of Morphology* 88,1951, pp. 49-92.

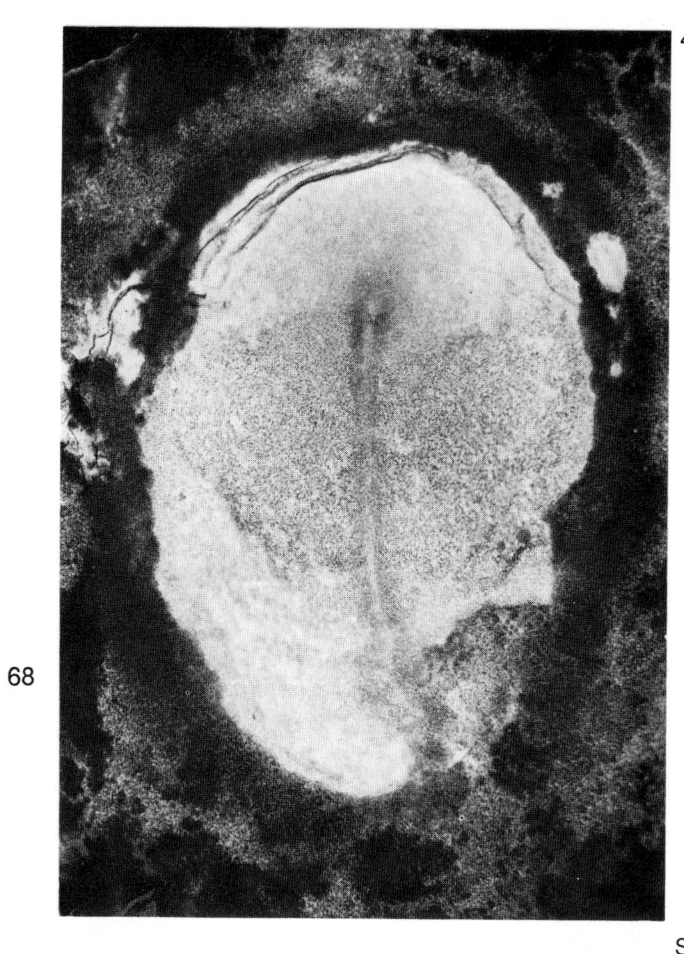

4

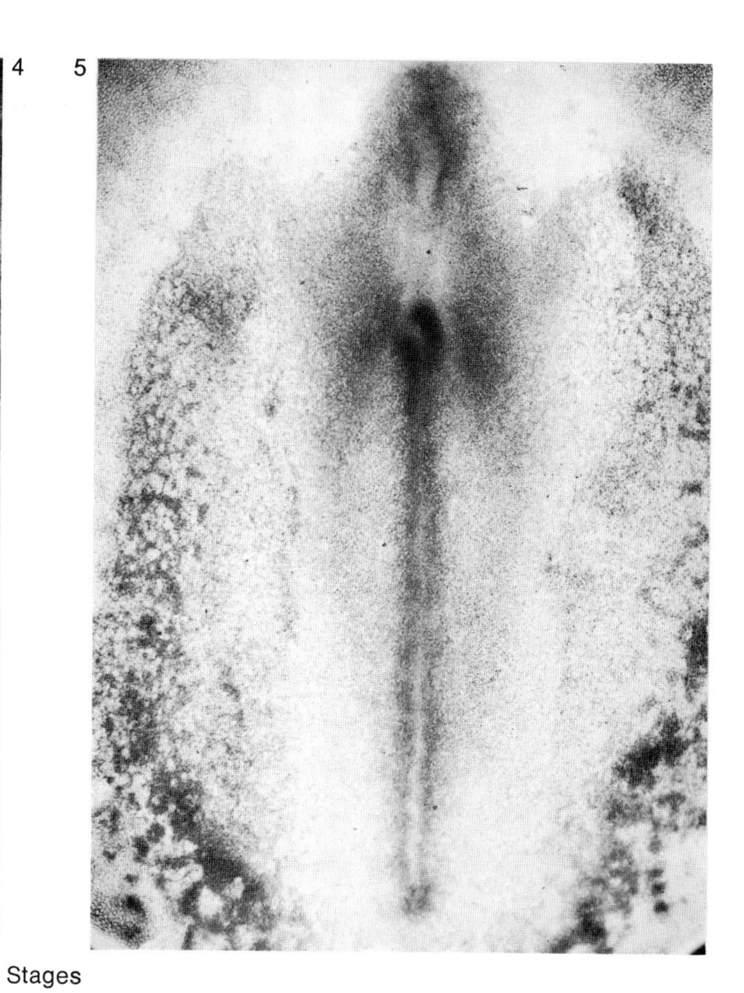

5

Stages

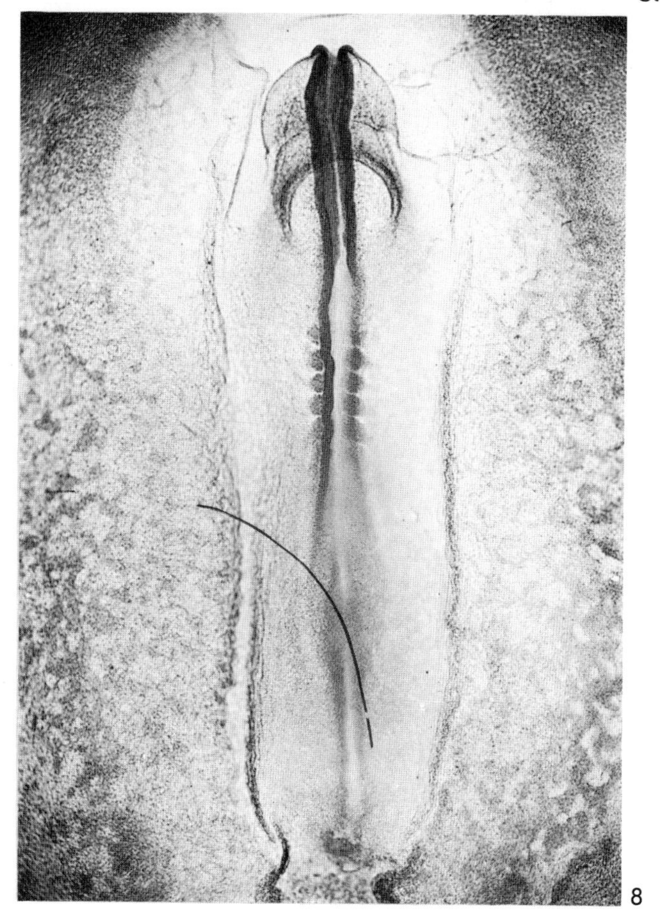

8

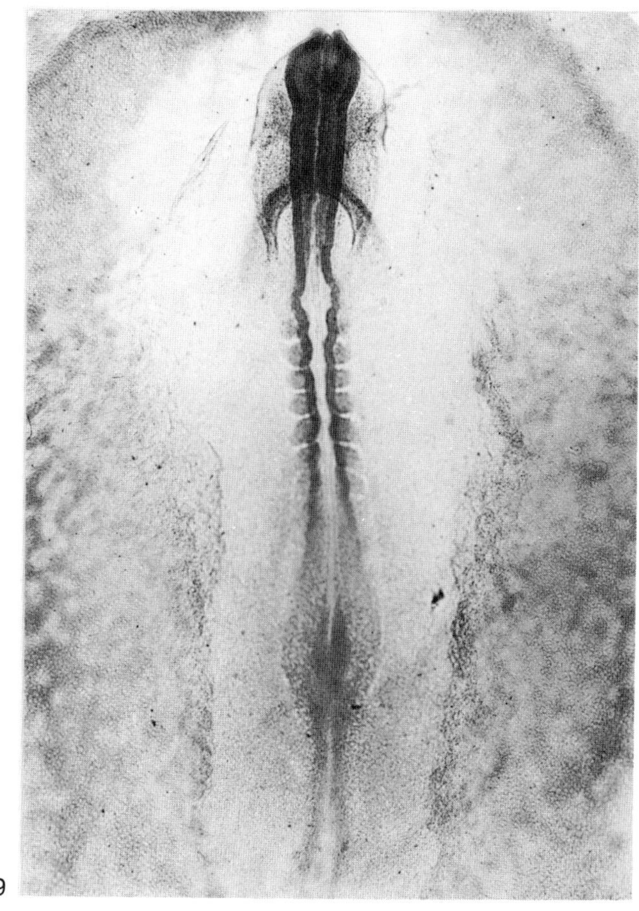

9

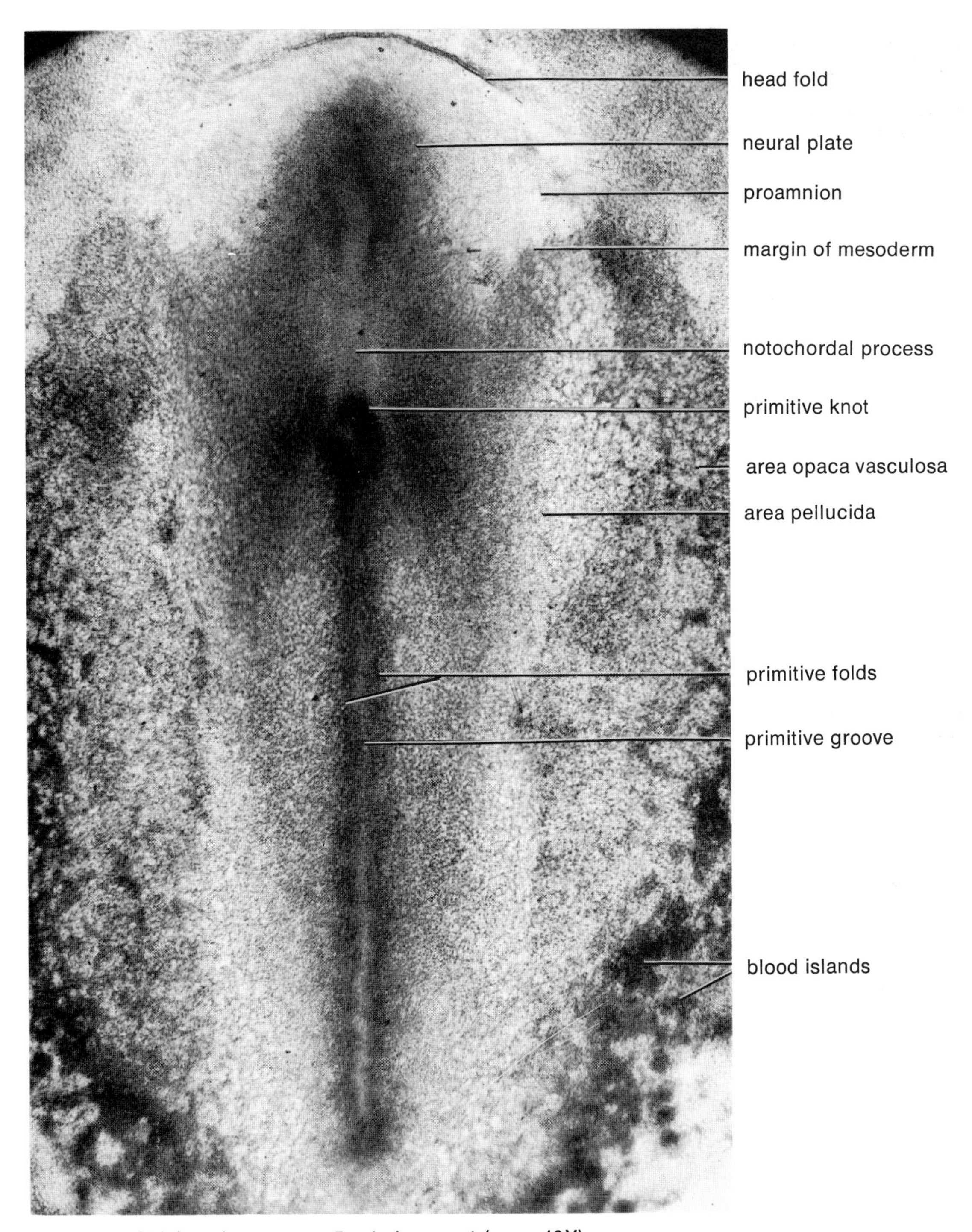

Figure 80 Chick embryo, stage 5, whole mount (mag. 40X)

Figure 79 Chick embryos, stages 4, 5, 8, and 9, whole mounts (mag. 27.5X)

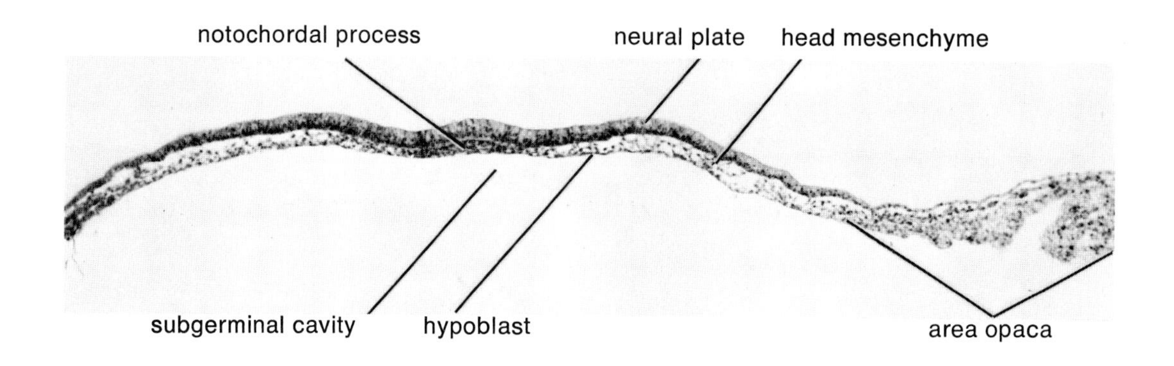
notochordal process
neural plate
head mesenchyme
subgerminal cavity
hypoblast
area opaca

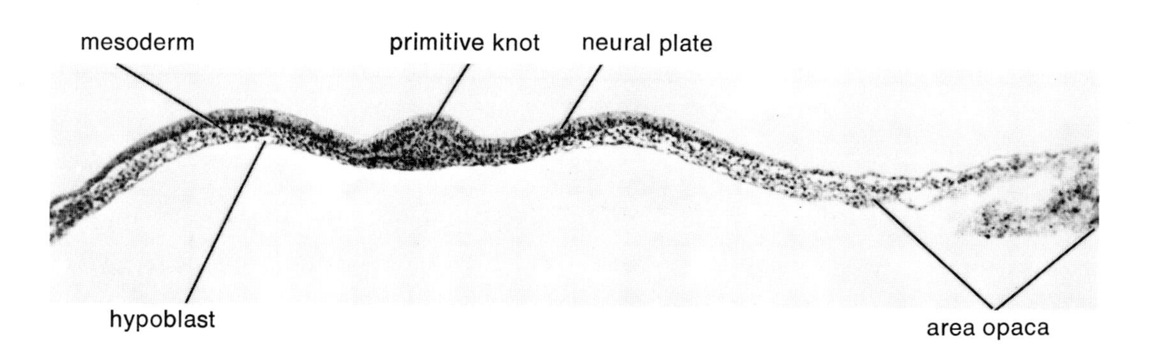
mesoderm
primitive knot
neural plate
hypoblast
area opaca

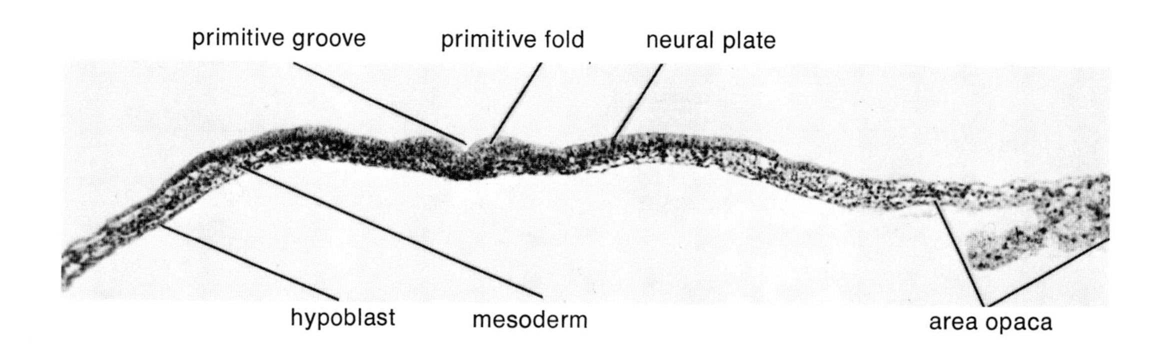
primitive groove
primitive fold
neural plate
hypoblast
mesoderm
area opaca

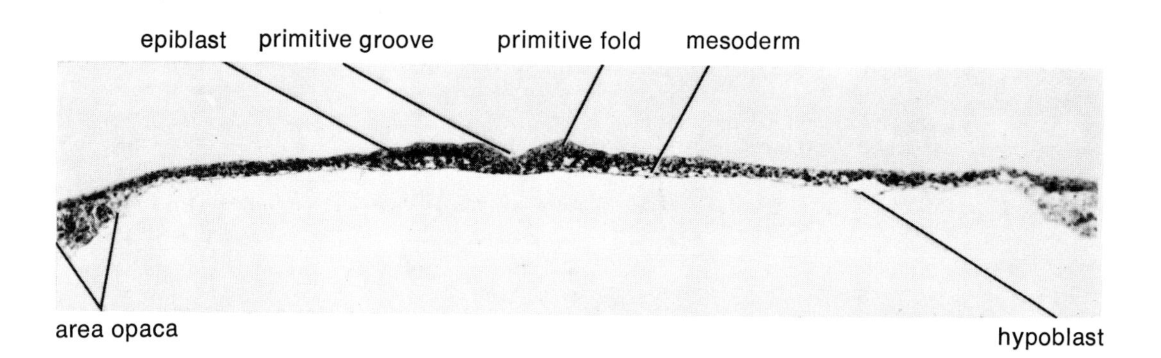
epiblast
primitive groove
primitive fold
mesoderm
area opaca
hypoblast

Figure 81 Chick embryo, stage 5, transverse section through notochordal process (mag. 75X)

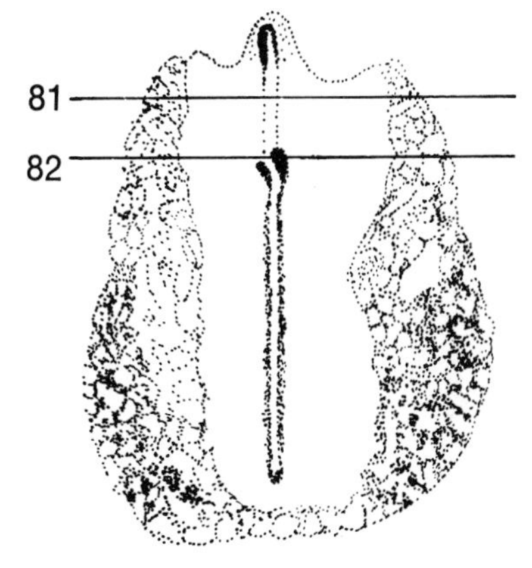

Figure 82 Chick embryo, stage 5, transverse section through primitive knot (mag. 75X)

Figure 83 Chick embryo, stage 5, transverse section through anterior primitive groove (mag. 75X)

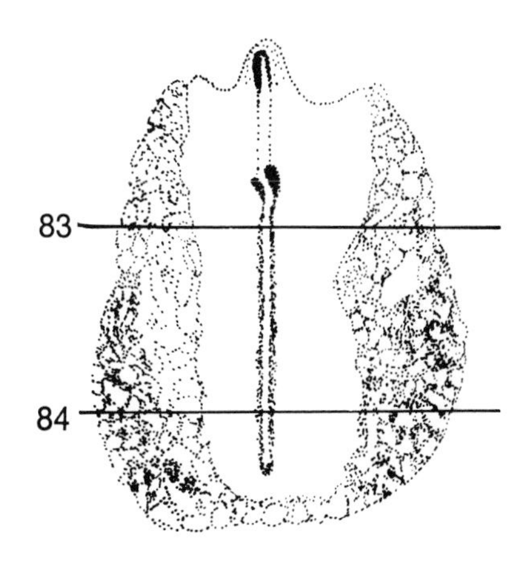

Figure 84 Chick embryo, stage 5, transverse section through posterior primitive groove (mag. 75X)

10. The Stage 8 Chick Embryo (26-29 hours incubation)

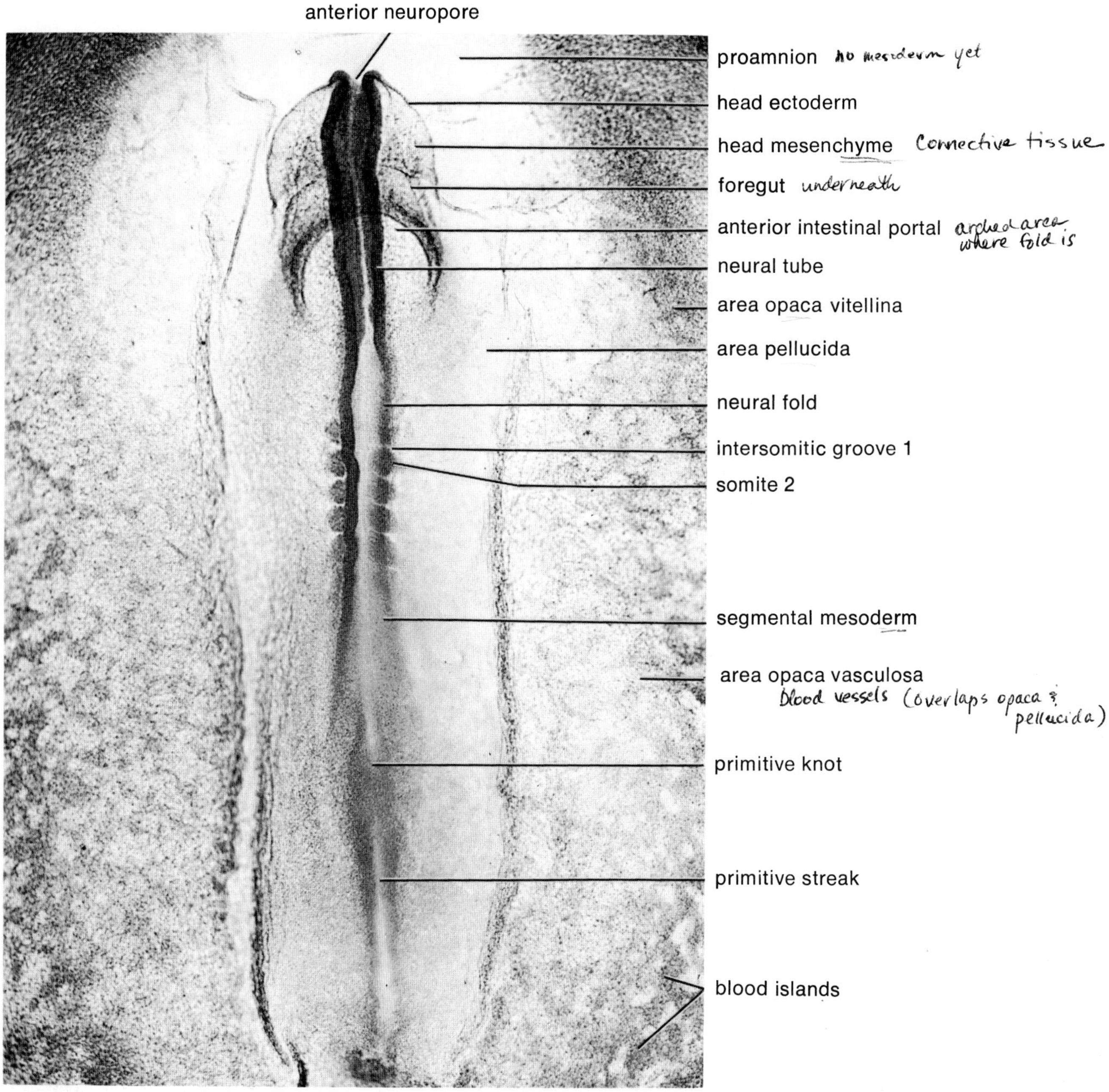

anterior neuropore

proamnion no mesoderm yet

head ectoderm

head mesenchyme Connective tissue

foregut underneath

anterior intestinal portal arched area where fold is

neural tube

area opaca vitellina

area pellucida

neural fold

intersomitic groove 1

somite 2

segmental mesoderm

area opaca vasculosa
blood vessels (overlaps opaca & pellucida)

primitive knot

primitive streak

blood islands

Figure 85 Whole mount (mag. 40X)

Figure 86
Chick embryo, stage 8, opaque whole mount, incident illumination (mag. 40X)

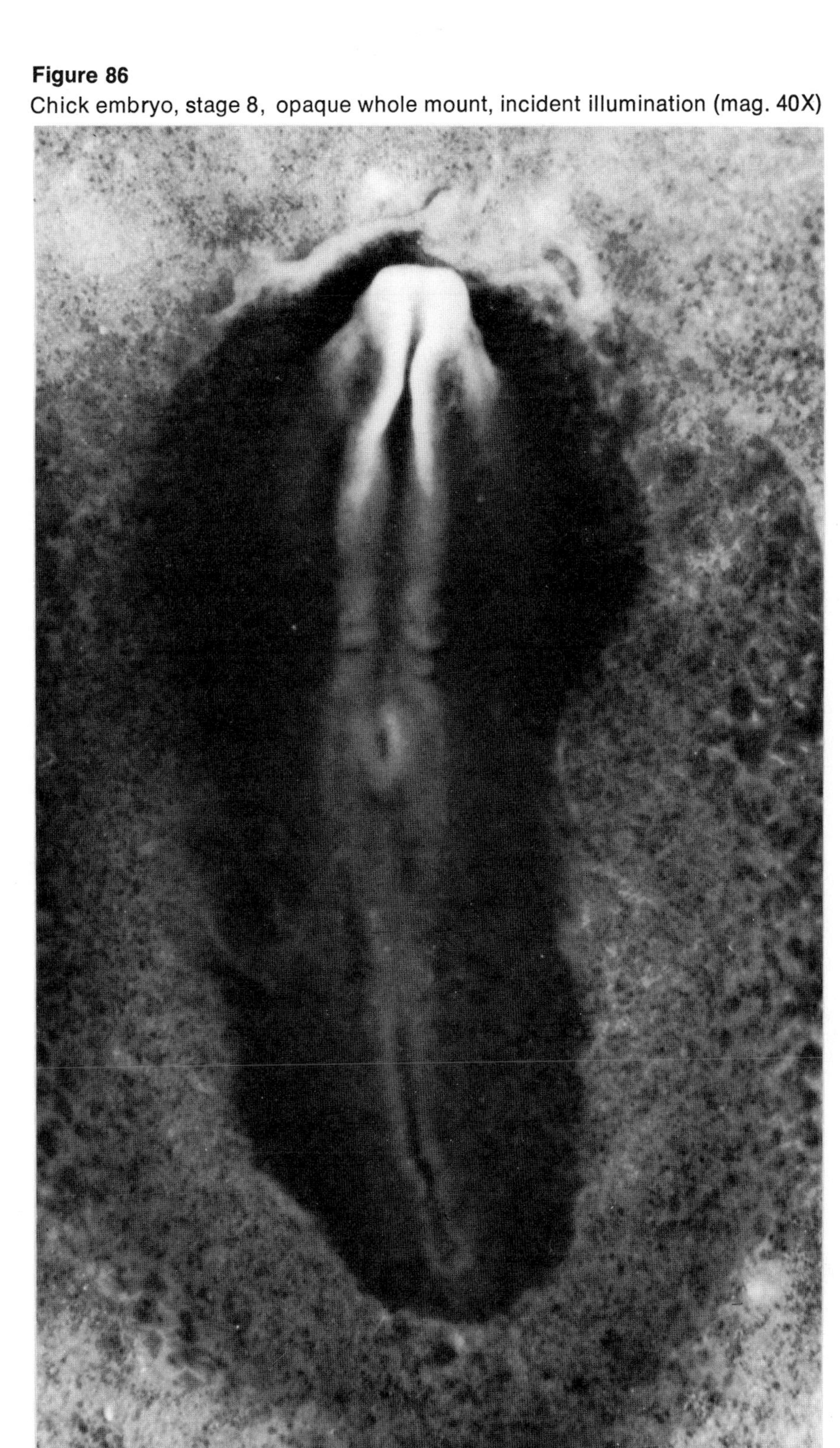

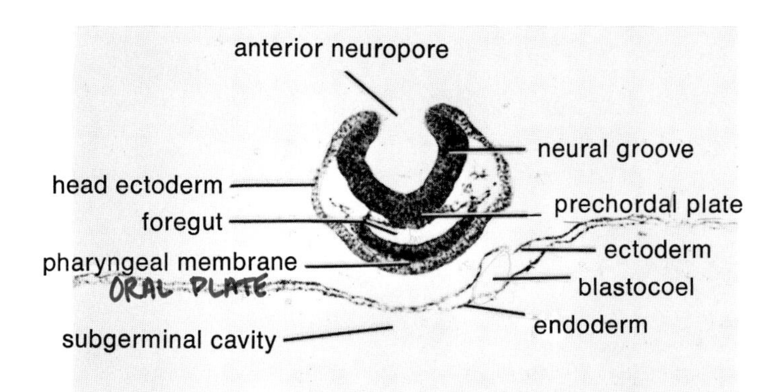

Figure 87 Chick embryo, stage 8, transverse section through pharyngeal membrane (mag. 75X)

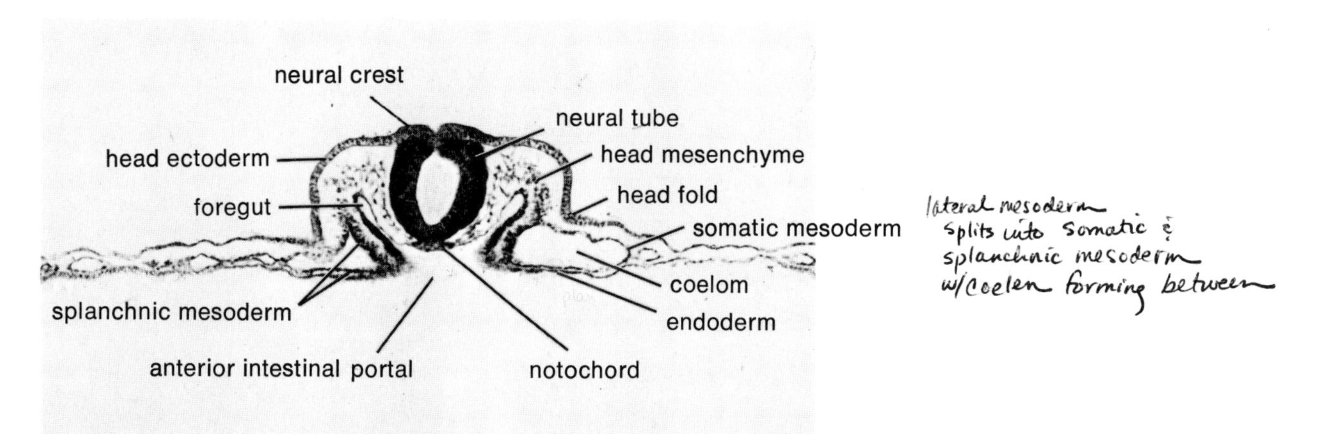

Figure 88 Chick embryo, stage 8, transverse section through anterior intestinal portal (mag. 75X)

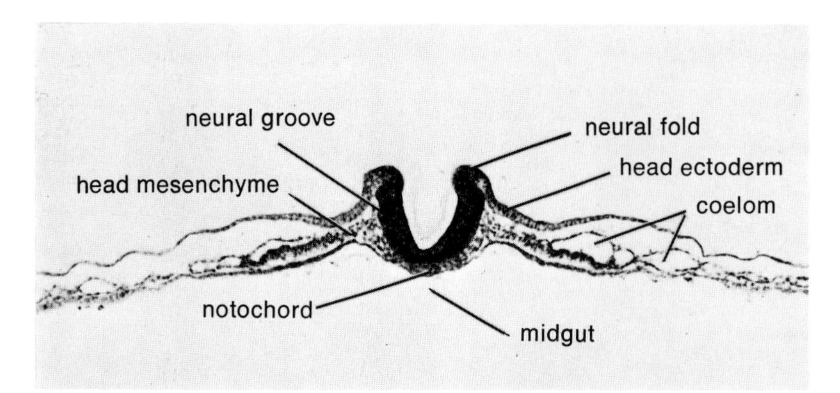

Figure 89 Chick embryo, stage 8, transverse section through neural groove (mag. 75X)

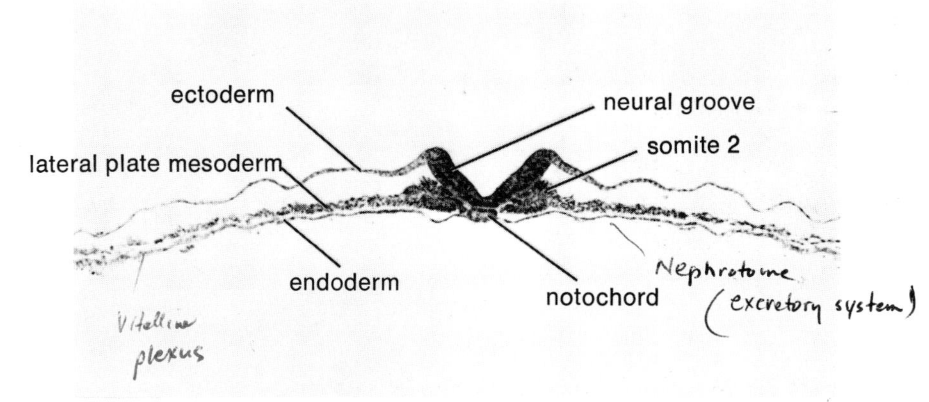

Figure 90 Chick embryo, stage 8, transverse section through somites (mag. 75X)

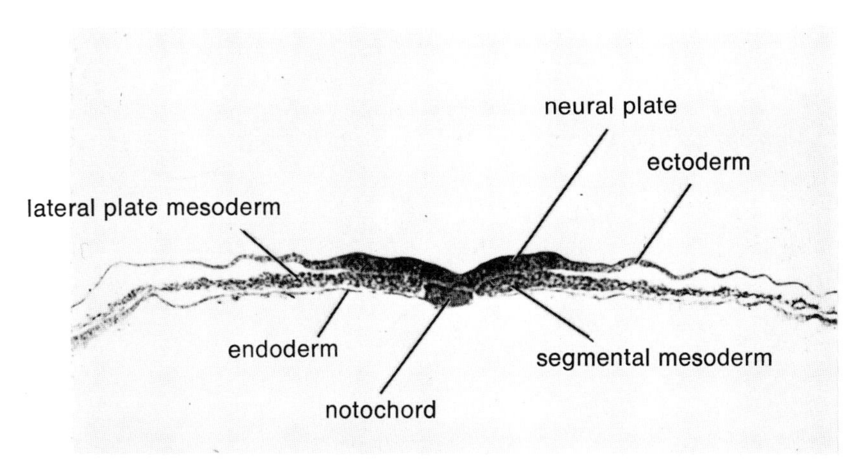

Figure 91 Chick embryo, stage 8, transverse section through neural plate (mag. 75X)

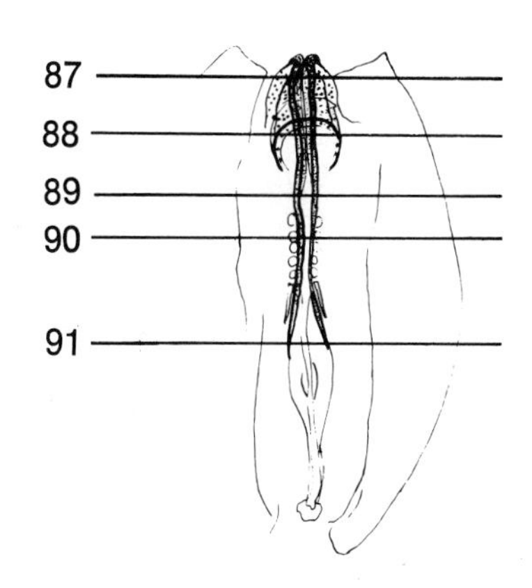

11. The Stage 11 Chick Embryo (40-45 hours incubation)

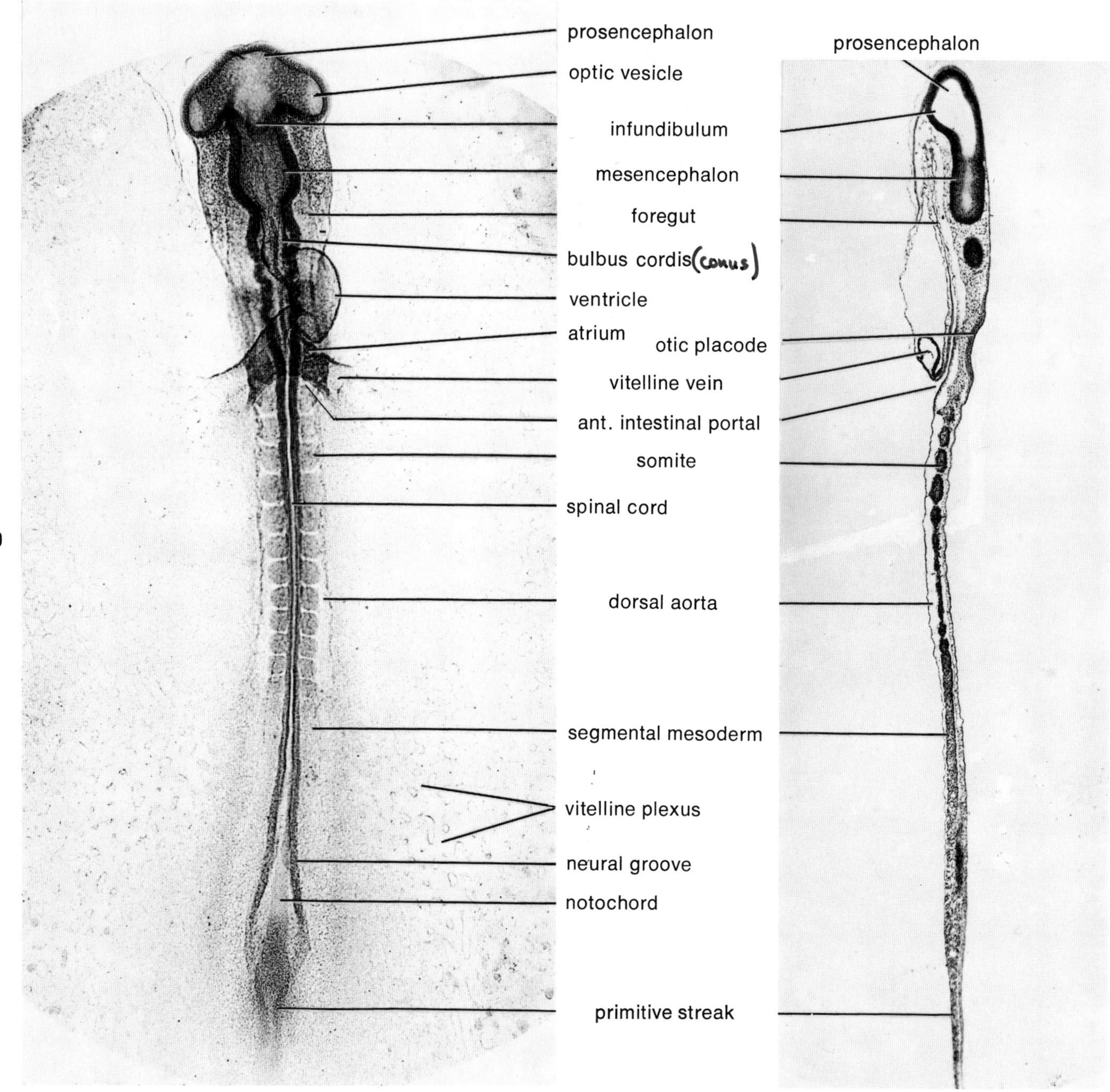

Figure 92 Chick embryo, stage 11, whole mount (mag. 30X)

Figure 93 Chick embryo, stage 11, sagittal section (mag. 30X)

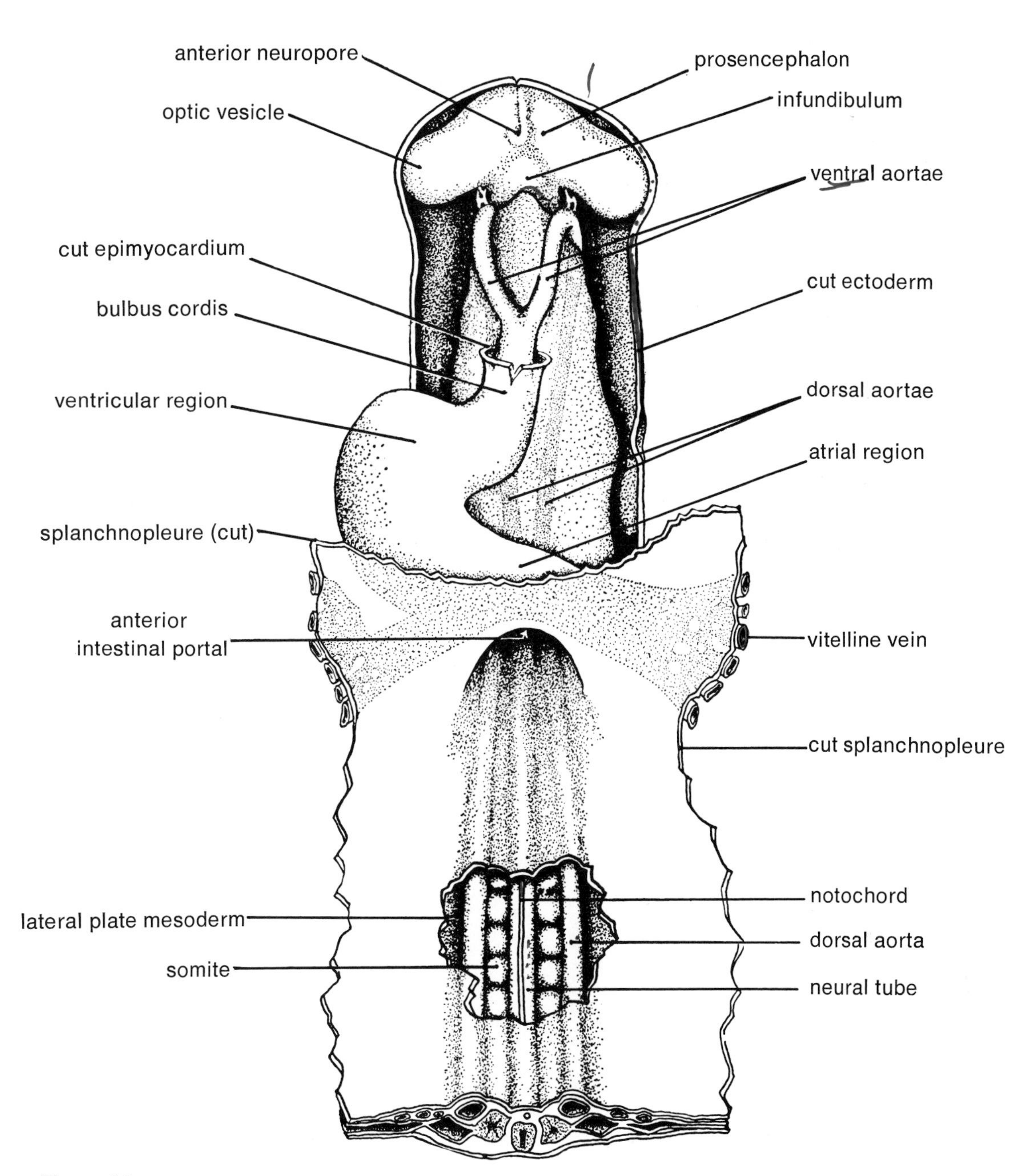

Figure 94

Diagrammatic ventral view of dissection of a stage 11 chick embryo. (Modified from Prentiss.) The splanchnopleure of the yolk sac cephalic to the anterior intestinal portal, the ectoderm of the ventral surface of the head, and the mesoderm of the pericardial region, have been removed to show the underlying structures. (From *Early Embryology of the Chick*, 4th ed., by Bradley M. Patten. Copyright 1957 by B. M. Patten. Used with permission of the McGraw-Hill Book Co.)

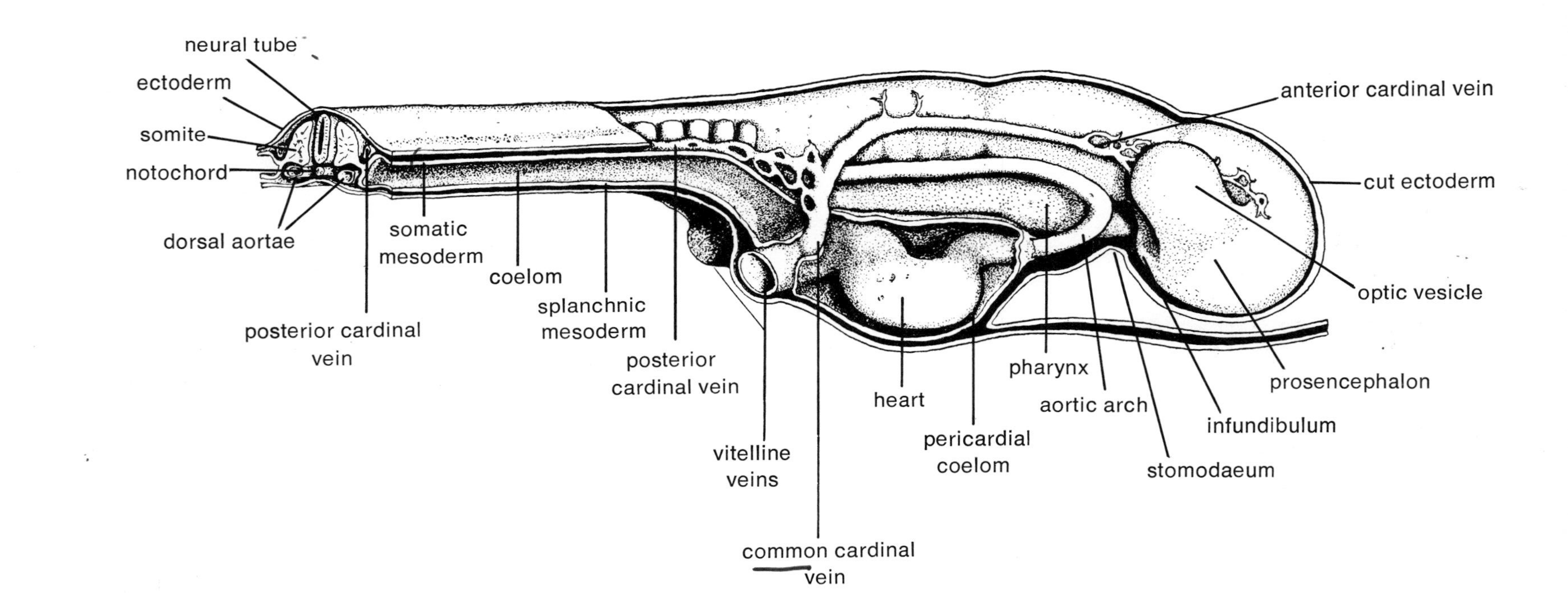

Figure 95

Diagrammatic lateral view of dissection of a stage 12 chick. The lateral body wall on the right side has been removed to show the internal structures. Note especially the relations of the pericardial region to that part of the coelom which lies farther caudally, and the small anastomosing channels of the developing posterior cardinal vein from which a single main vessel is later derived. (From *Early Embryology of the Chick*, 3rd ed., by Bradley M. Patten. Copyright 1929 by P. Blakiston's Sons & Co. Inc. Used with permission of the McGraw-Hill Book Co.)

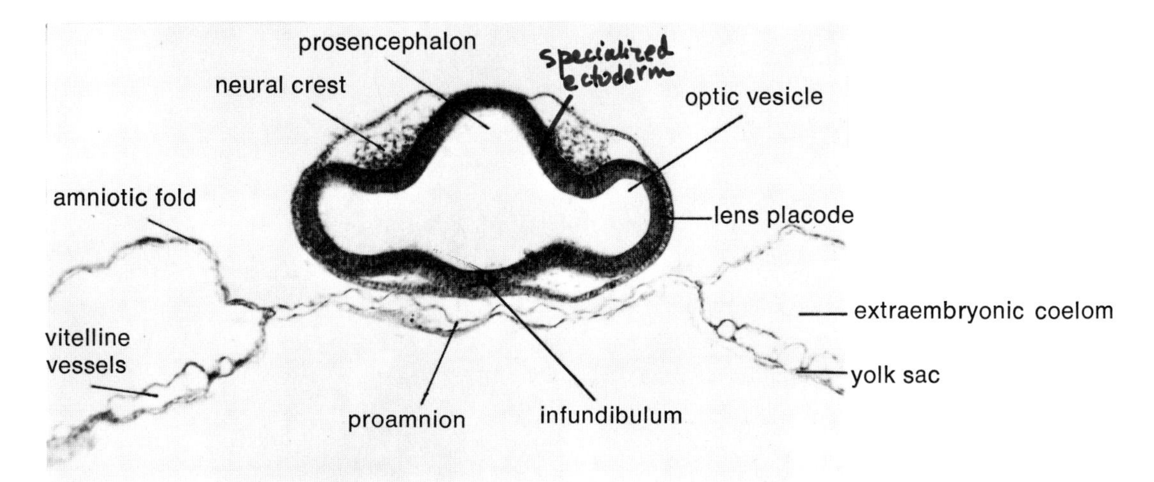

Figure 96 Chick embryo, stage 11, transverse section through optic vesicles (mag. 75X)

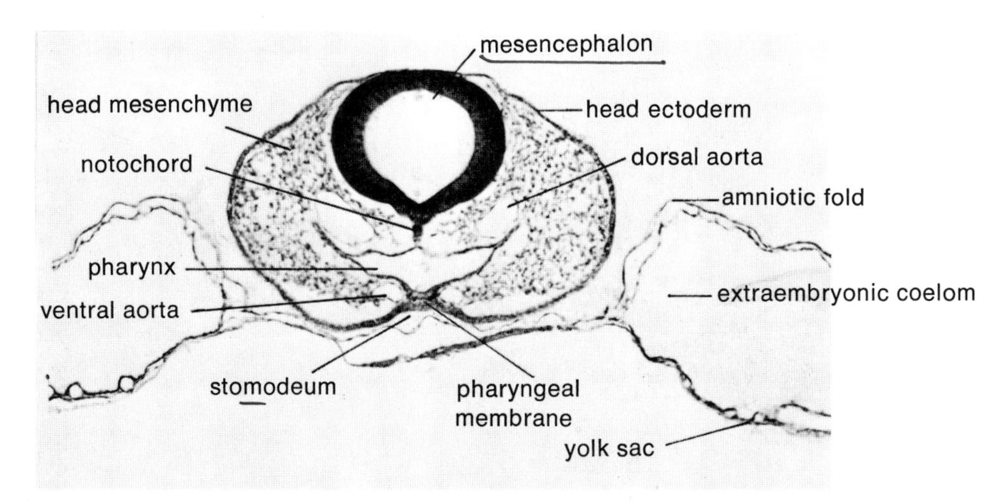

Figure 97 Chick embryo, stage 11, transverse section through pharyngeal membrane (mag. 75X)

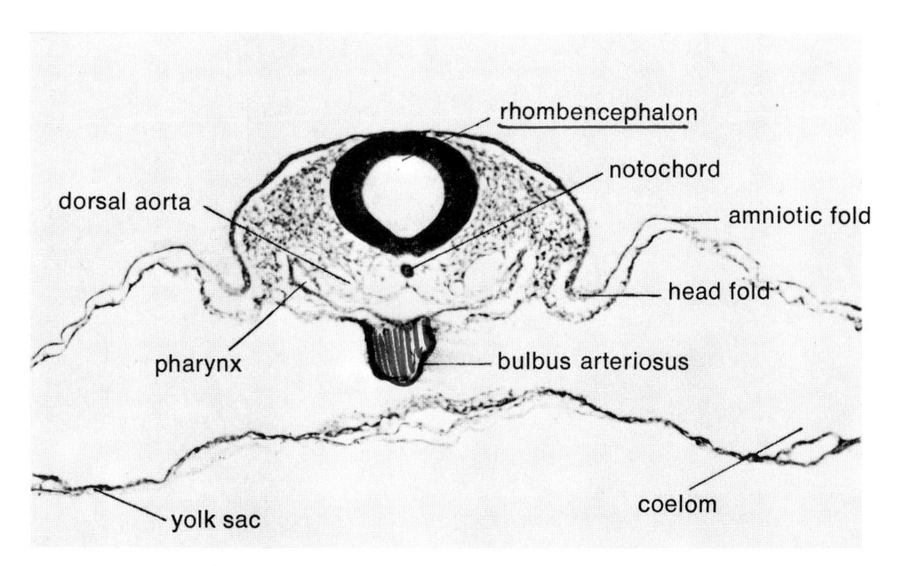

96
97
98

Figure 98 Chick embryo, stage 11, transverse section through bulbus arteriosus (cordis) (mag. 75X)

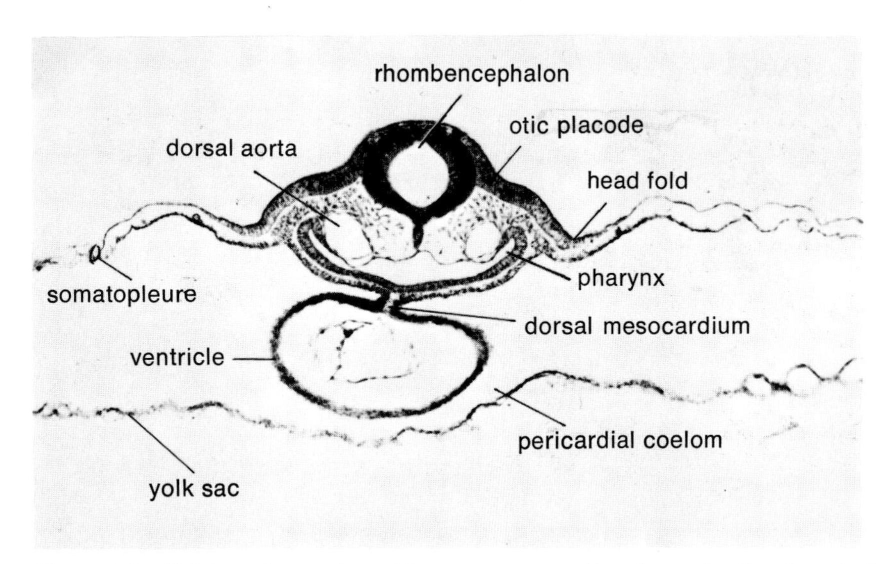

Figure 99 Chick embryo, stage 11, transverse section through otic placode (mag. 75X)

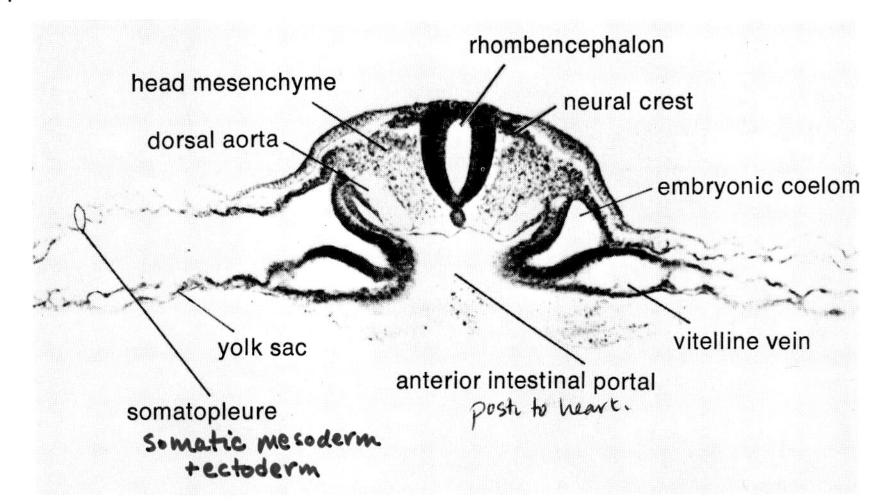

Figure 100 Chick embryo, stage 11, transverse section through anterior intestinal portal (mag. 75X)

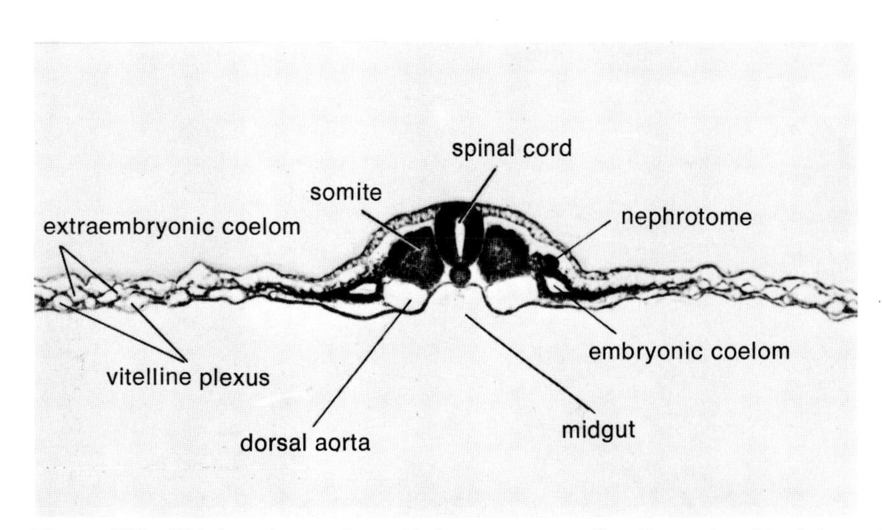

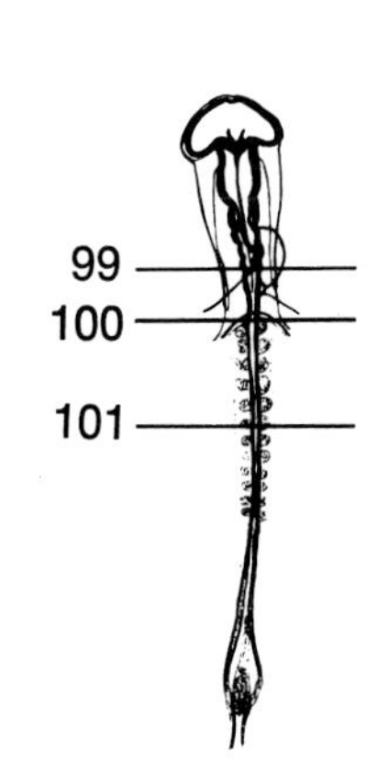

Figure 101 Chick embryo, stage 11, transverse section through midgut (mag. 75X)

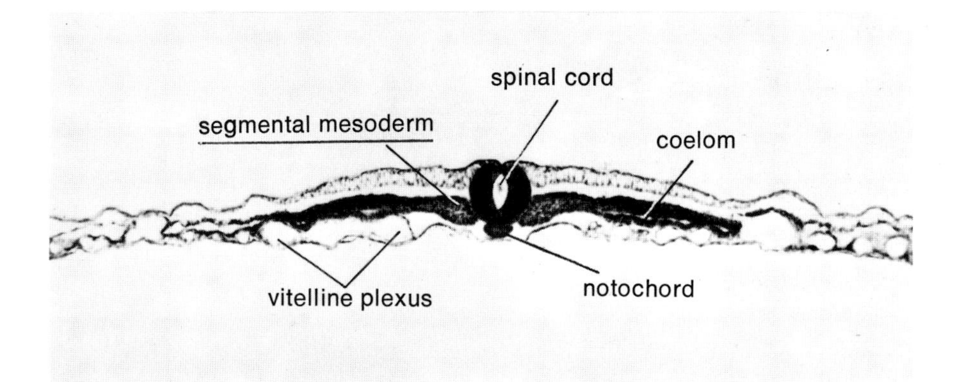

Figure 102 Chick embryo, stage 11, transverse section through segmental mesoderm (mag. 75X)

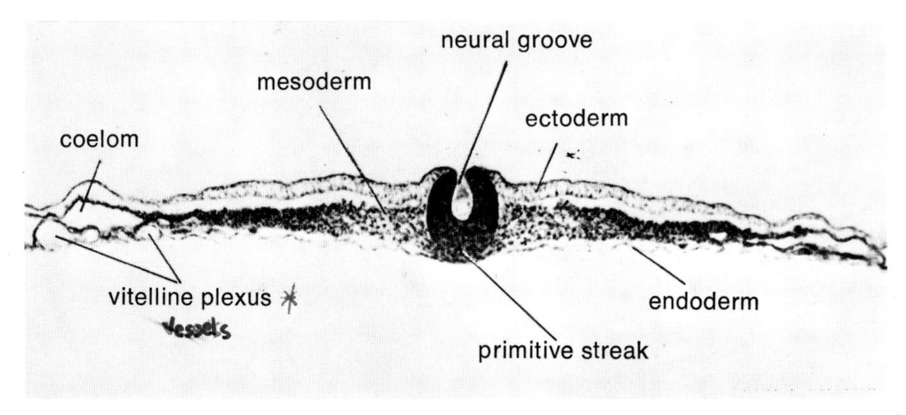

Figure 103 Chick embryo, stage 11, transverse section through neural groove (mag. 75X)

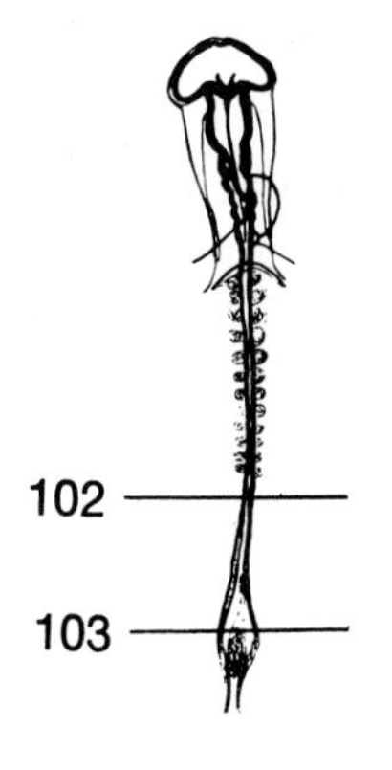

12. The Stage 15 Chick Embryo (50-55 hours incubation)

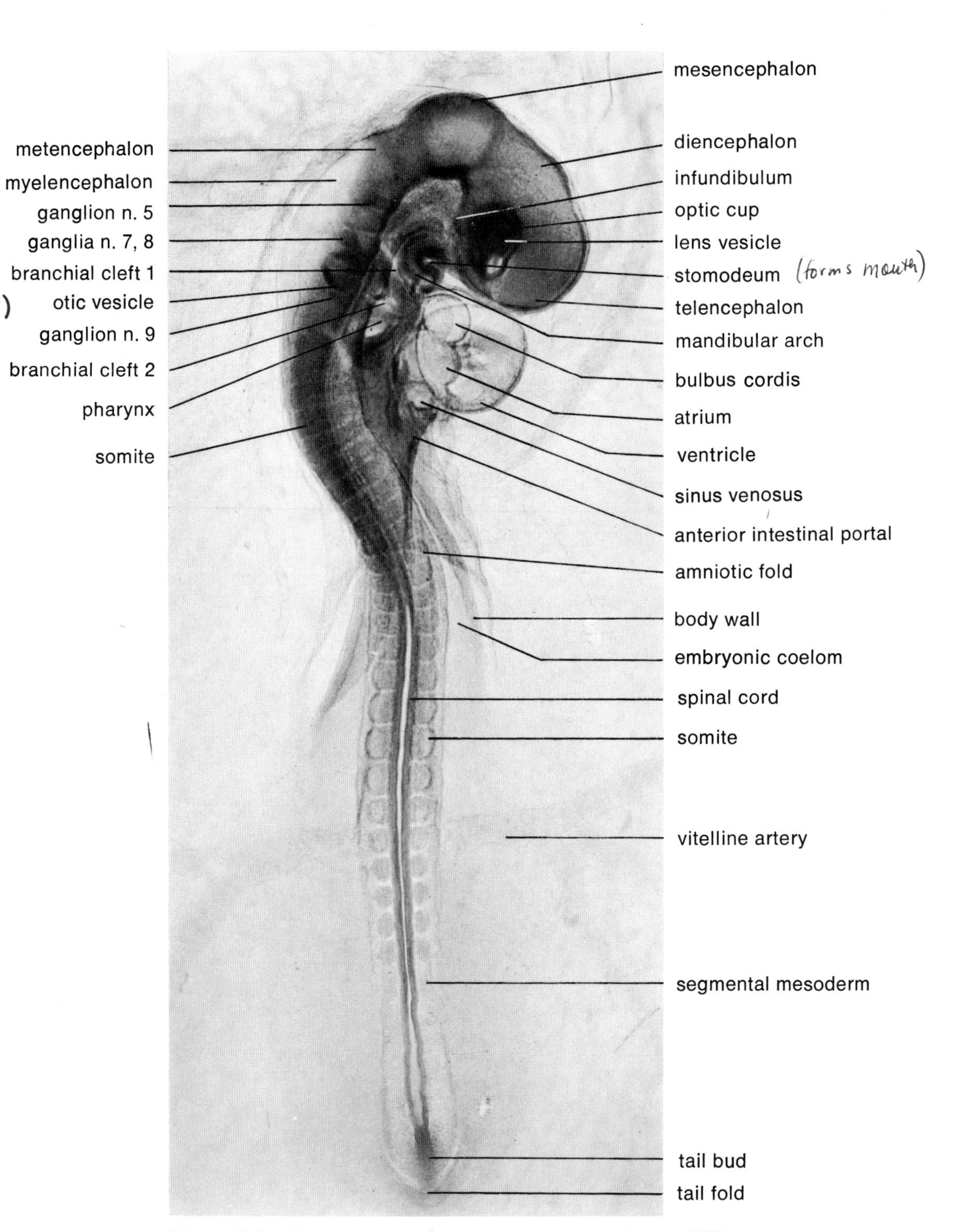

Figure 104 Chick embryo, stage 15, whole mount (mag. 25X)

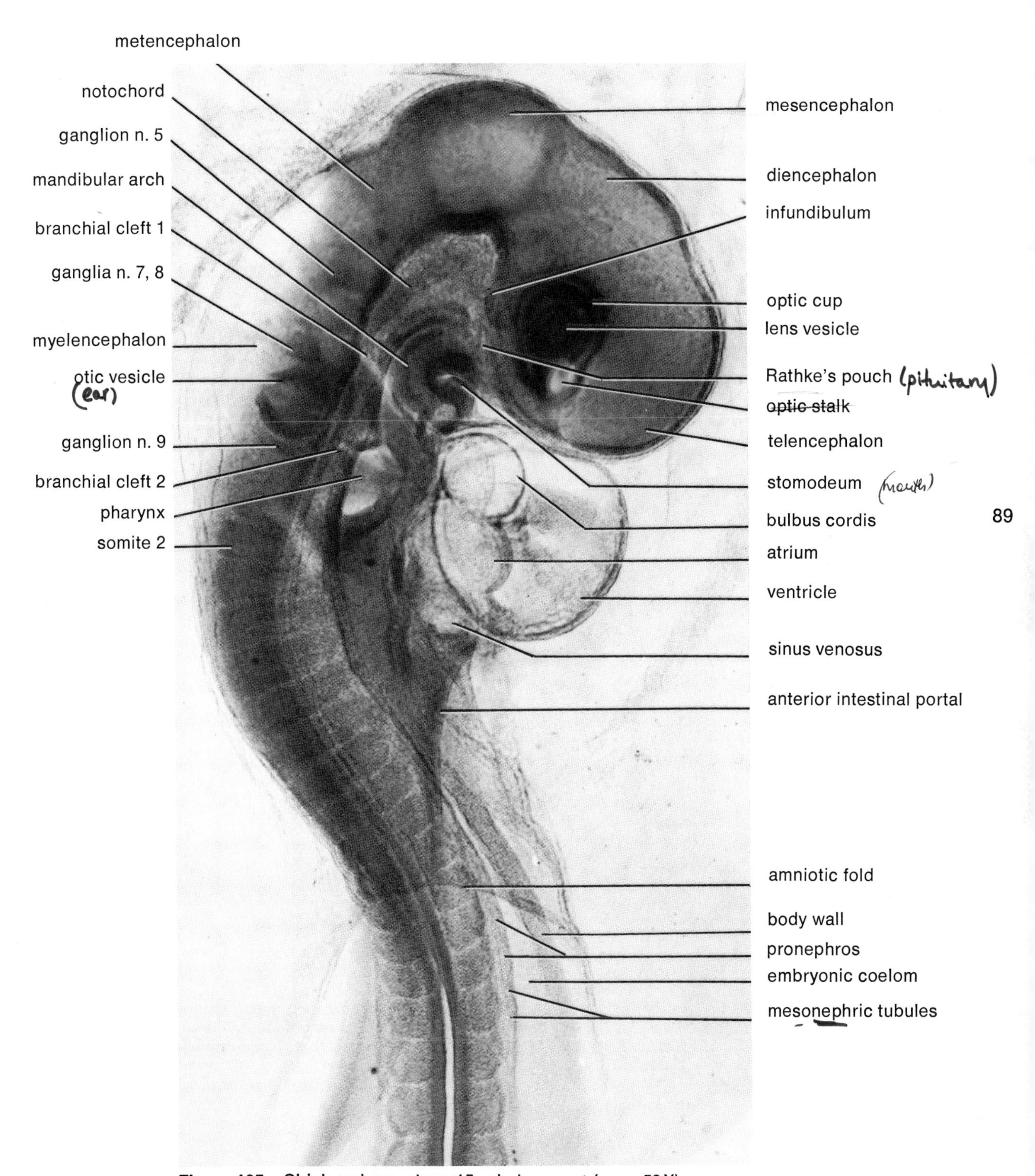

Figure 105 Chick embryo, stage 15, whole mount (mag. 50 X)

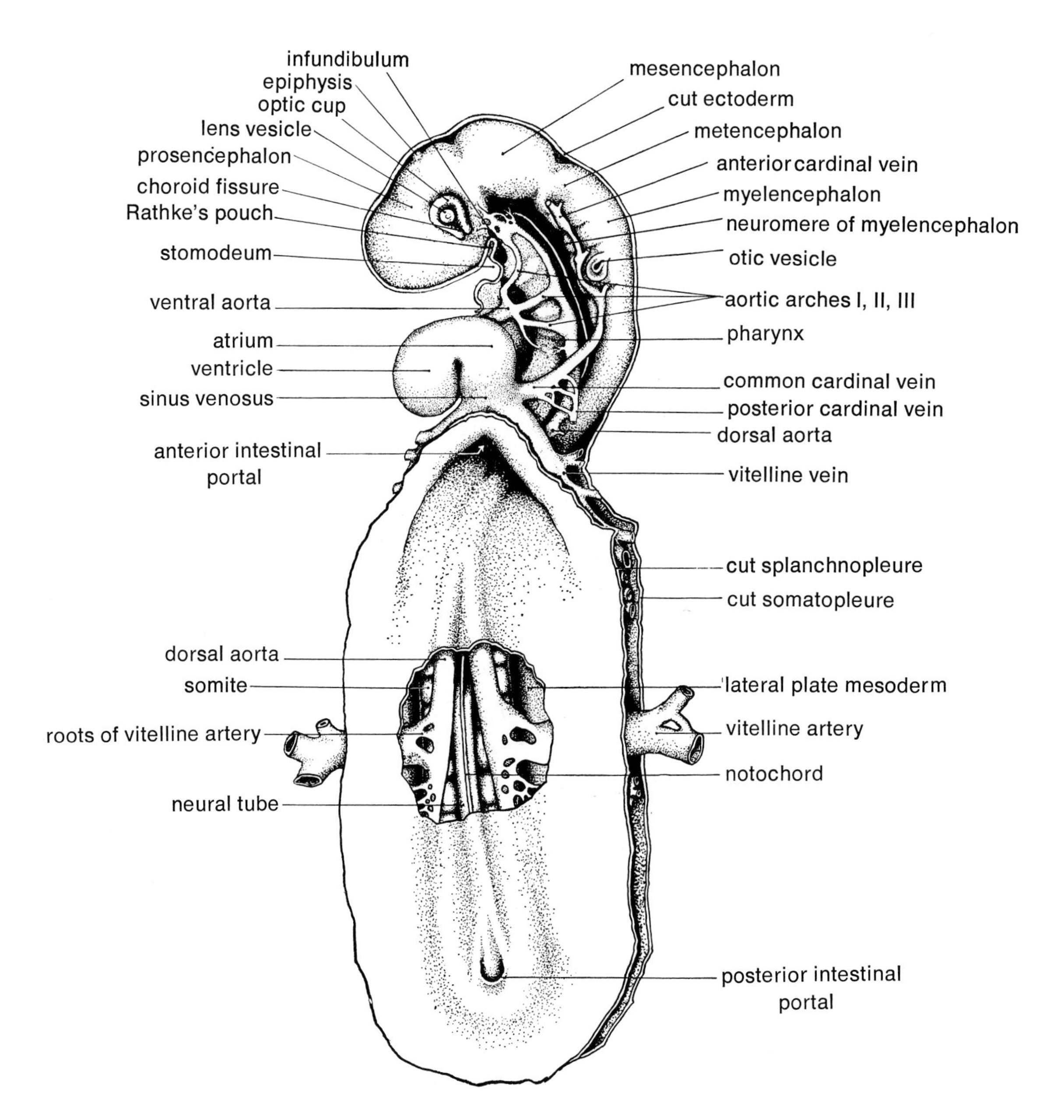

Figure 106

Diagram of dissection of chick of about 50 hours. (Modified from Prentiss.) The splanchnopleure of the yolk sac cephalic to the anterior intestinal portal, the ectoderm of the left side of the head, and the mesoderm in the pericardial region have been dissected away. A window has been cut in the splanchnopleure of the dorsal wall of the mid gut to show the origin of the vitelline artery. (From *Early Embryology of the Chick*, 4th ed., by Bradley M. Patten. Copyright 1957 by B. M. Patten. Used with permission of the McGraw-Hill Book Co.)

Figure 107 Chick embryo, stage 15, whole mount, blood vessels injected (mag. 25X)

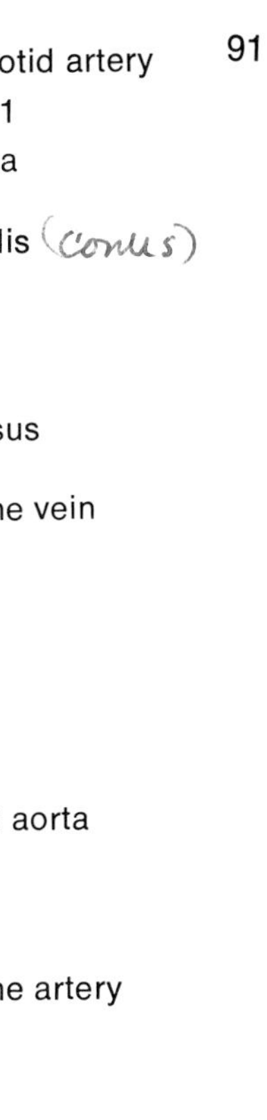

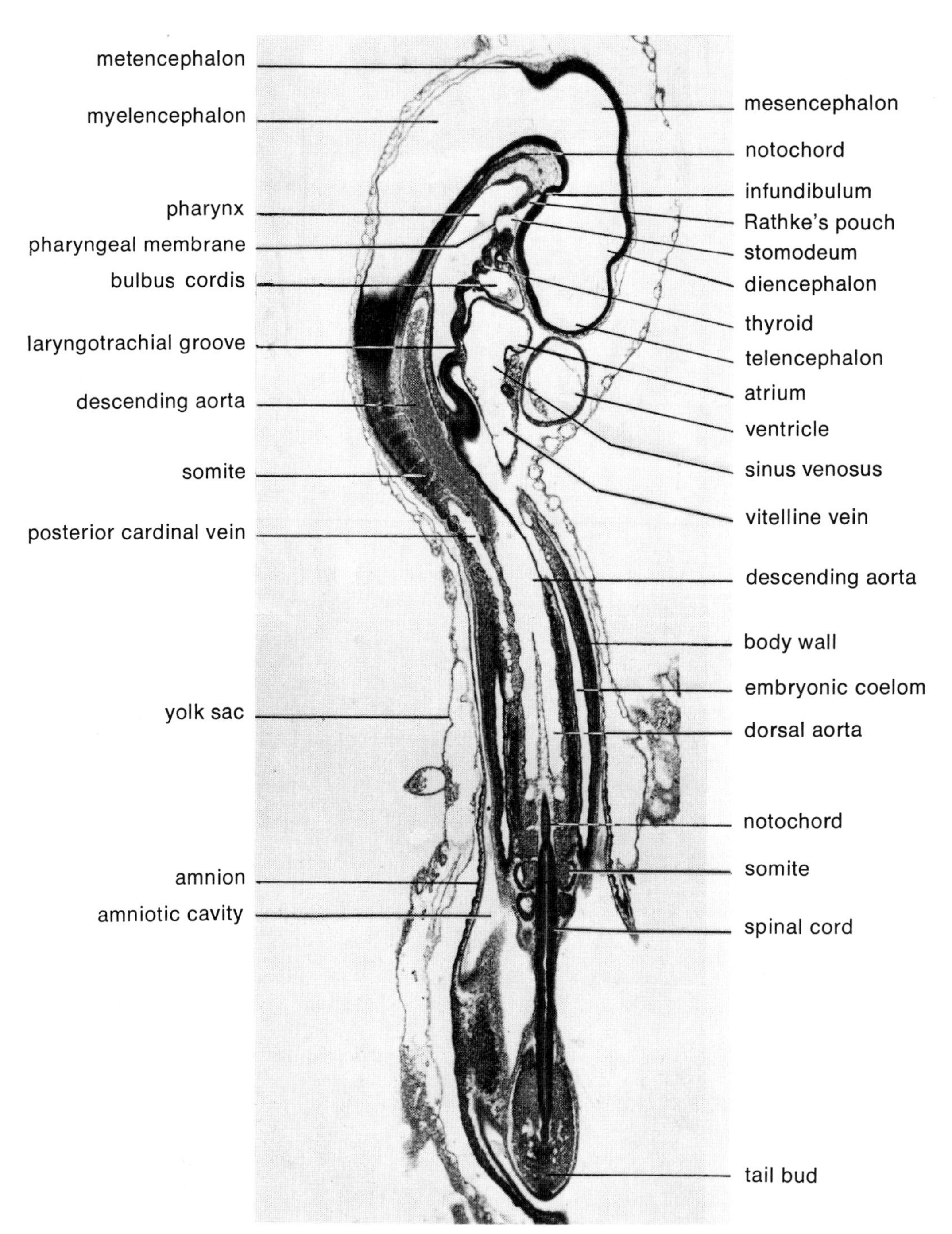

Figure 108 Chick embryo, stage 15, sagittal section (mag. 25X)

If see vitelline vesicles: left side w/yolk hanging

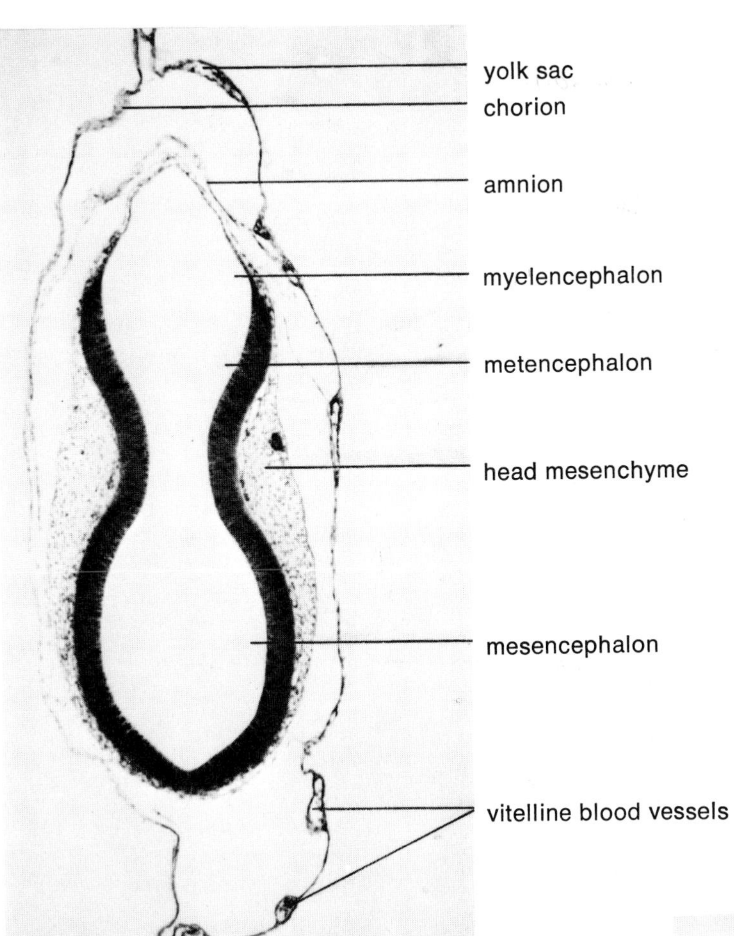

Figure 109 Chick embryo, stage 15, transverse section through mesencephalon (mag. 60X)

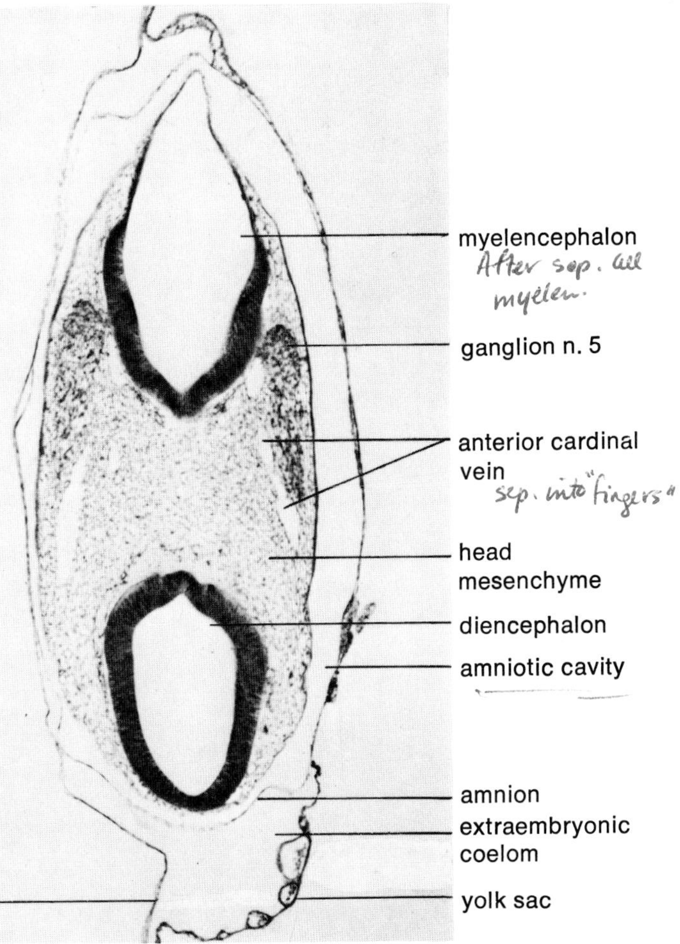

Figure 110
Chick embryo, stage 15, transverse section through ganglion of cranial nerve 5 (mag. 60X)

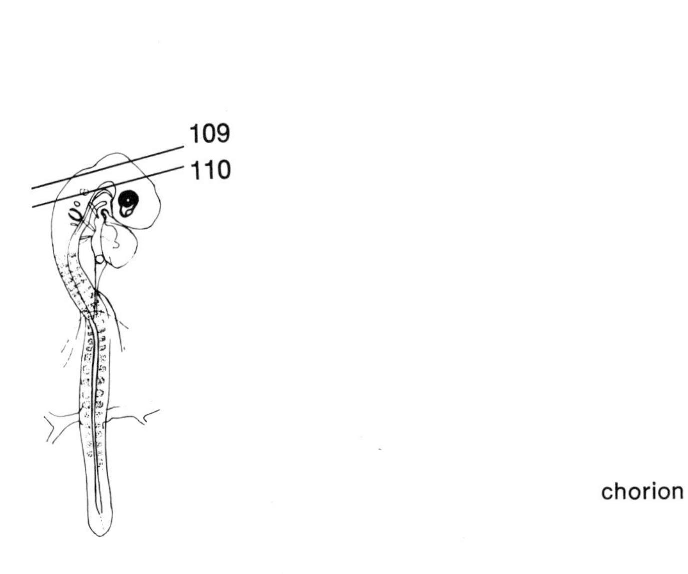

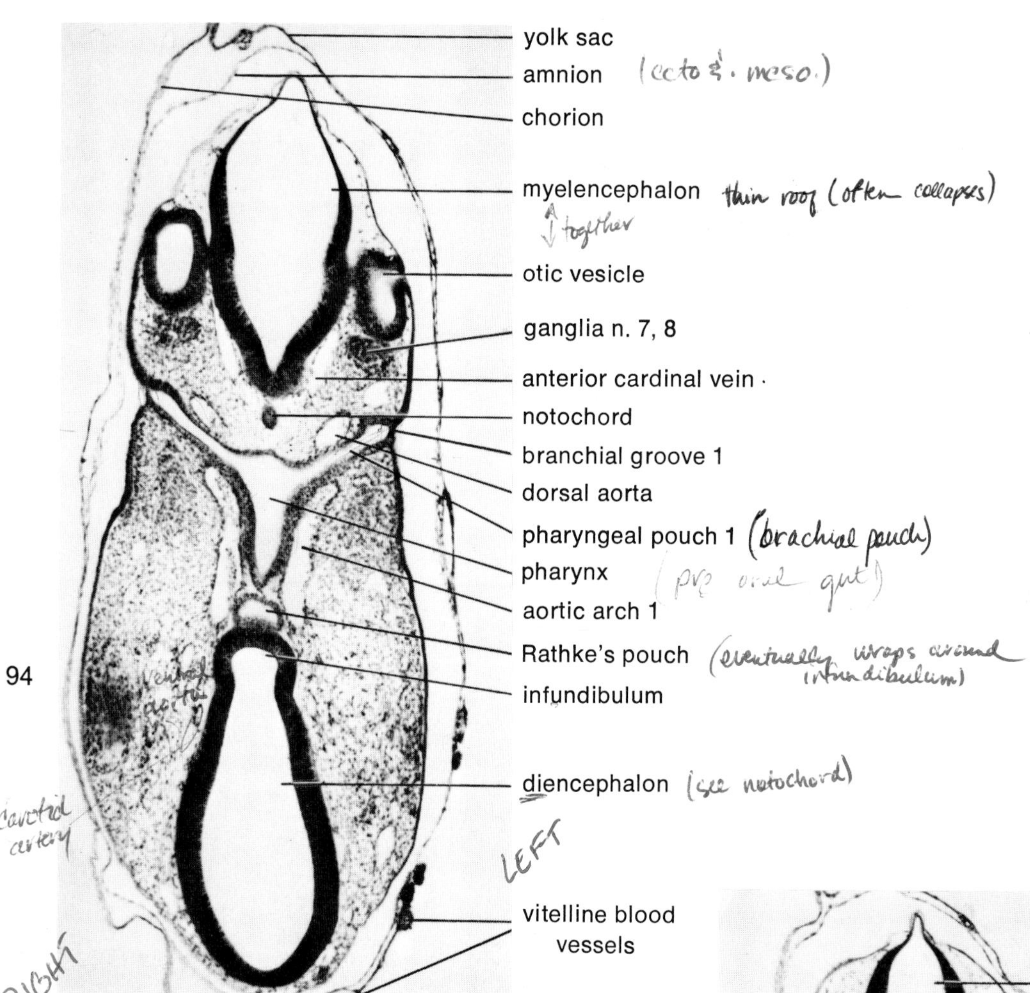

Figure 111 Chick embryo, stage 15, transverse section through otic vesicle (mag. 60X)

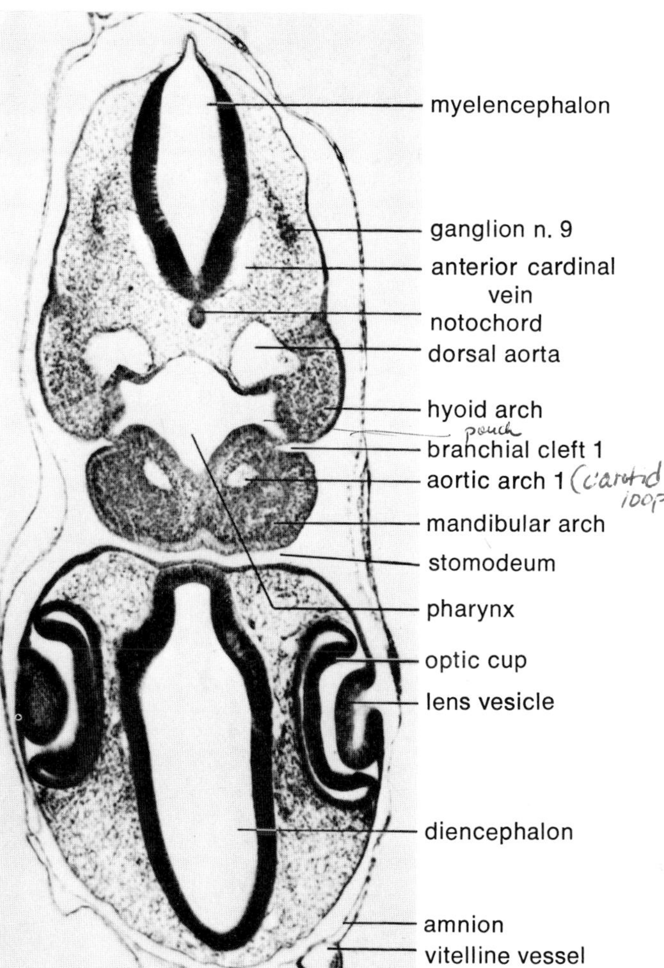

Figure 112
Chick embryo, stage 15, transverse section through optic cups (mag. 60X)

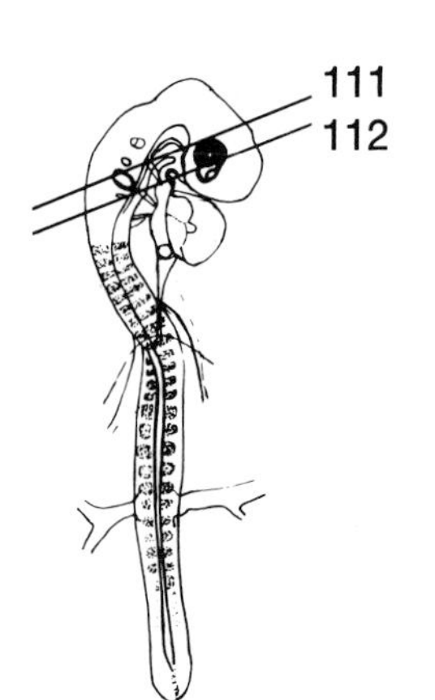

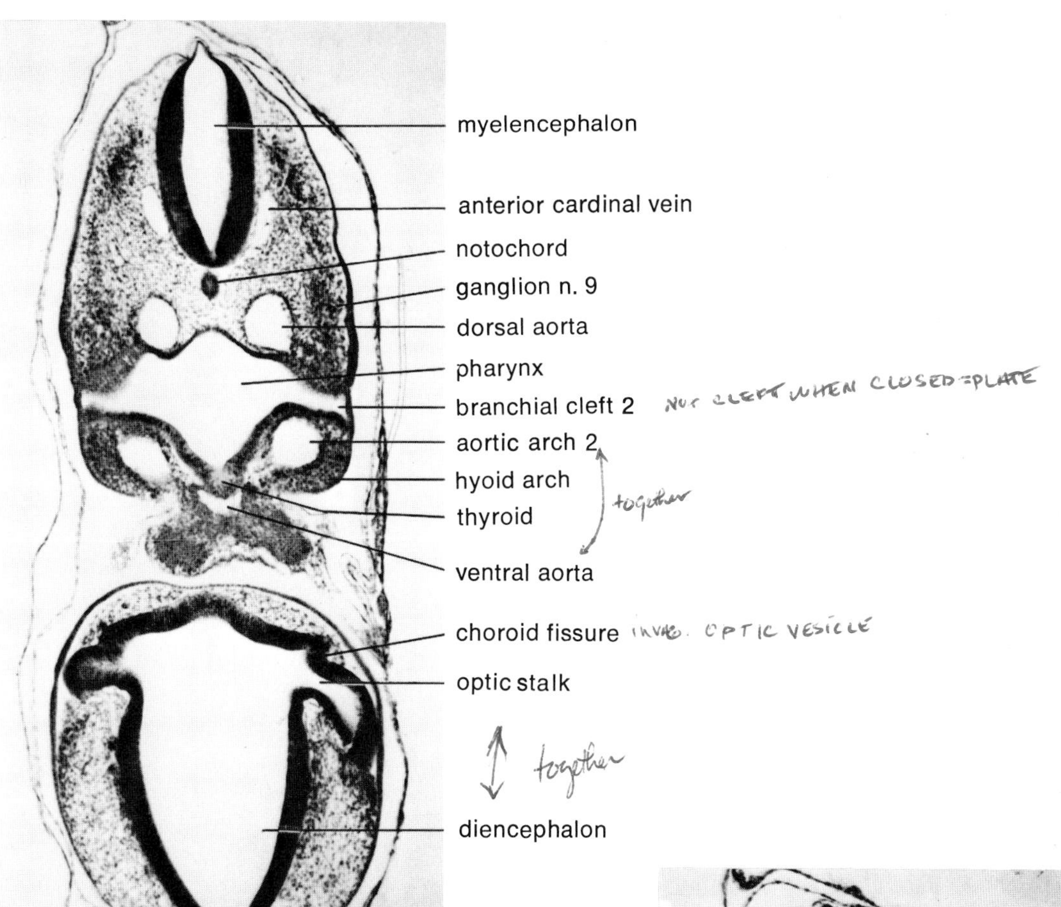

Figure 113 Chick embryo, stage 15, transverse section through thyroid (mag. 60X)

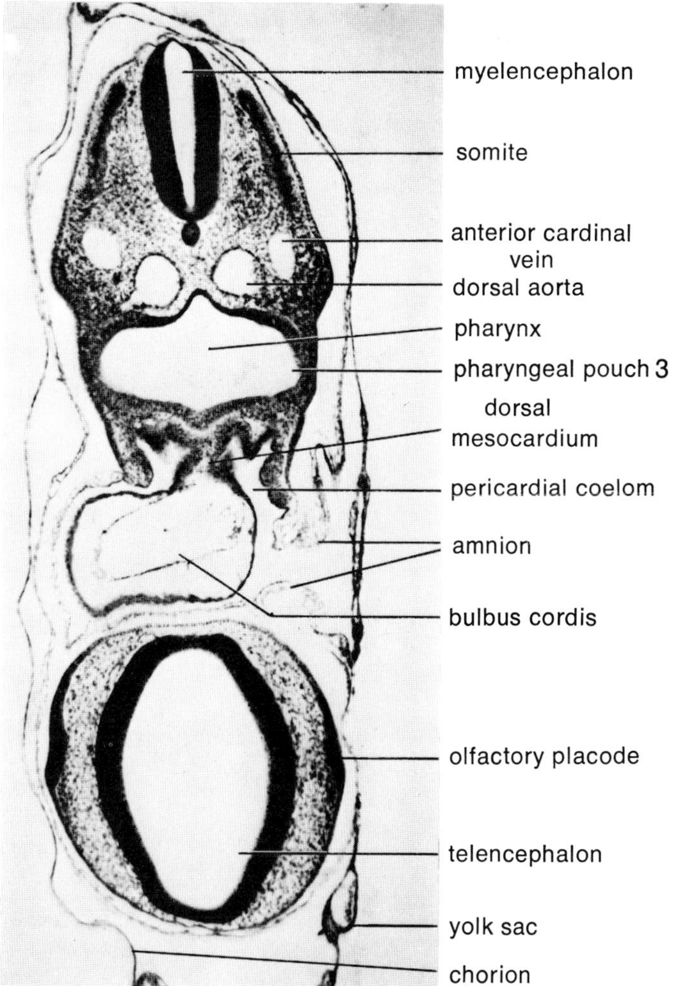

Figure 114
Chick embryo, stage 15, transverse section through olfactory placodes (mag. 60X)

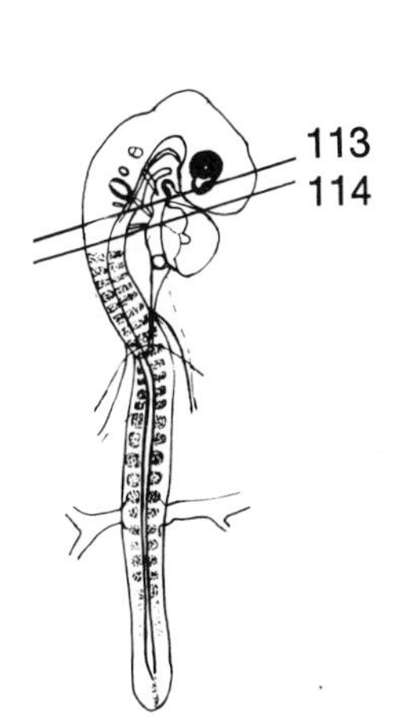

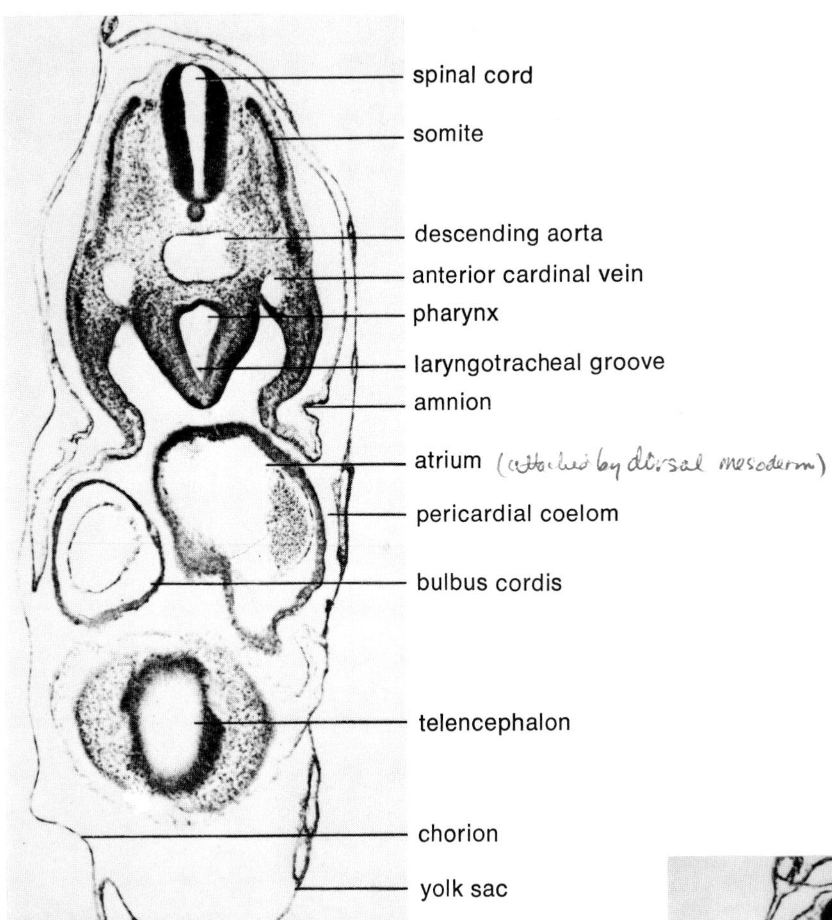

Figure 115 Chick embryo, stage 15, transverse section through atrium (mag. 60X)

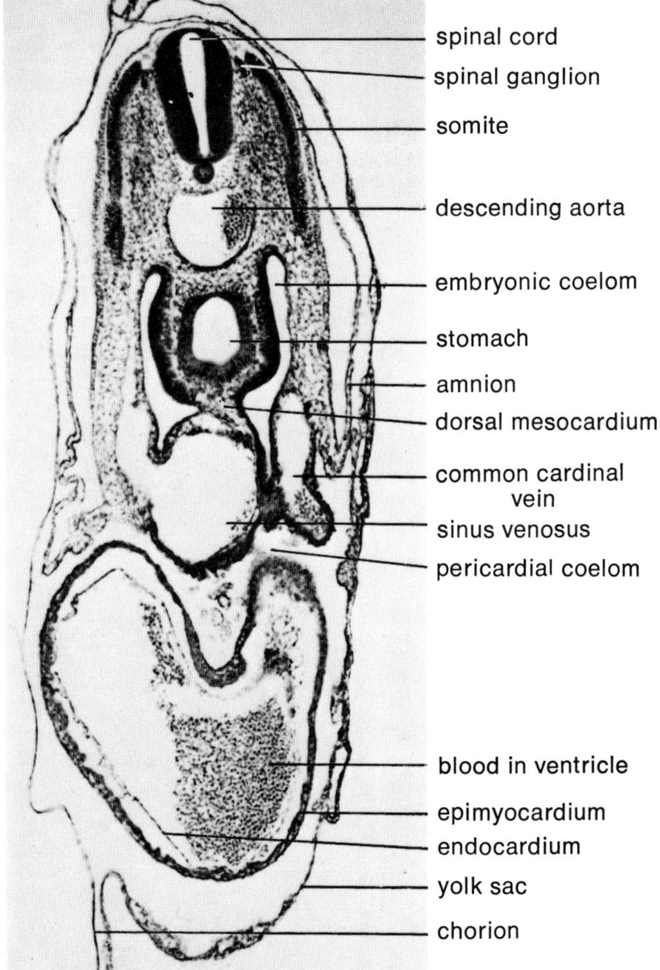

Figure 116
Chick embryo, stage 15, transverse section through sinus venosus (mag. 60X)

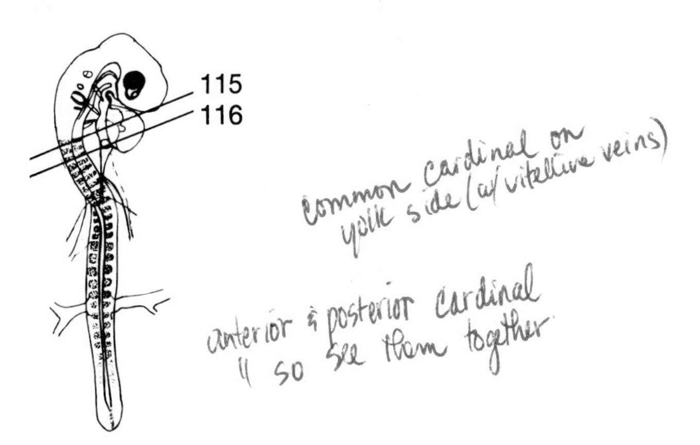

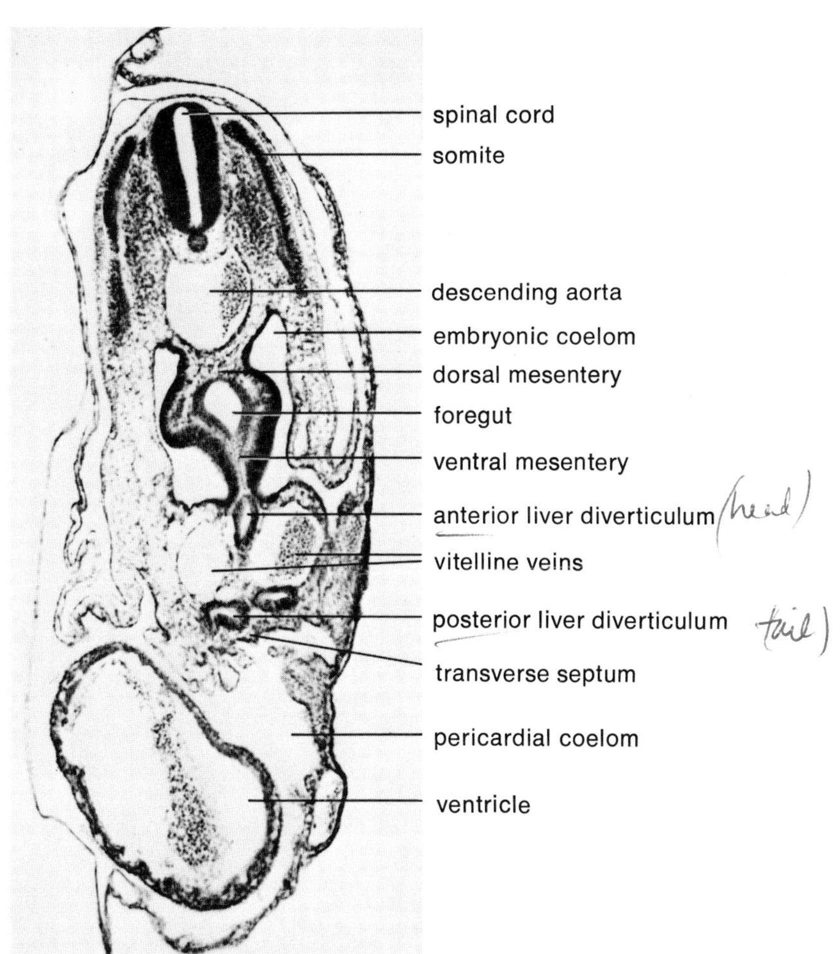

Figure 117 Chick embryo, stage 15, transverse section through liver diverticula (mag. 60X)

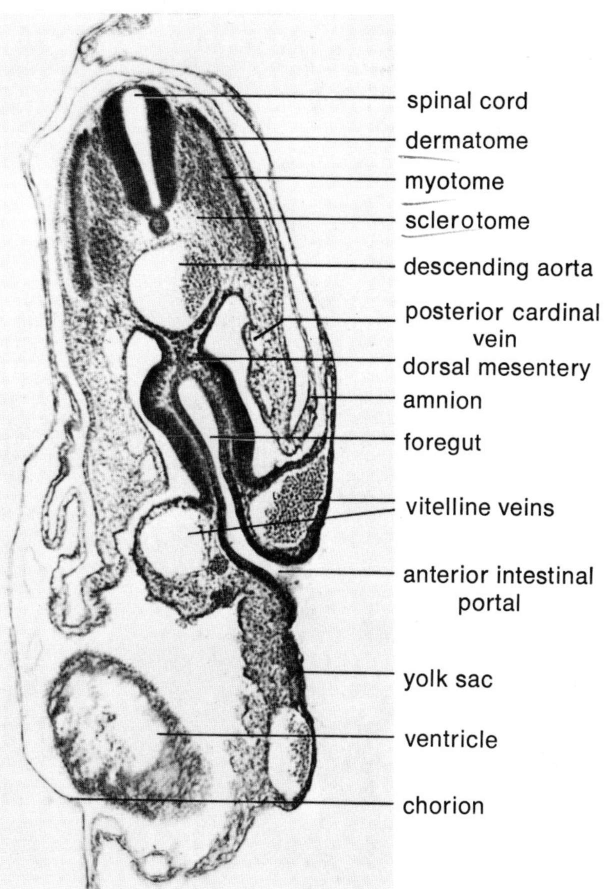

Figure 118
Chick embryo, stage 15, transverse section through anterior intestinal portal (mag. 60X)

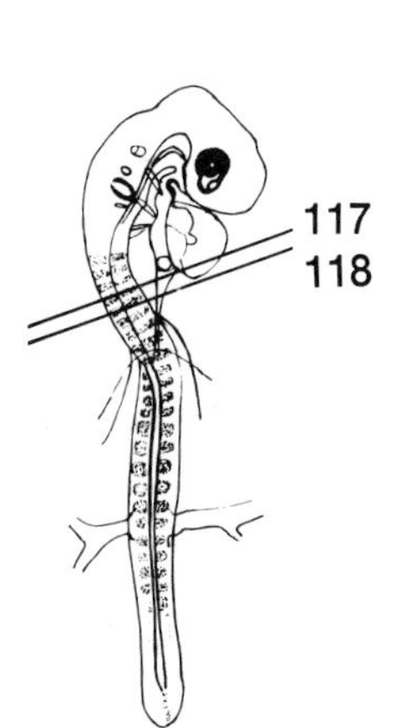

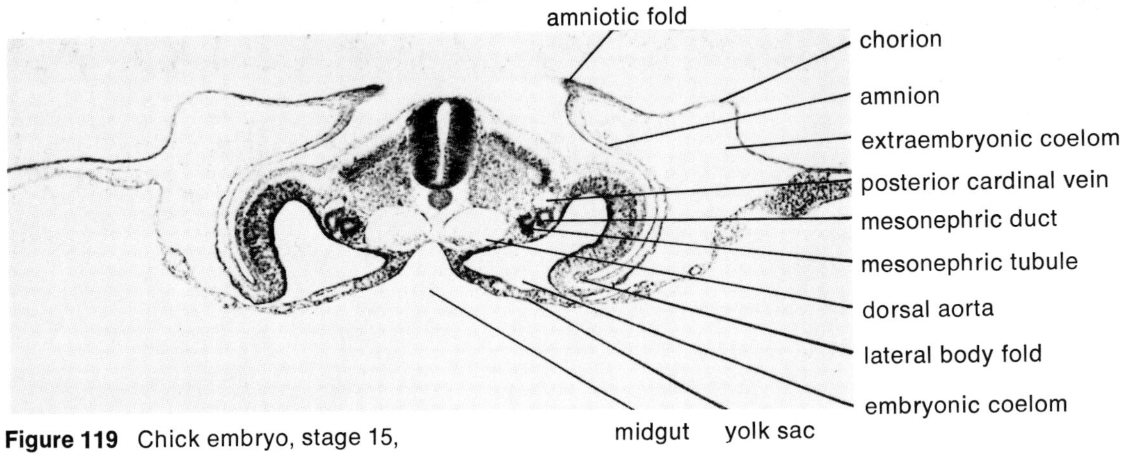

Figure 119 Chick embryo, stage 15, transverse section through mesonephros (mag. 60X)

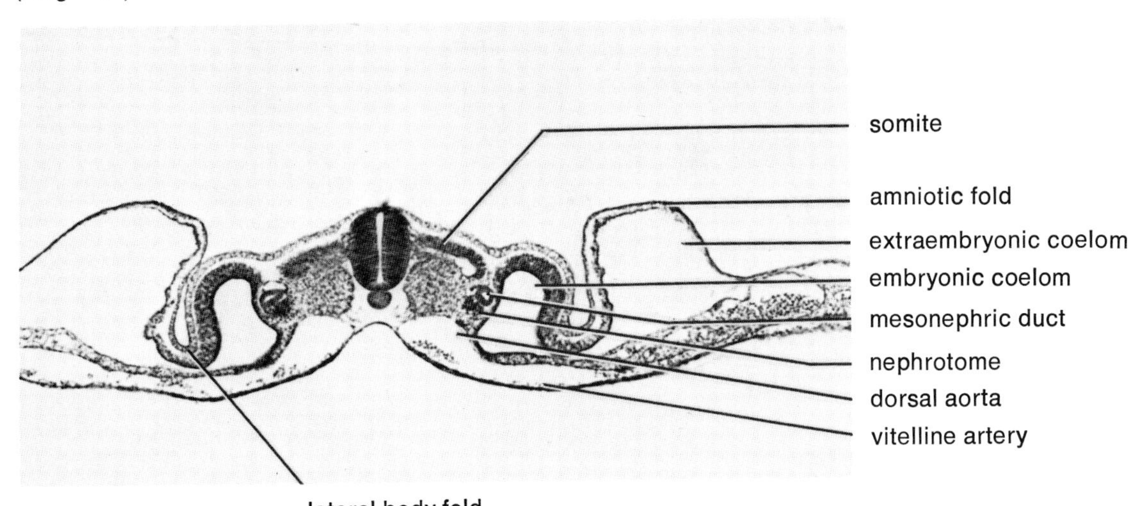

Figure 120 Chick embryo, stage 15, transverse section through vitelline artery (mag. 60 X)

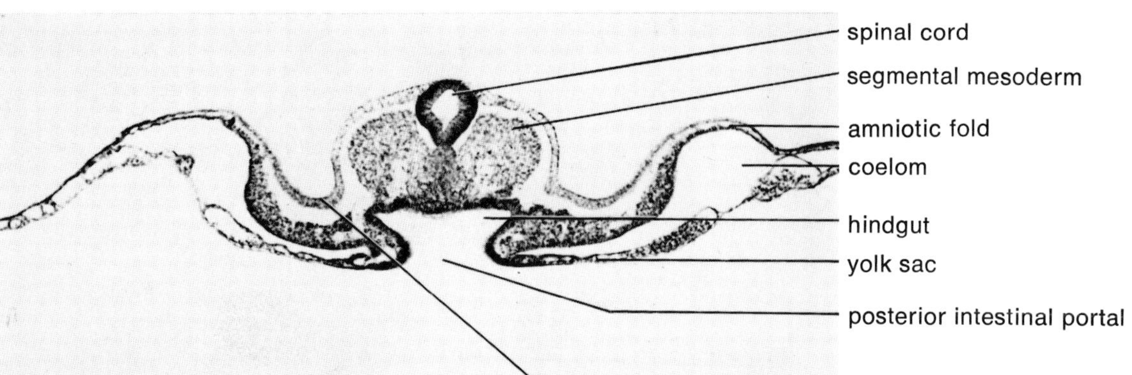

Figure 121 Chick embryo, stage 15, transverse section through posterior intestinal portal (mag. 60X)

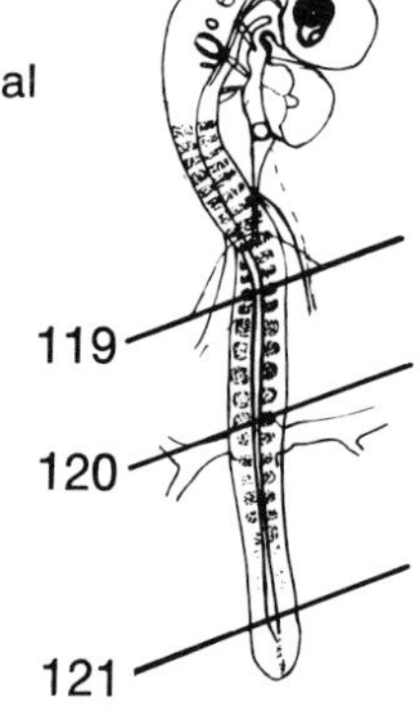

13. The Stage 18 Chick Embryo (3 days incubation)

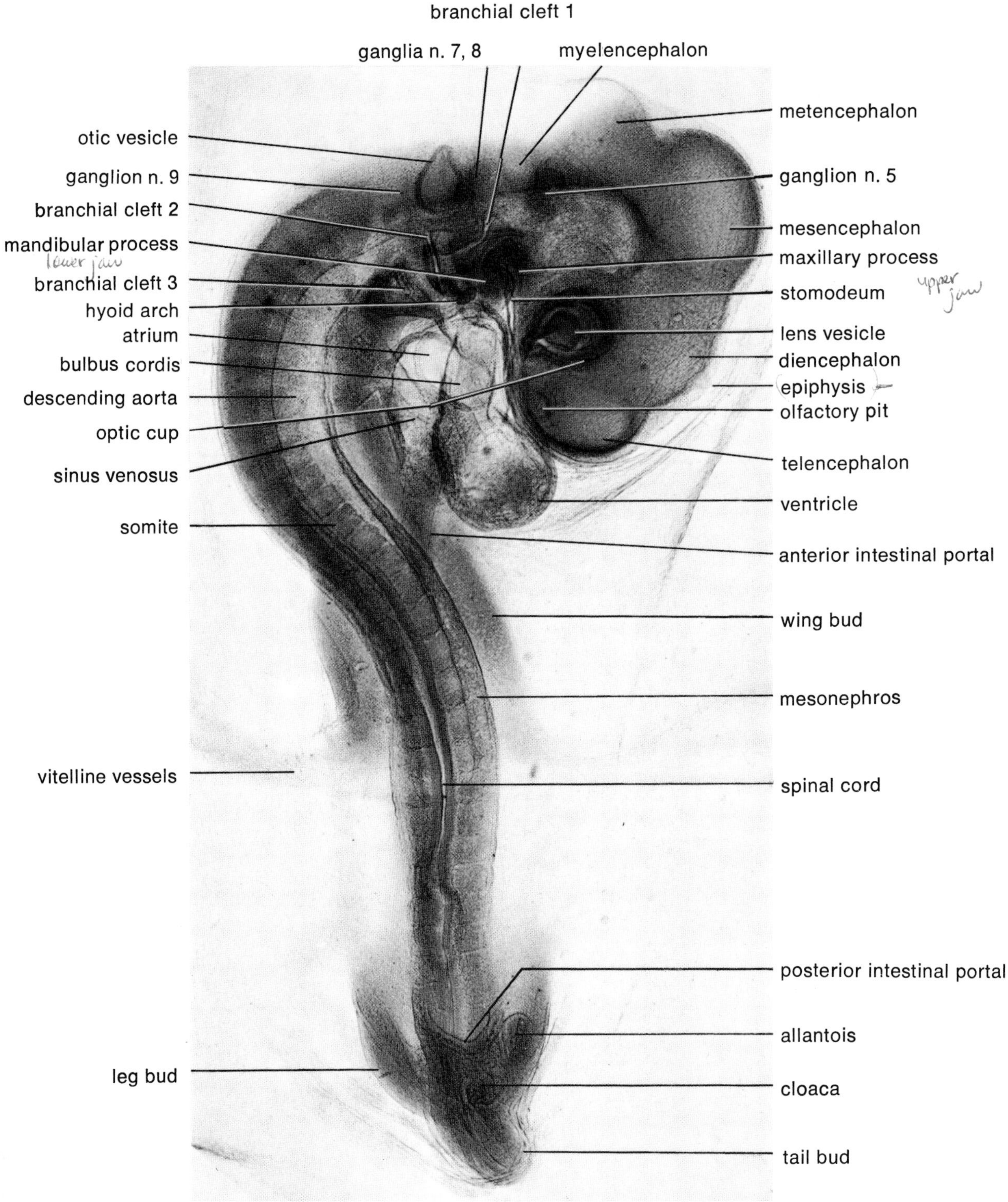

branchial cleft 1
ganglia n. 7, 8
myelencephalon
otic vesicle
ganglion n. 9
branchial cleft 2
mandibular process
branchial cleft 3
hyoid arch
atrium
bulbus cordis
descending aorta
optic cup
sinus venosus
somite
vitelline vessels
leg bud
metencephalon
ganglion n. 5
mesencephalon
maxillary process
stomodeum
lens vesicle
diencephalon
epiphysis
olfactory pit
telencephalon
ventricle
anterior intestinal portal
wing bud
mesonephros
spinal cord
posterior intestinal portal
allantois
cloaca
tail bud

Figure 122 Chick embryo, stage 18, whole mount (mag. 25X; transmitted illumination)

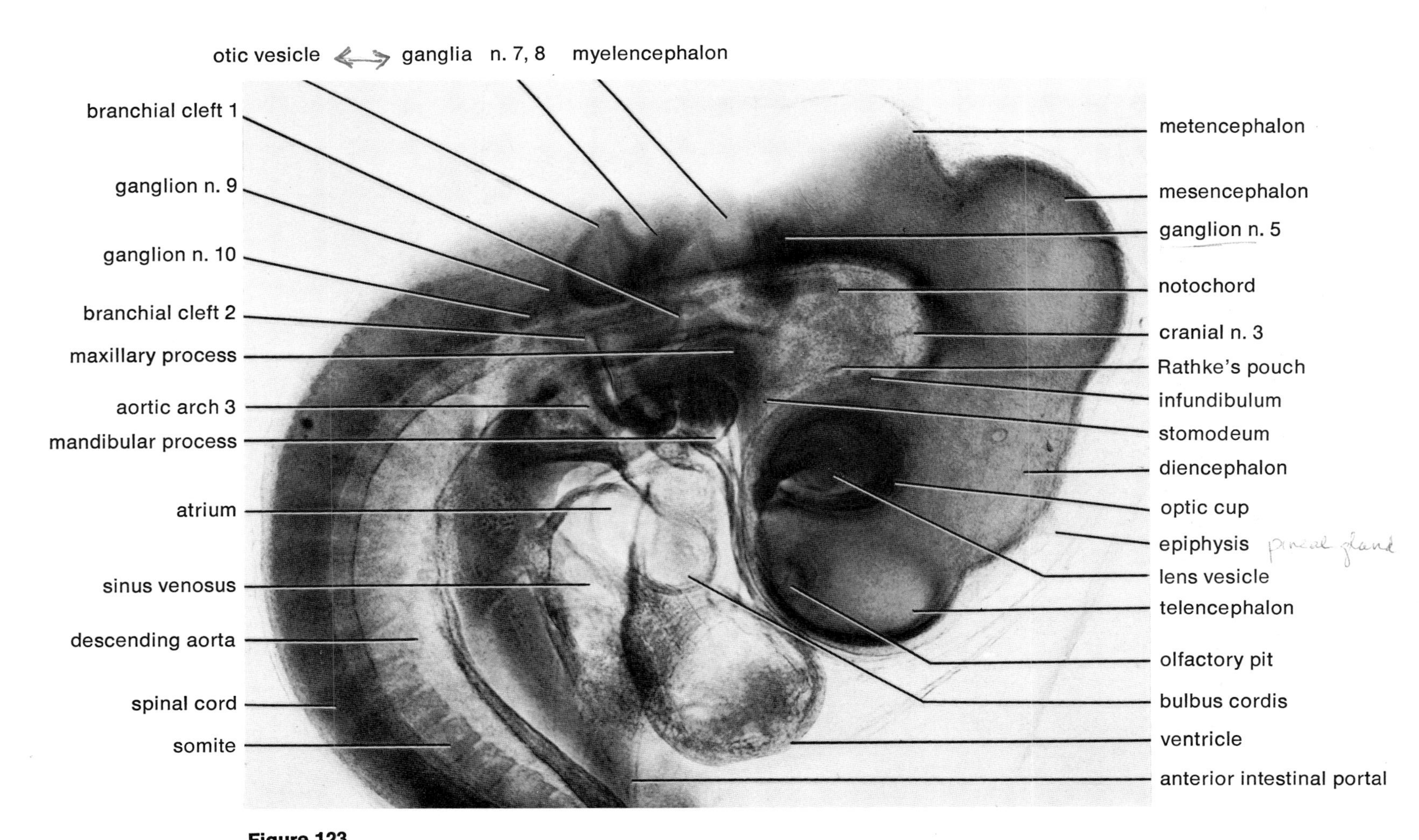

Figure 123
Chick embryo, stage 18, whole mount (mag. 50×)

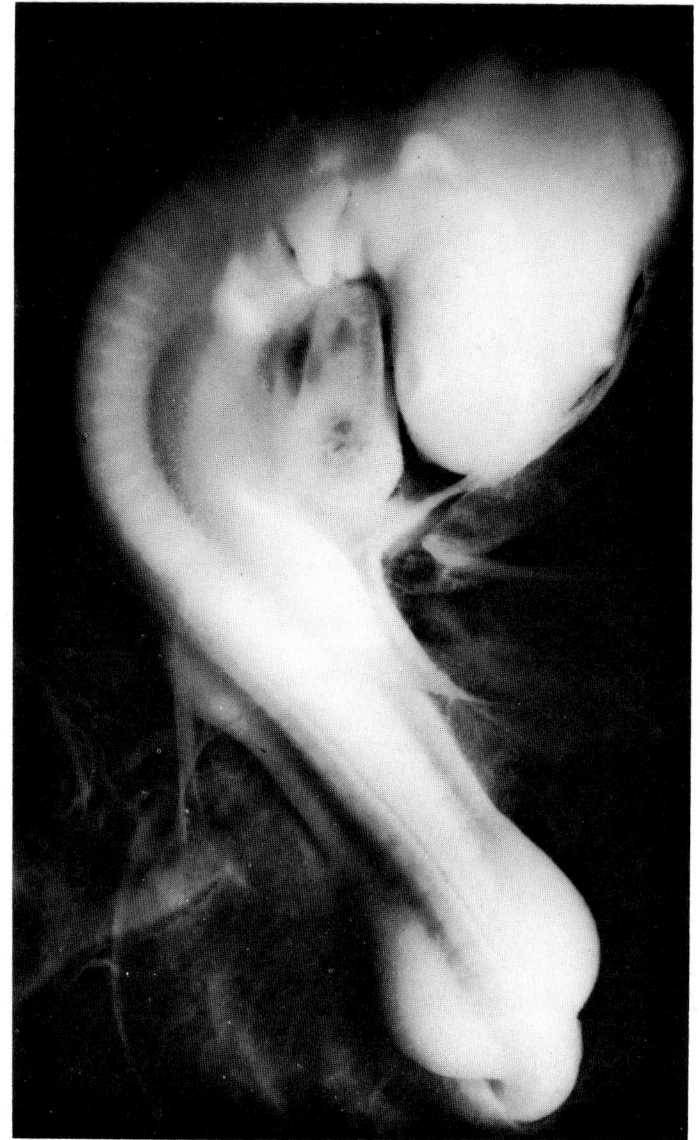

Figure 124 Chick embryo, stage 19, opaque mount, unstained, incident illumination

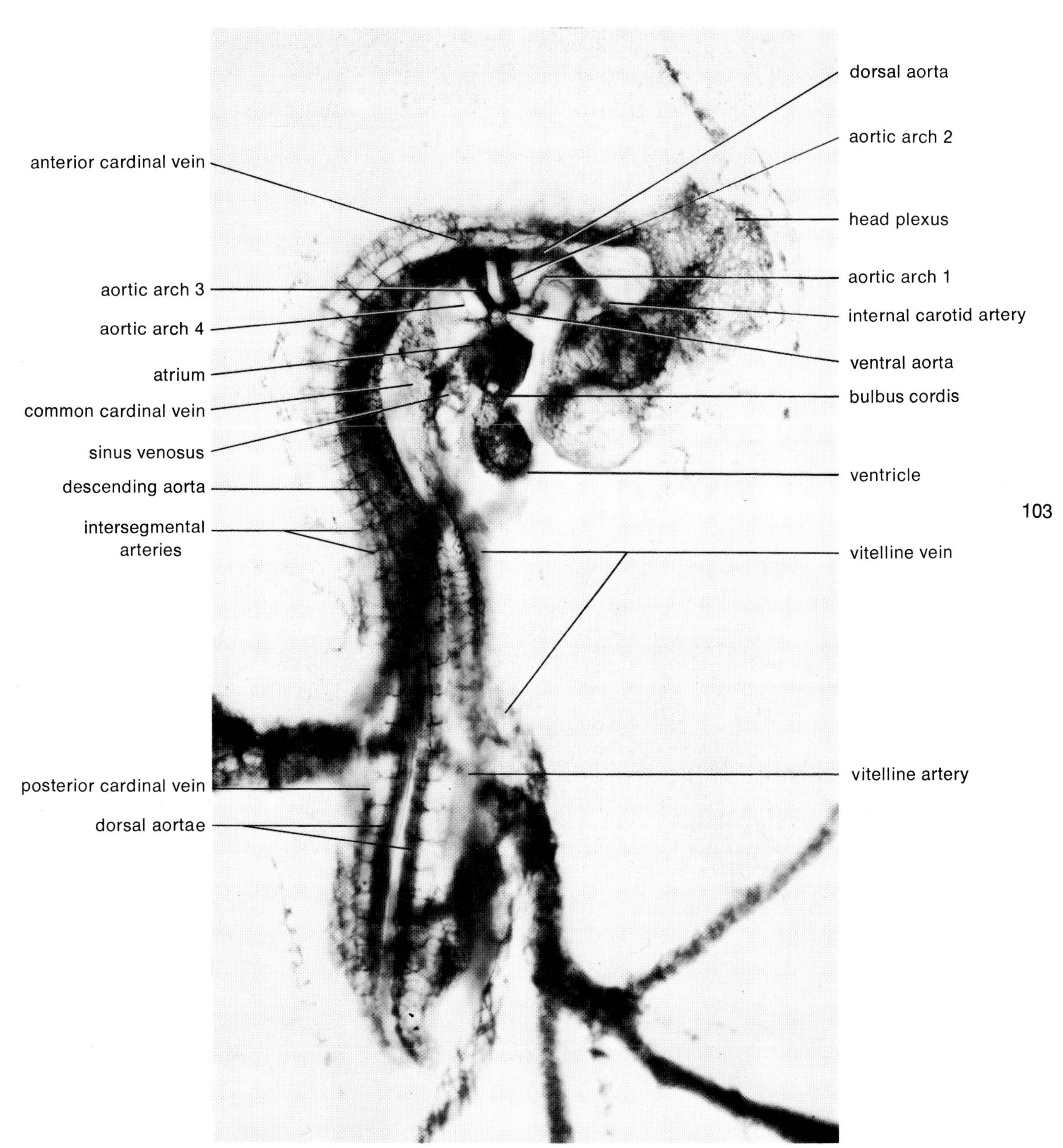

Figure 125 Chick embryo, stage 18, whole mount, blood vessels injected (mag. 25X; transmitted illumination)

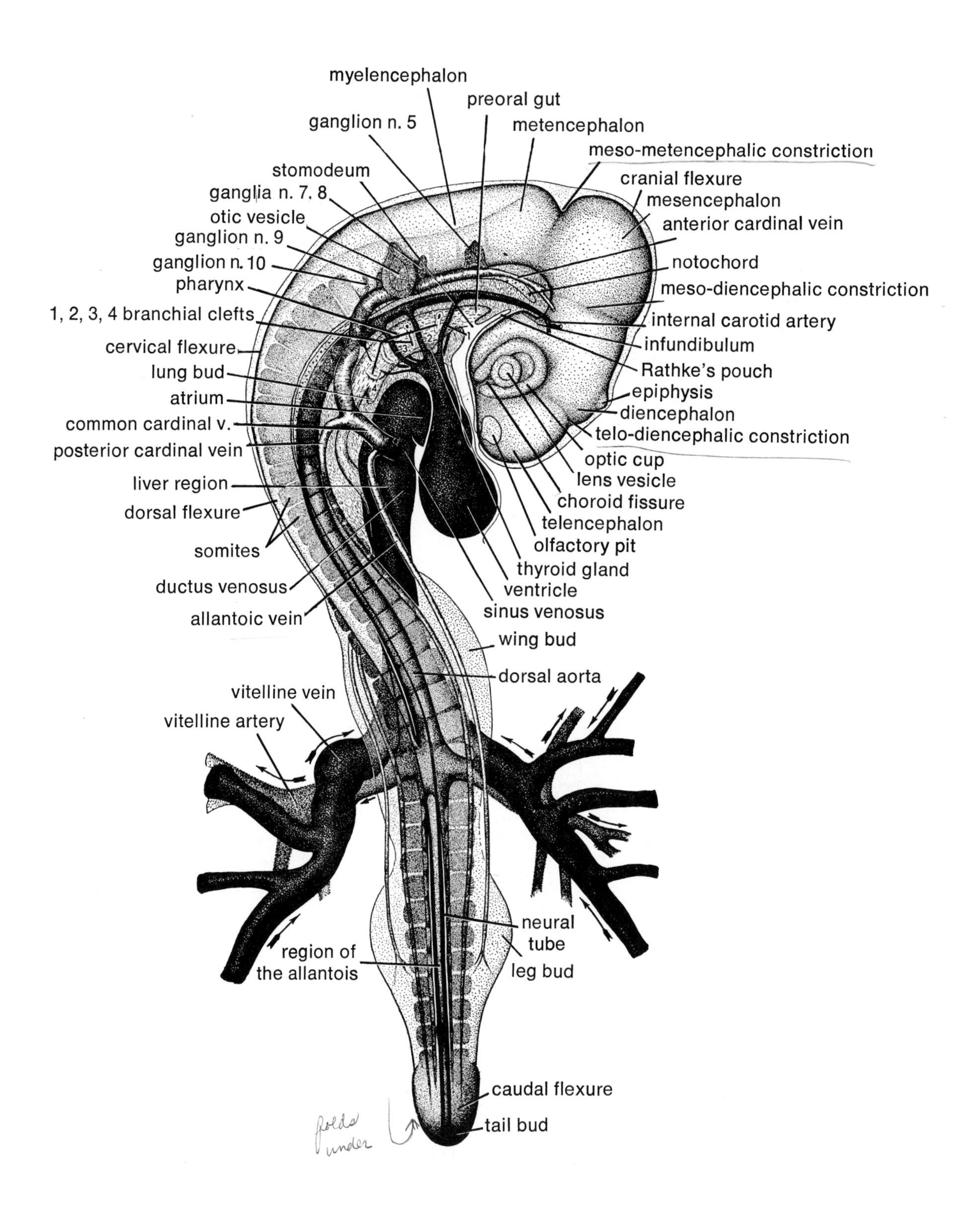

Figure 126

The 72-hour chick embryo with 35 somites, stage 18, dorsal view. (From *Fundamentals of Comparative Embryology of the Vertebrates* by Alfred F. Huettner. Copyright 1941 by Macmillan Publishing Co., Inc., New York. Used with permission of Macmillan Publishing Co.)

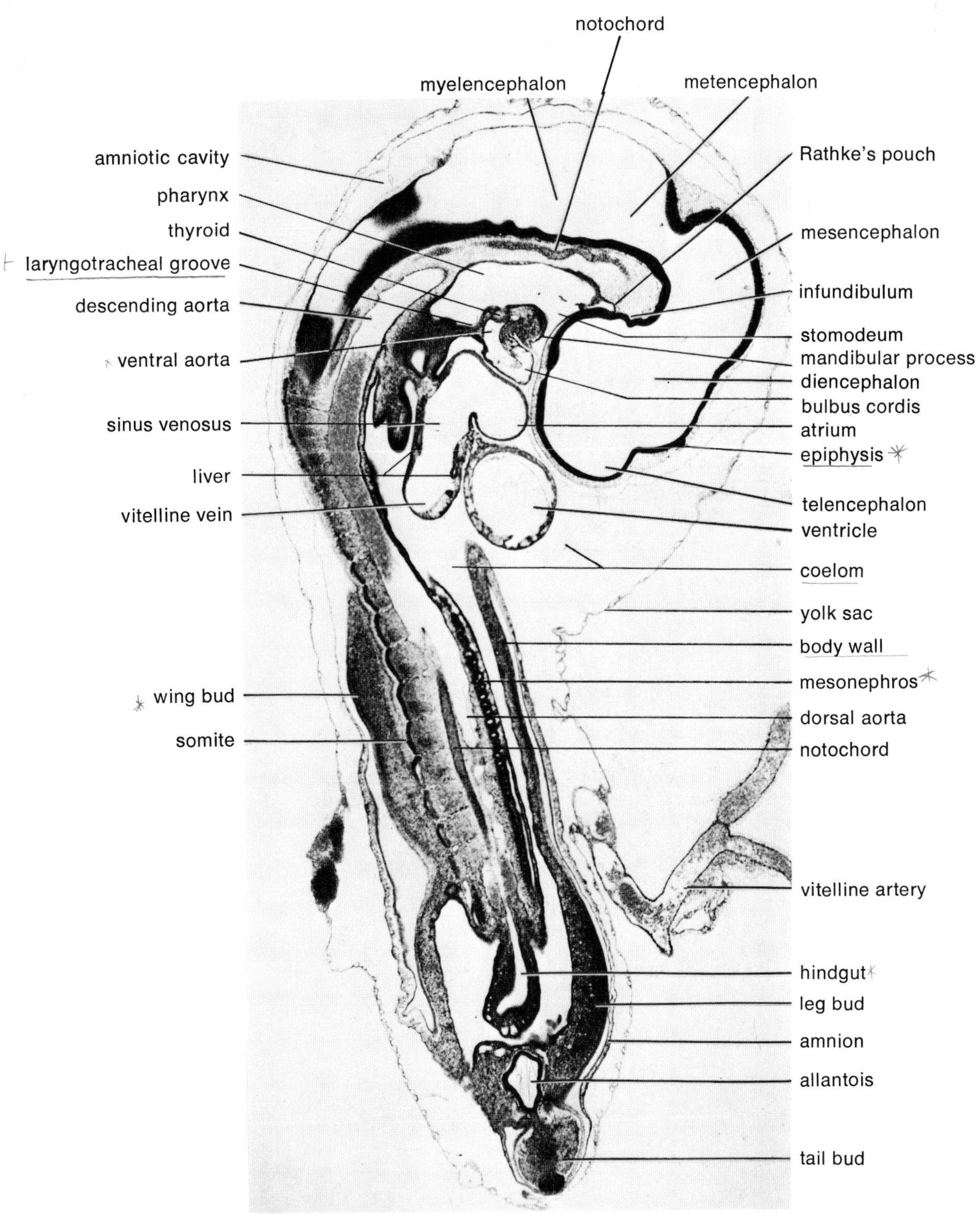

Figure 127 Chick embryo, stage 18, sagittal section (mag. 25X)

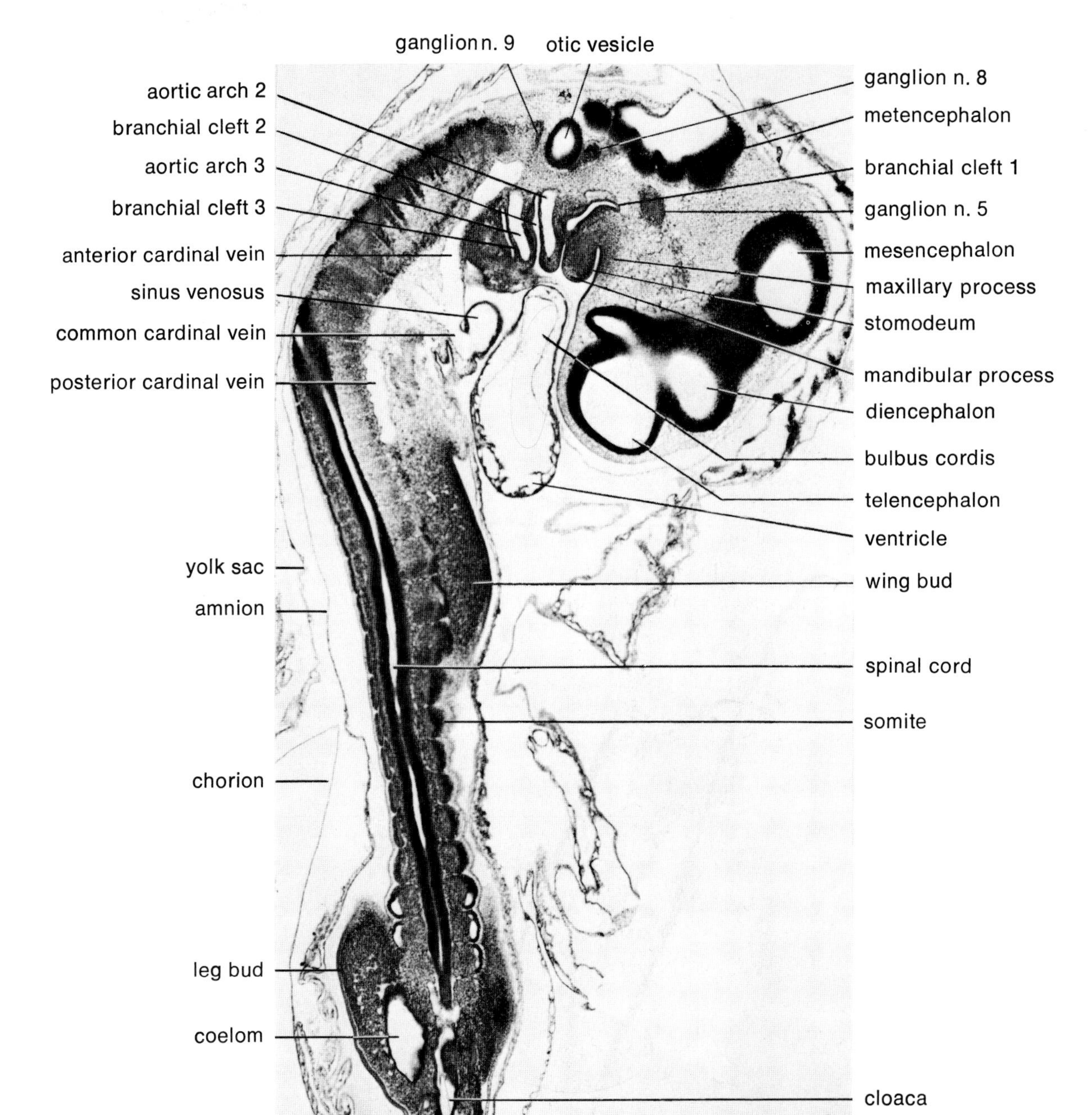

Figure 128 Chick embryo, stage 18, parasagittal section, right side (mag. 25X)

Collapsed → post. choridplexus

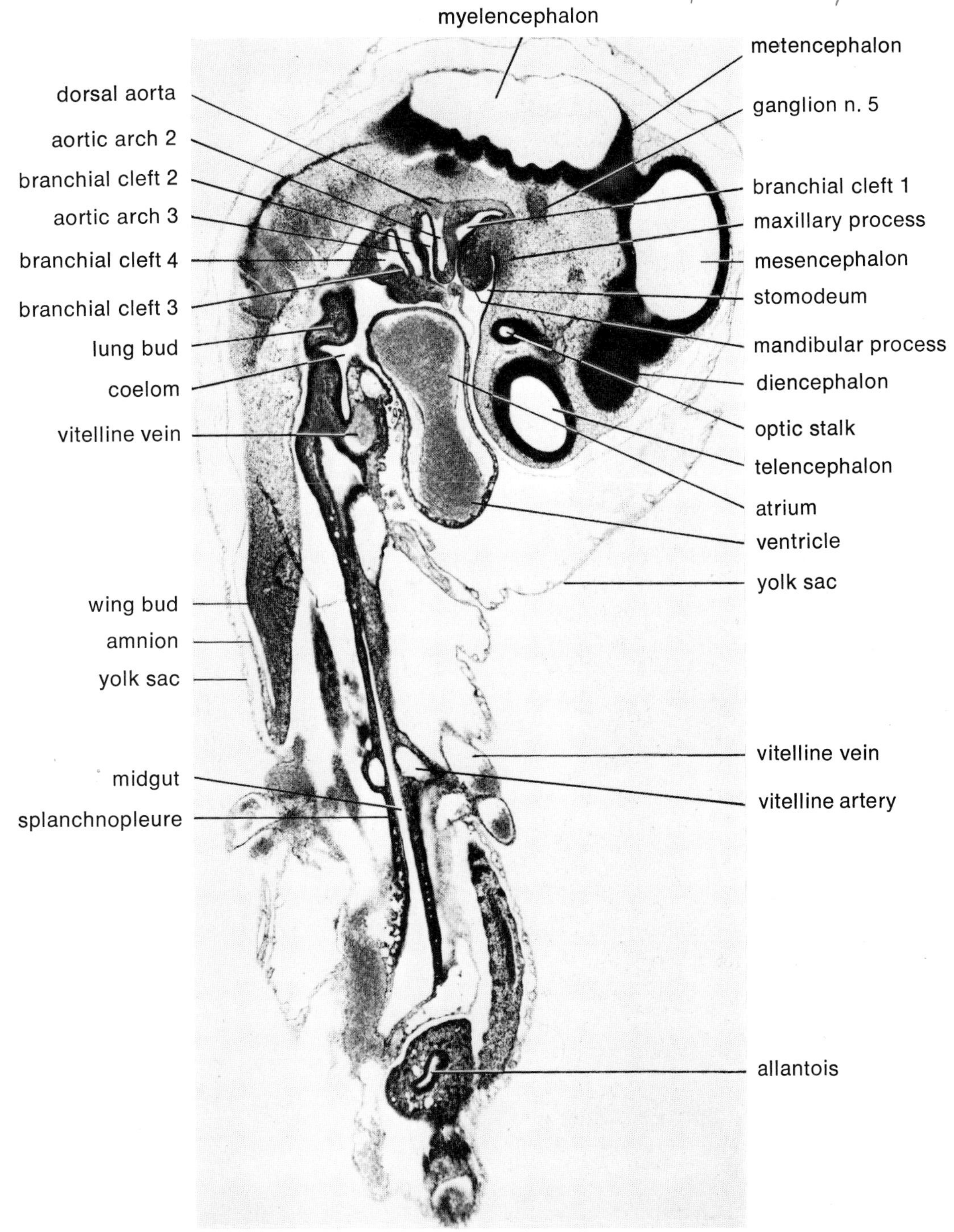

Figure 129 Chick embryo, stage 18, parasagittal section, left side (mag. 25X)

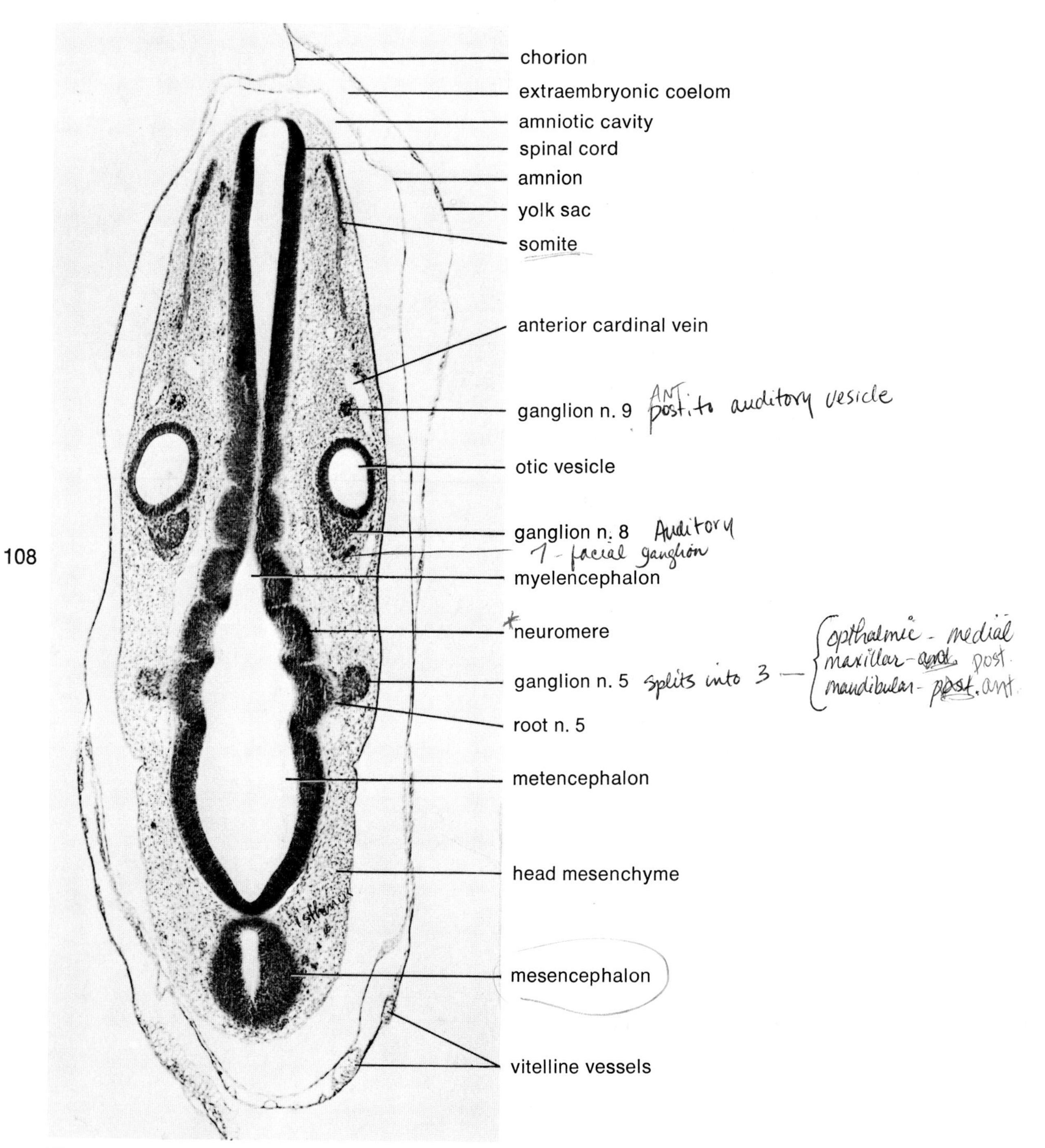

Figure 130 Chick embryo, stage 18, transverse section through otic vesicles (mag. 50X)

Ganglion

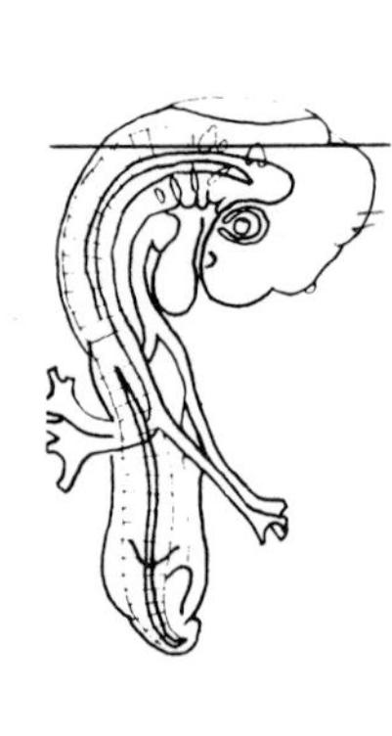

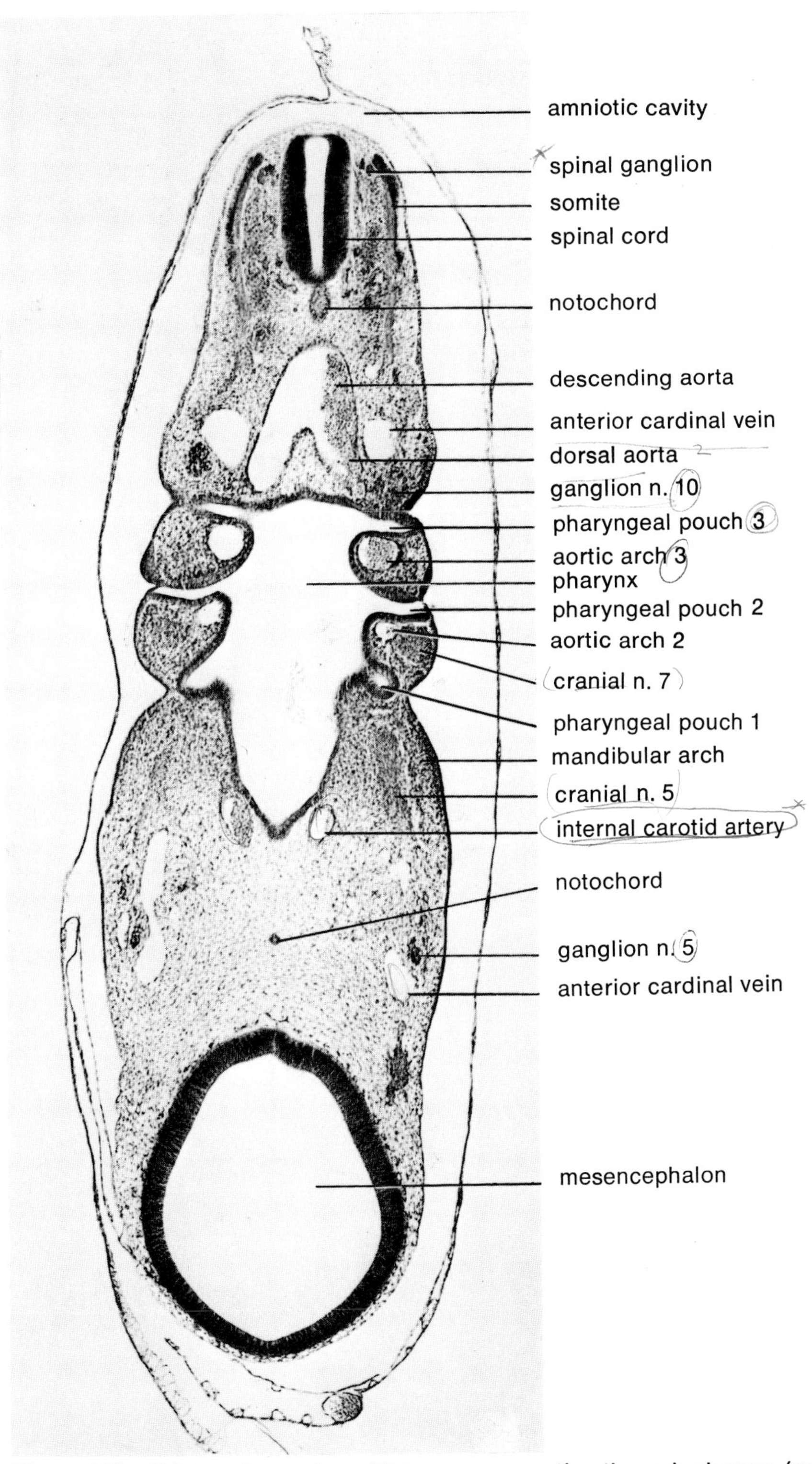

Figure 131 Chick embryo, stage 18, transverse section through pharynx (mag. 50X)

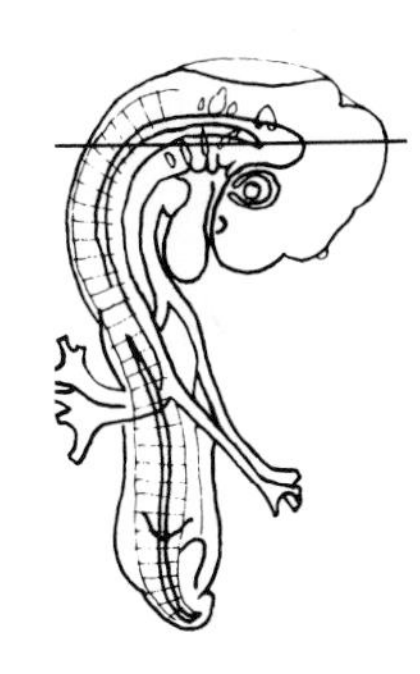

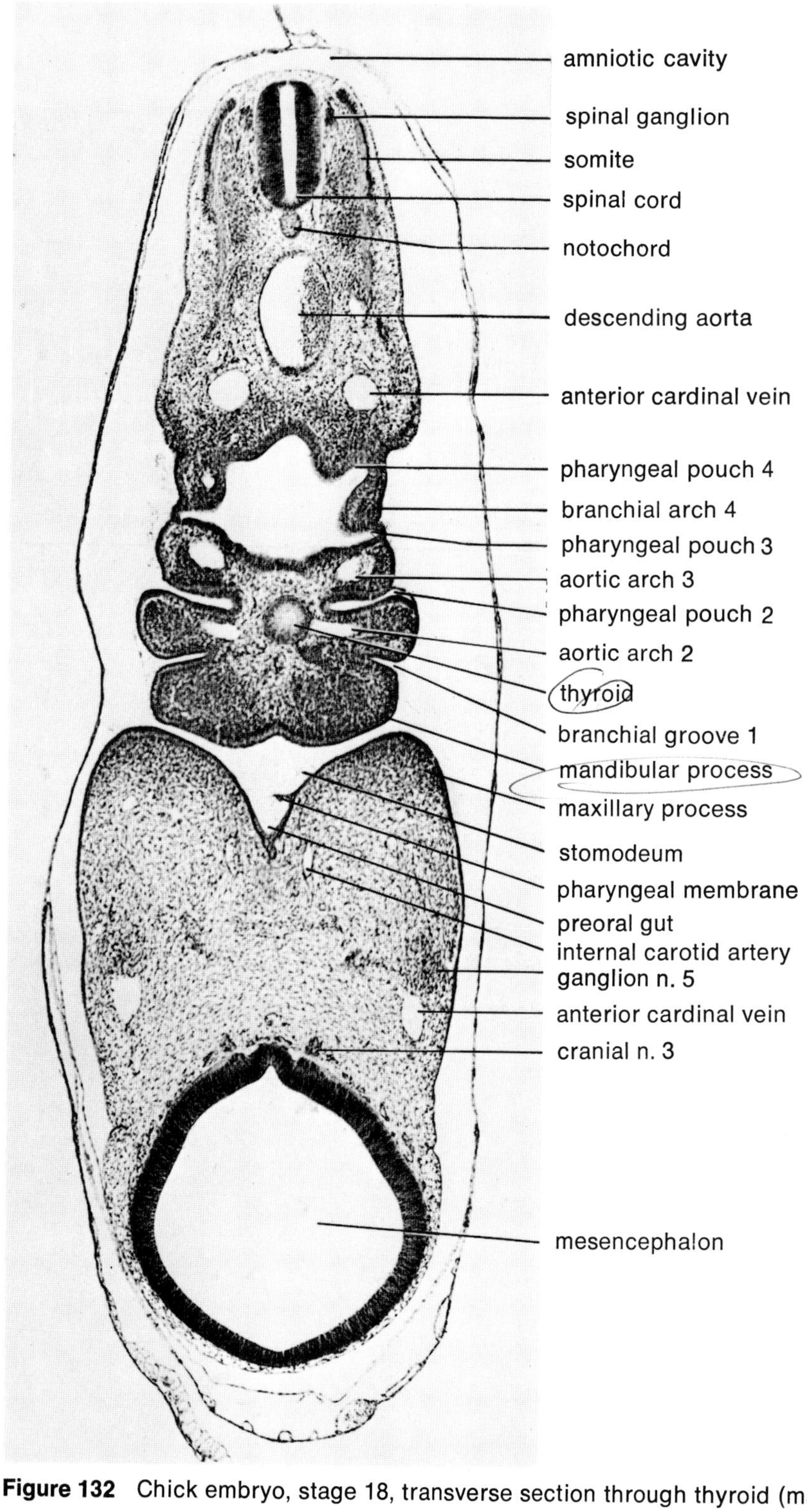

amniotic cavity
spinal ganglion
somite
spinal cord
notochord
descending aorta
anterior cardinal vein
pharyngeal pouch 4
branchial arch 4
pharyngeal pouch 3
aortic arch 3
pharyngeal pouch 2
aortic arch 2
thyroid
branchial groove 1
mandibular process
maxillary process
stomodeum
pharyngeal membrane
preoral gut
internal carotid artery
ganglion n. 5
anterior cardinal vein
cranial n. 3
mesencephalon

Figure 132 Chick embryo, stage 18, transverse section through thyroid (mag. 50X)

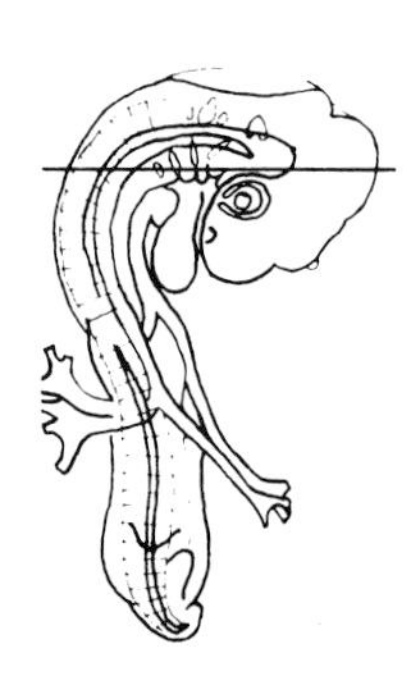

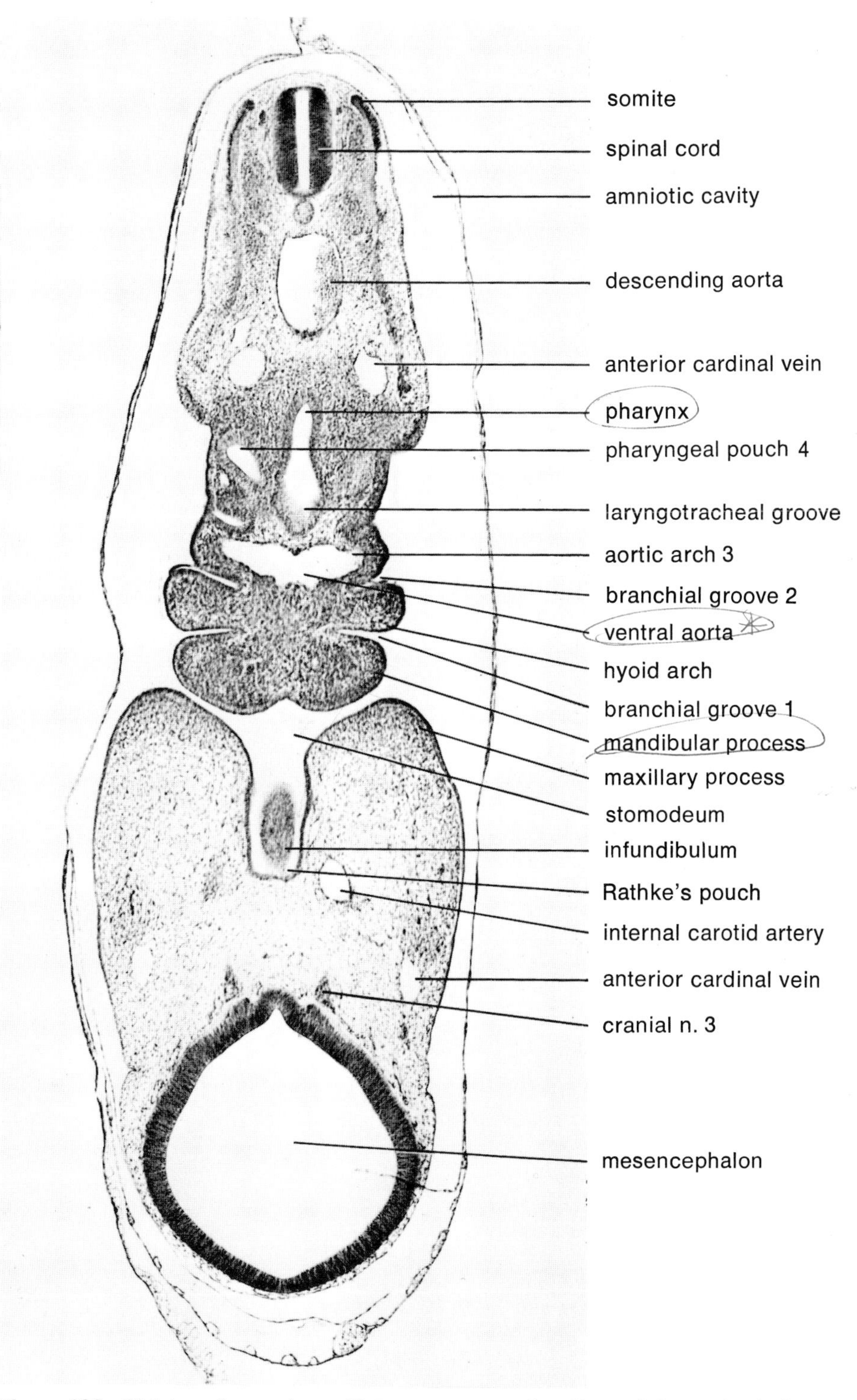

Figure 133 Chick embryo, stage 18, transverse section through hypophysis (mag. 50X)

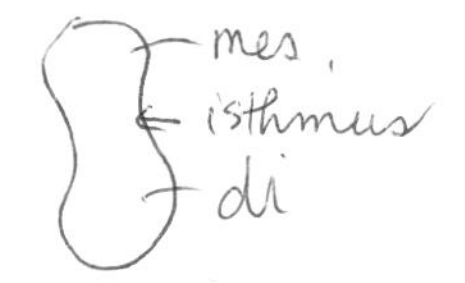

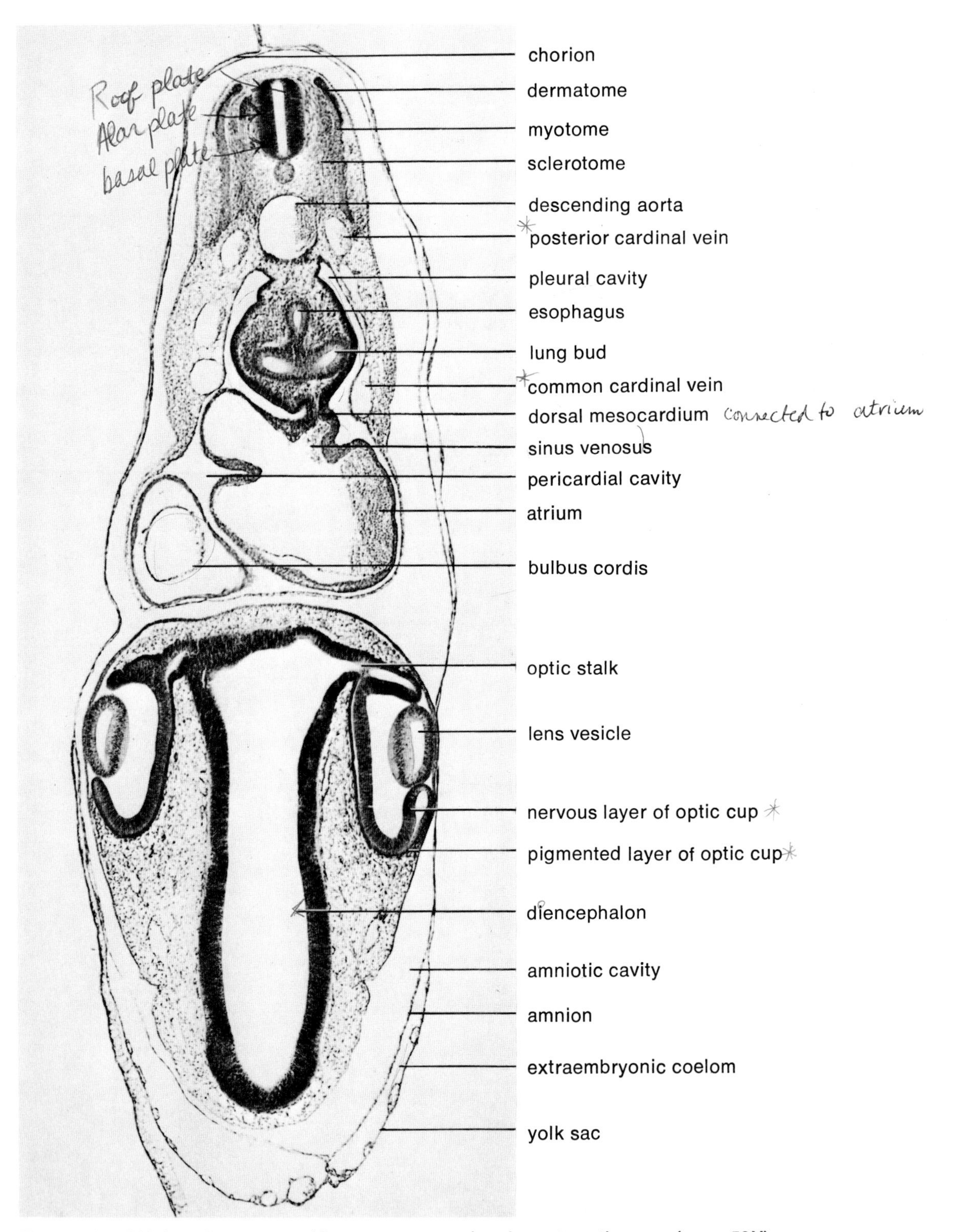

Figure 134 Chick embryo, stage 18, transverse section through optic cups (mag. 50X)

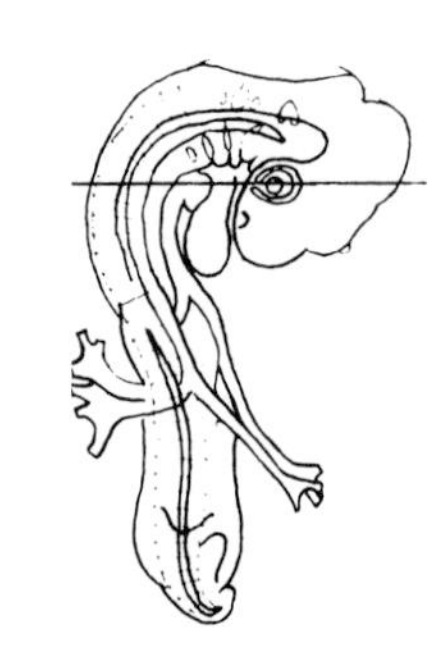

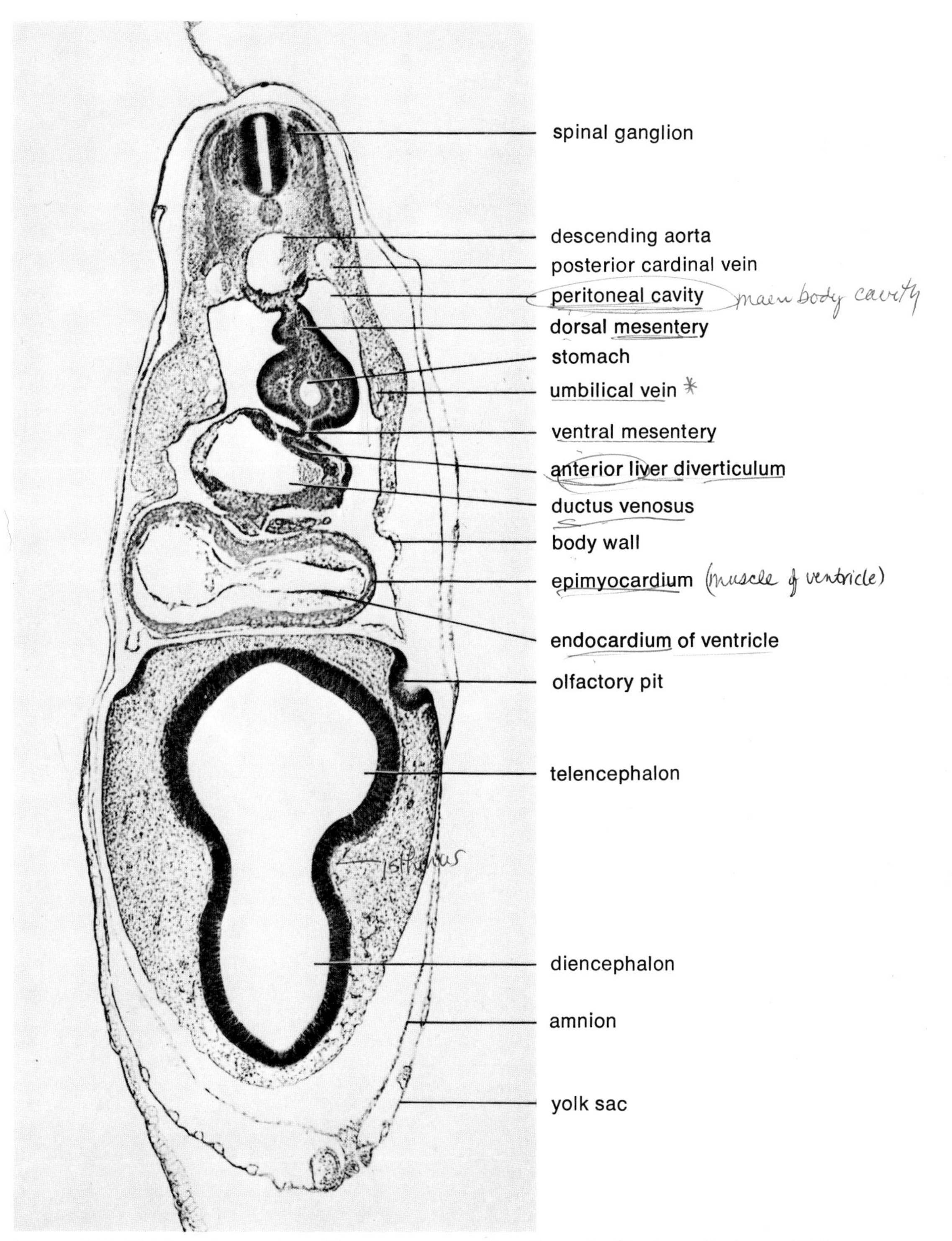

Figure 135 Chick embryo, stage 18, transverse section through olfactory pits (mag. 50X)

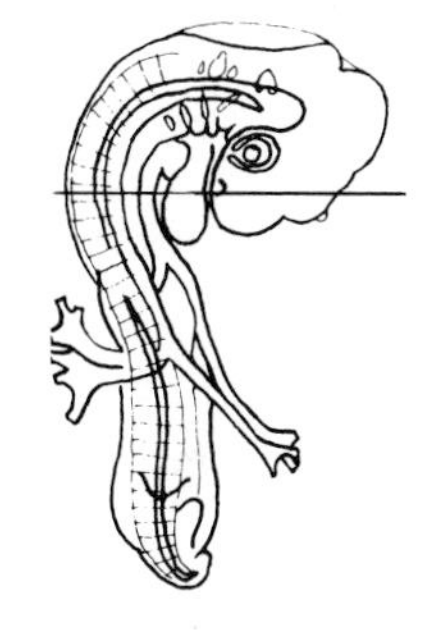

2ND slide

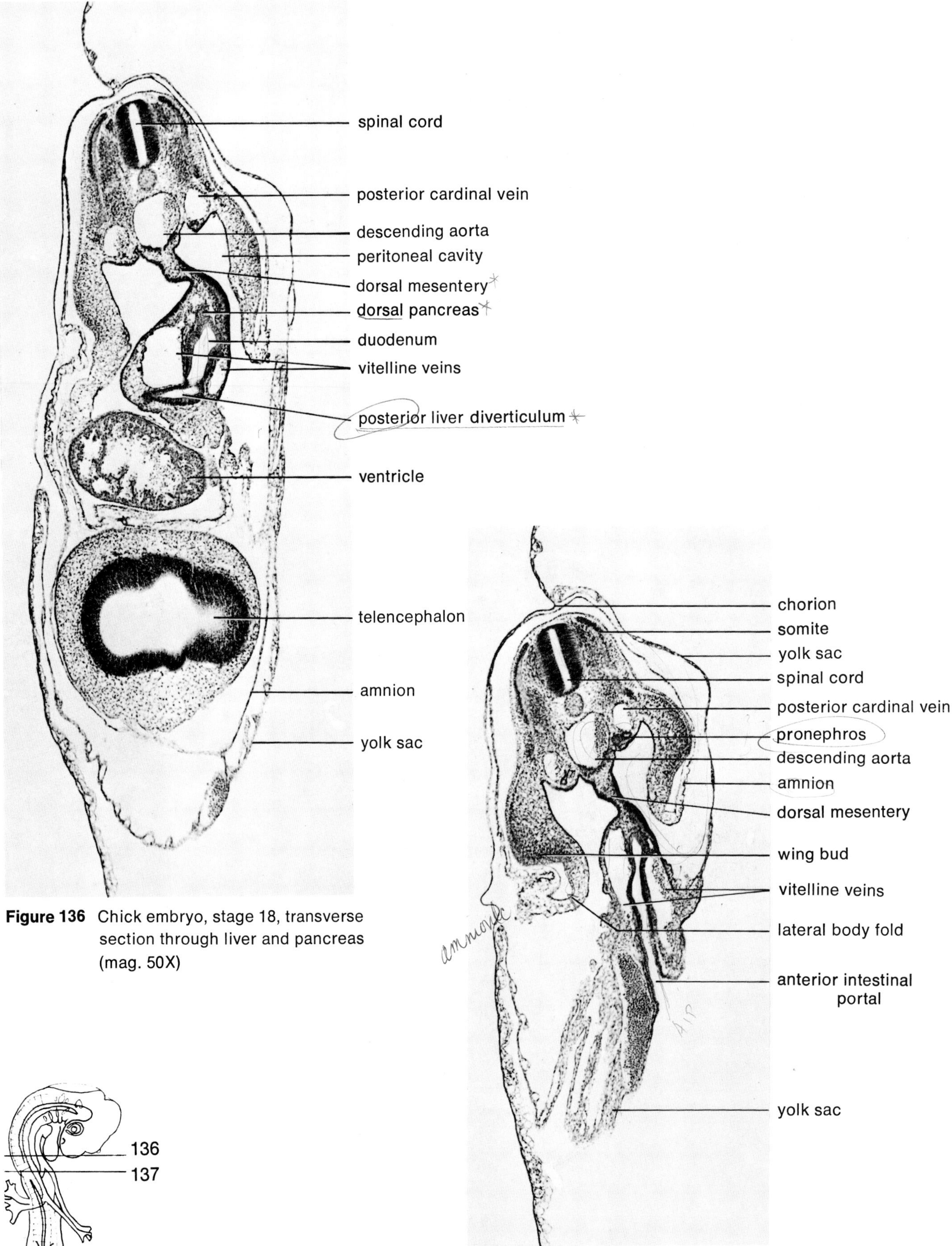

Figure 136 Chick embryo, stage 18, transverse section through liver and pancreas (mag. 50X)

Figure 137 Chick embryo, stage 18, transverse section through anterior intestinal portal (mag. 50X)

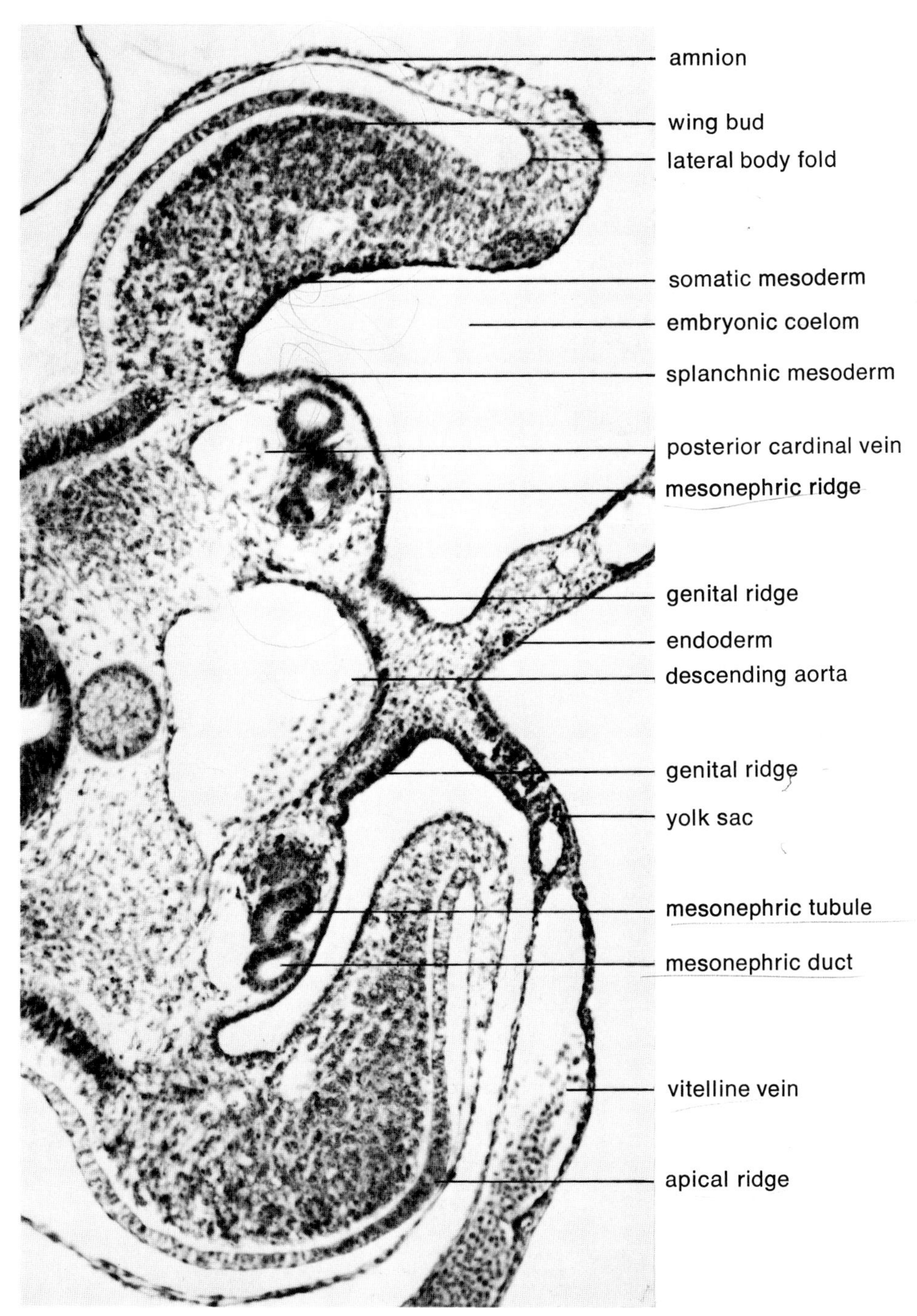

Figure 138 Chick embryo, stage 18, transverse section through genital ridge (mag. 140X)

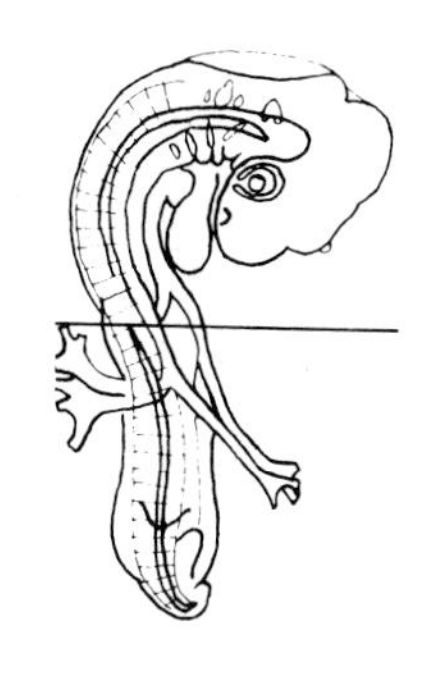

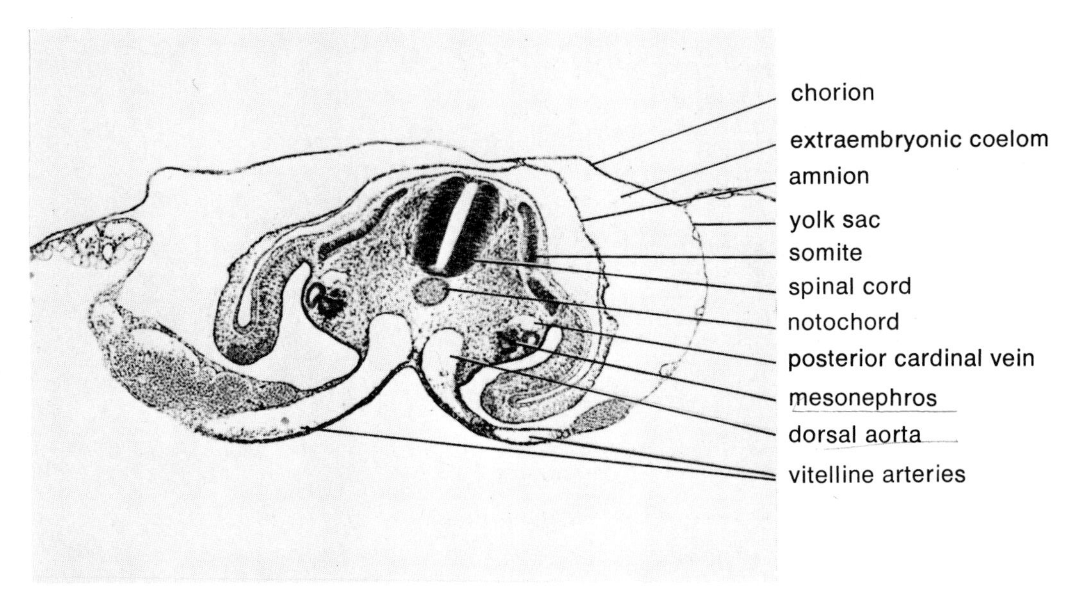

Figure 139 Chick embryo, stage 18, transverse section through vitelline arteries (mag. 50X)

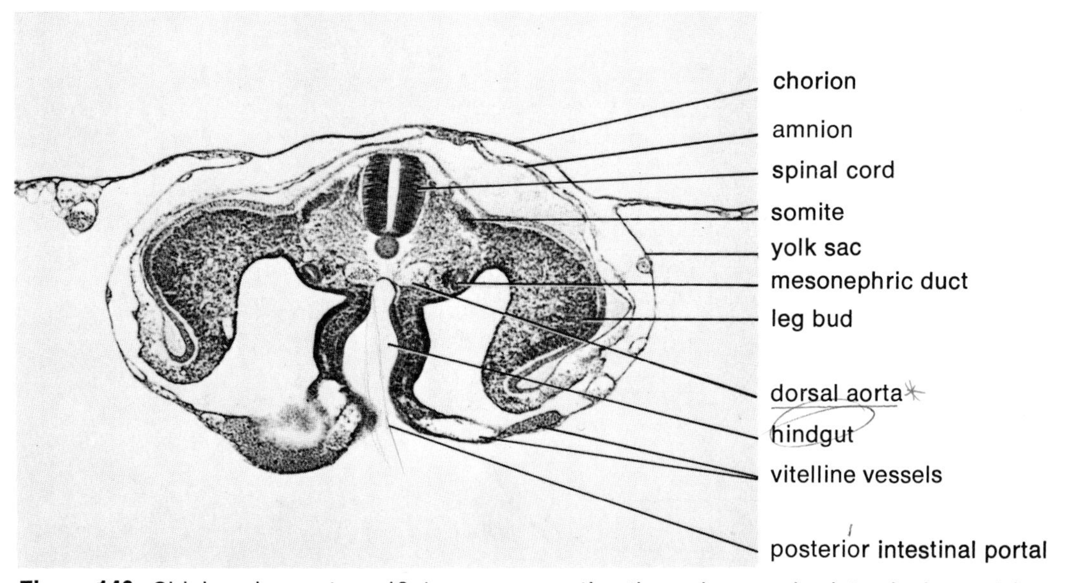

Figure 140 Chick embryo, stage 18, transverse section through posterior intestinal portal (mag. 50X)

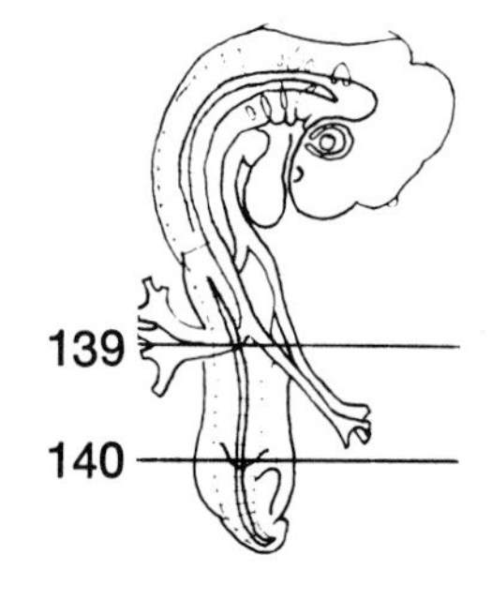

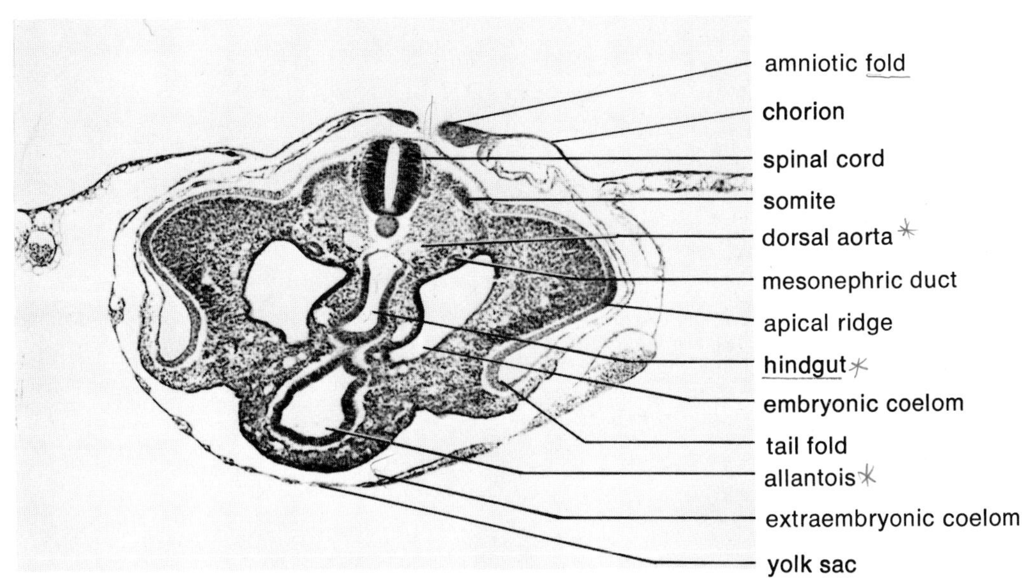

Figure 141 Chick embryo, stage 18, transverse section through allantois (mag. 50X)

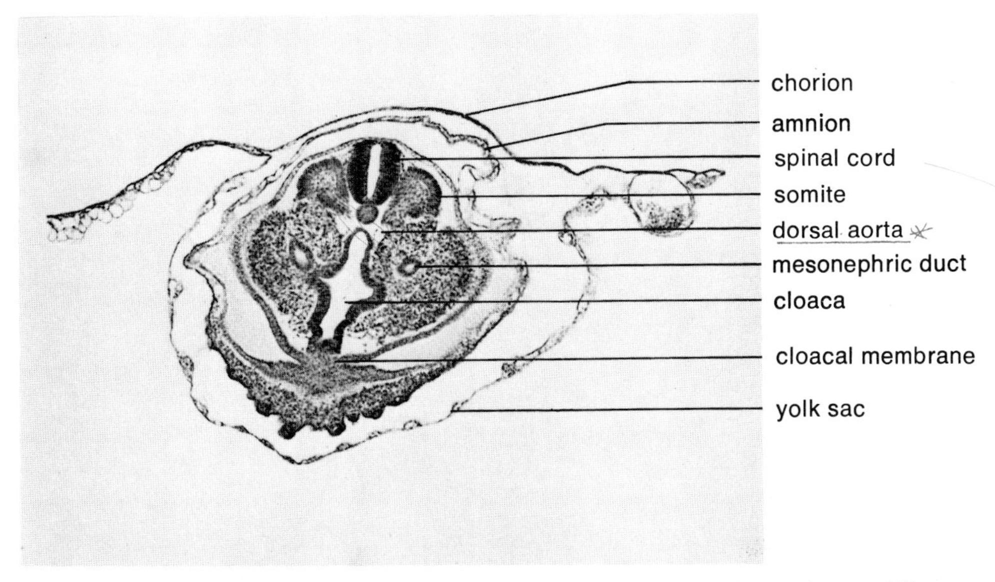

Figure 142 Chick embryo, stage 18, transverse section through cloaca (mag. 50X)

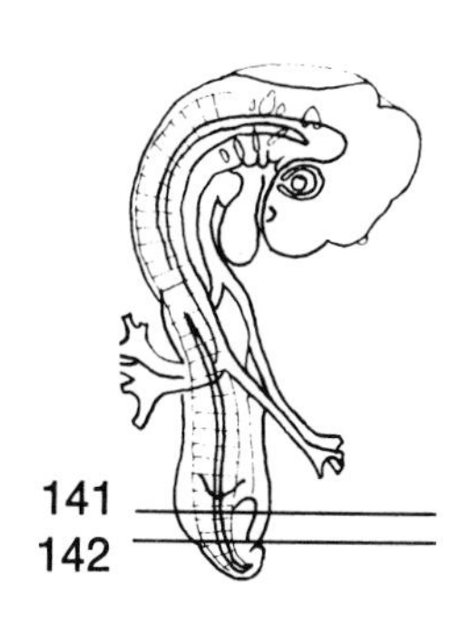

14. The Stage 21 Chick Embryo (3½ days incubation)

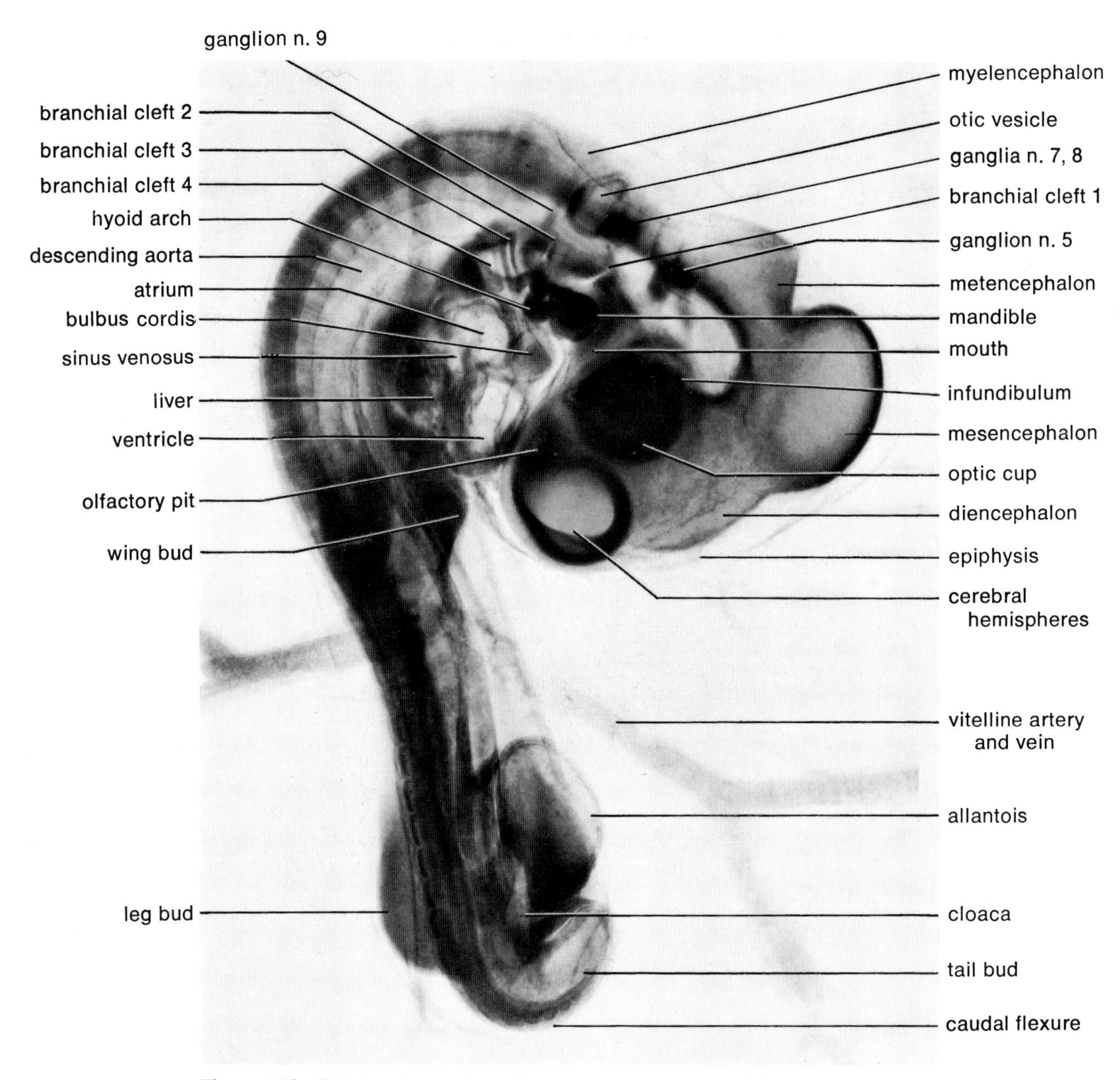

Figure 143 Chick embryo, stage 21, whole mount (mag. 20X)

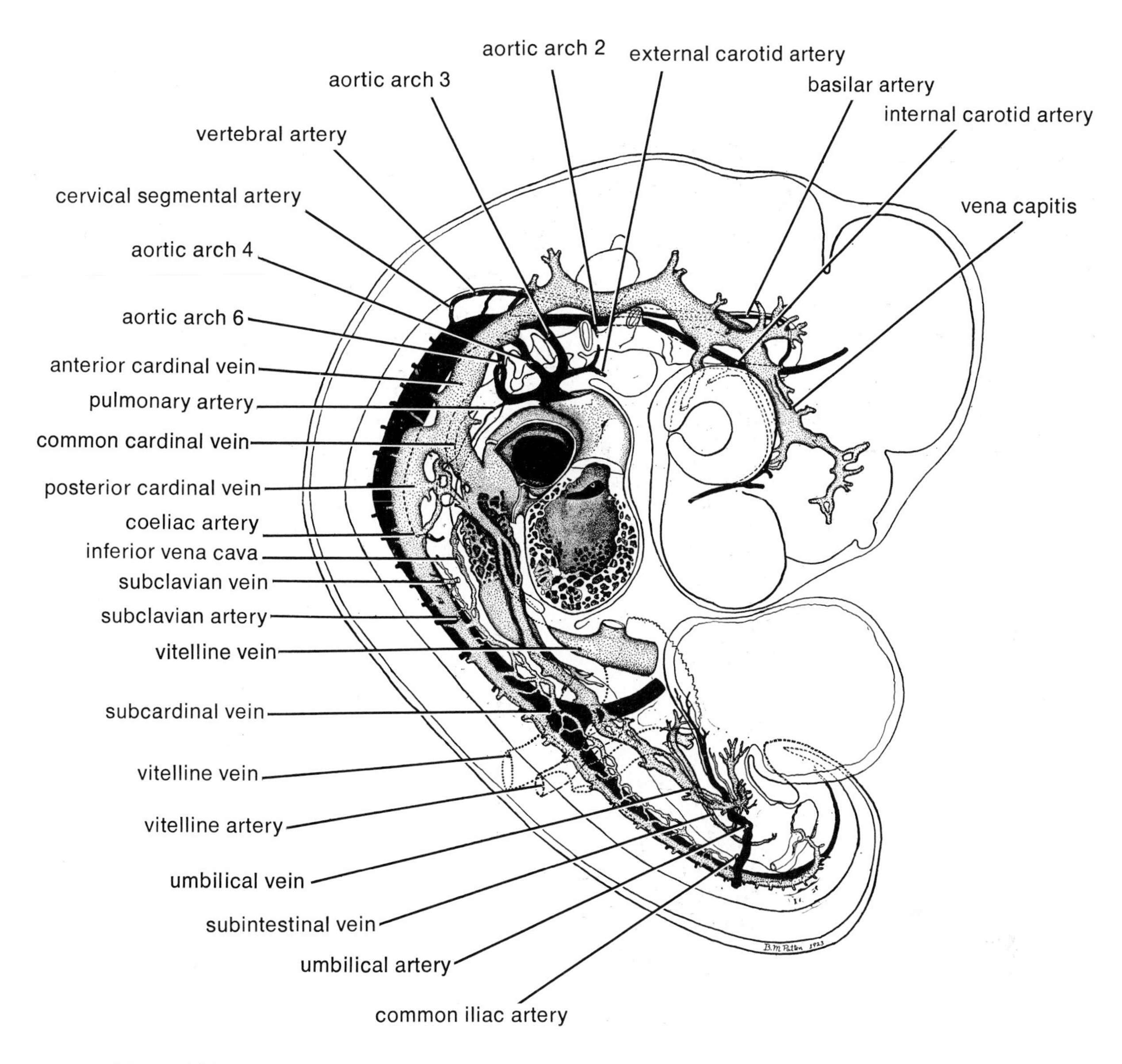

Figure 144

Reconstruction of circulatory system of stage 21 chick (mag. 18×). (From *Early Embryology of the Chick*, 4th ed., by Bradley M. Patten. Copyright 1957 by B. M. Patten. Used with permission of the McGraw-Hill Book Co.)

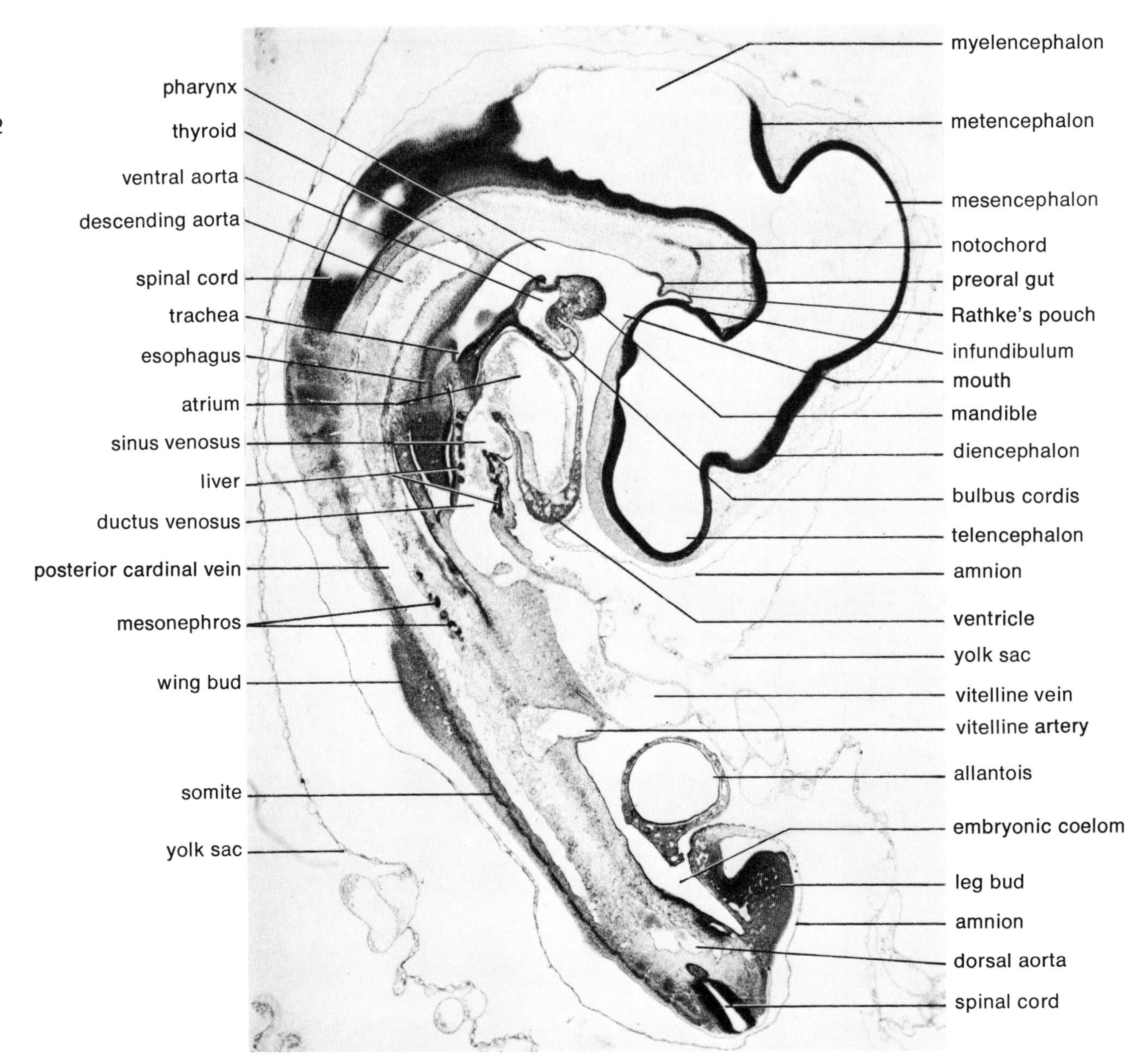

Figure 145 Chick embryo, stage 21, sagittal section (mag. 25X)

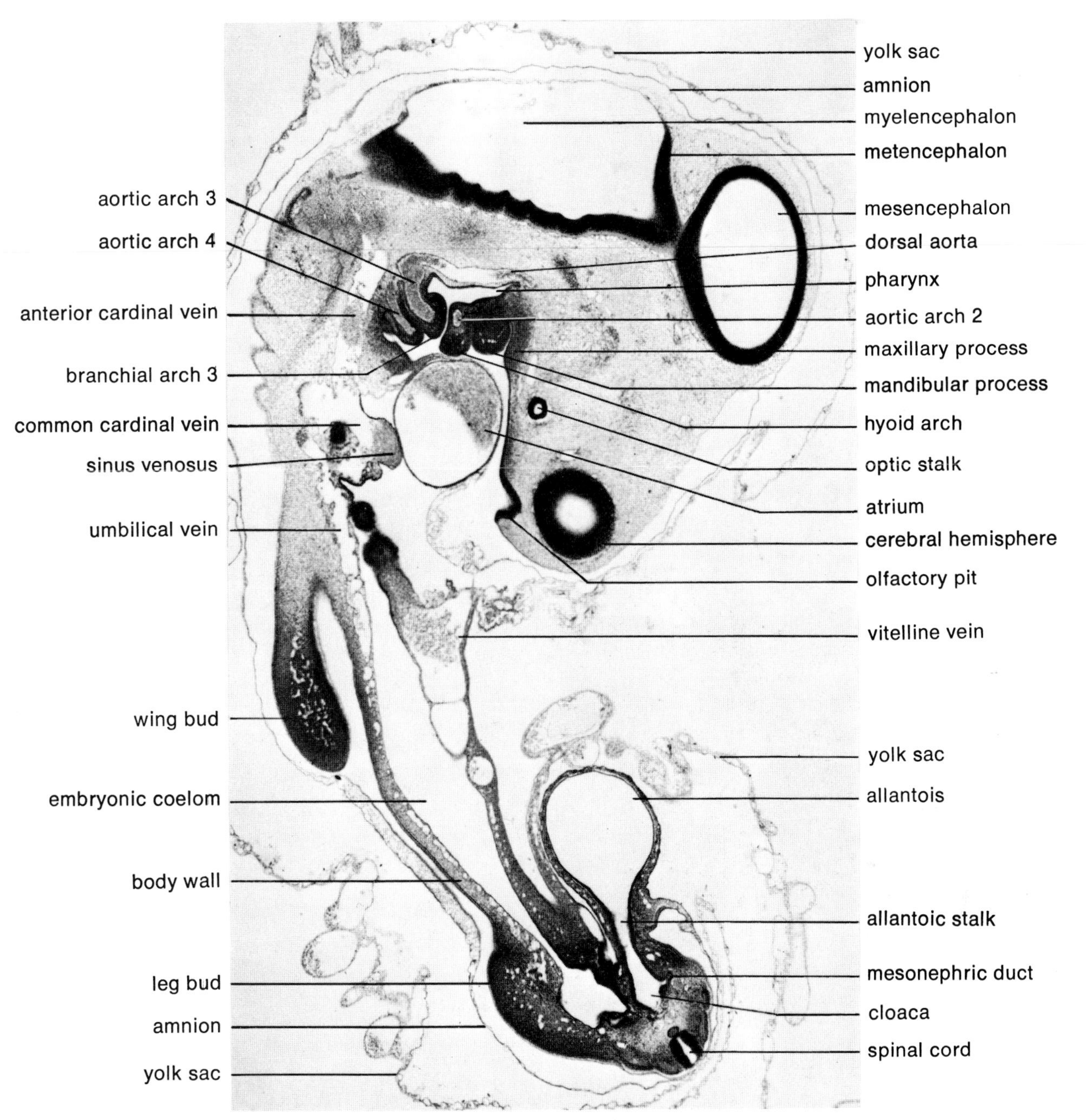

Figure 146 Chick embryo, stage 21, parasagittal section (mag. 25X)

15. The 10-mm Pig Embryo

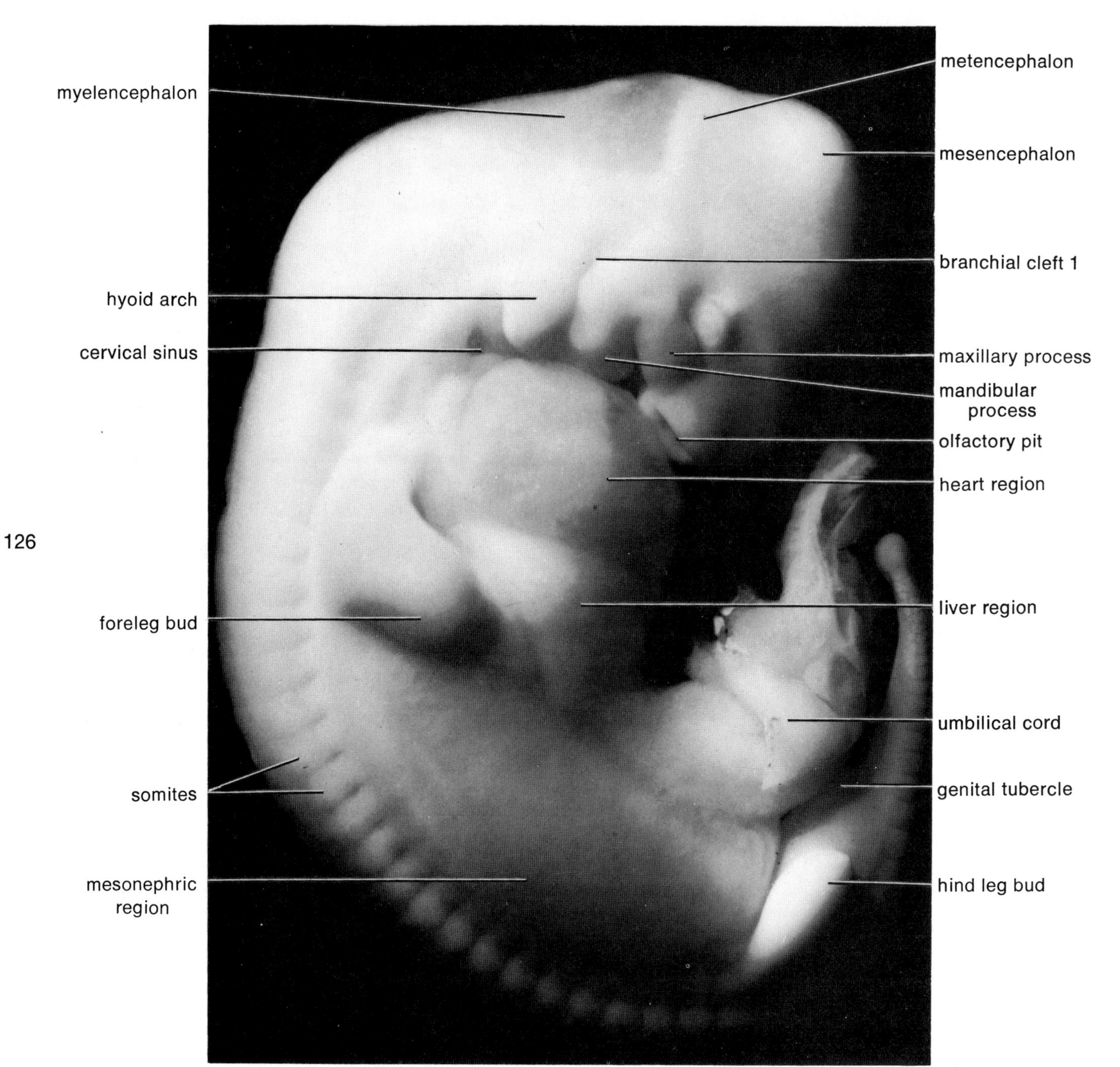

Figure 147 10-mm pig embryo, opaque mount (mag. 17X; incident illumination)

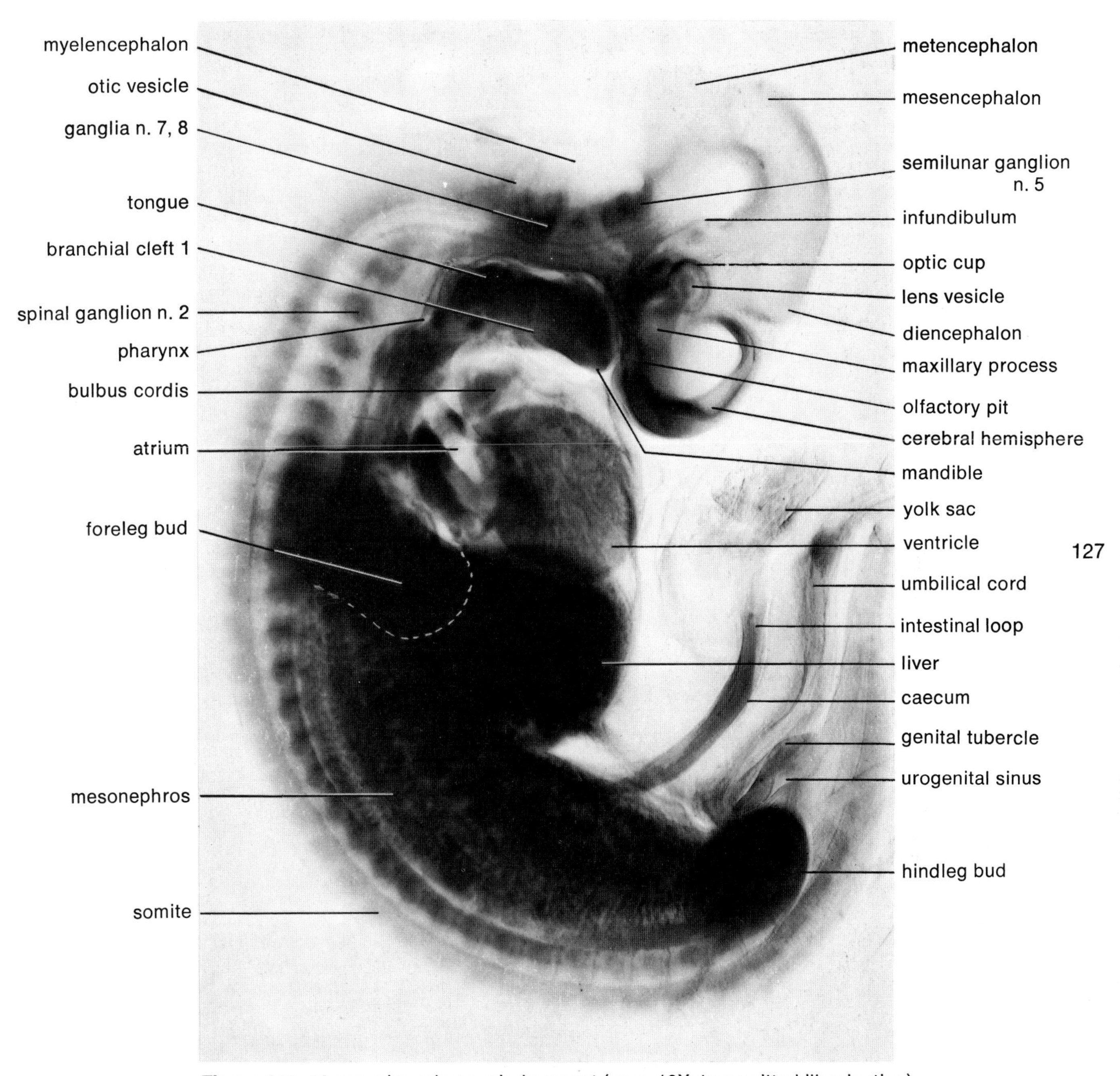

Figure 148 10-mm pig embryo, whole mount (mag. 16X; transmitted illumination)

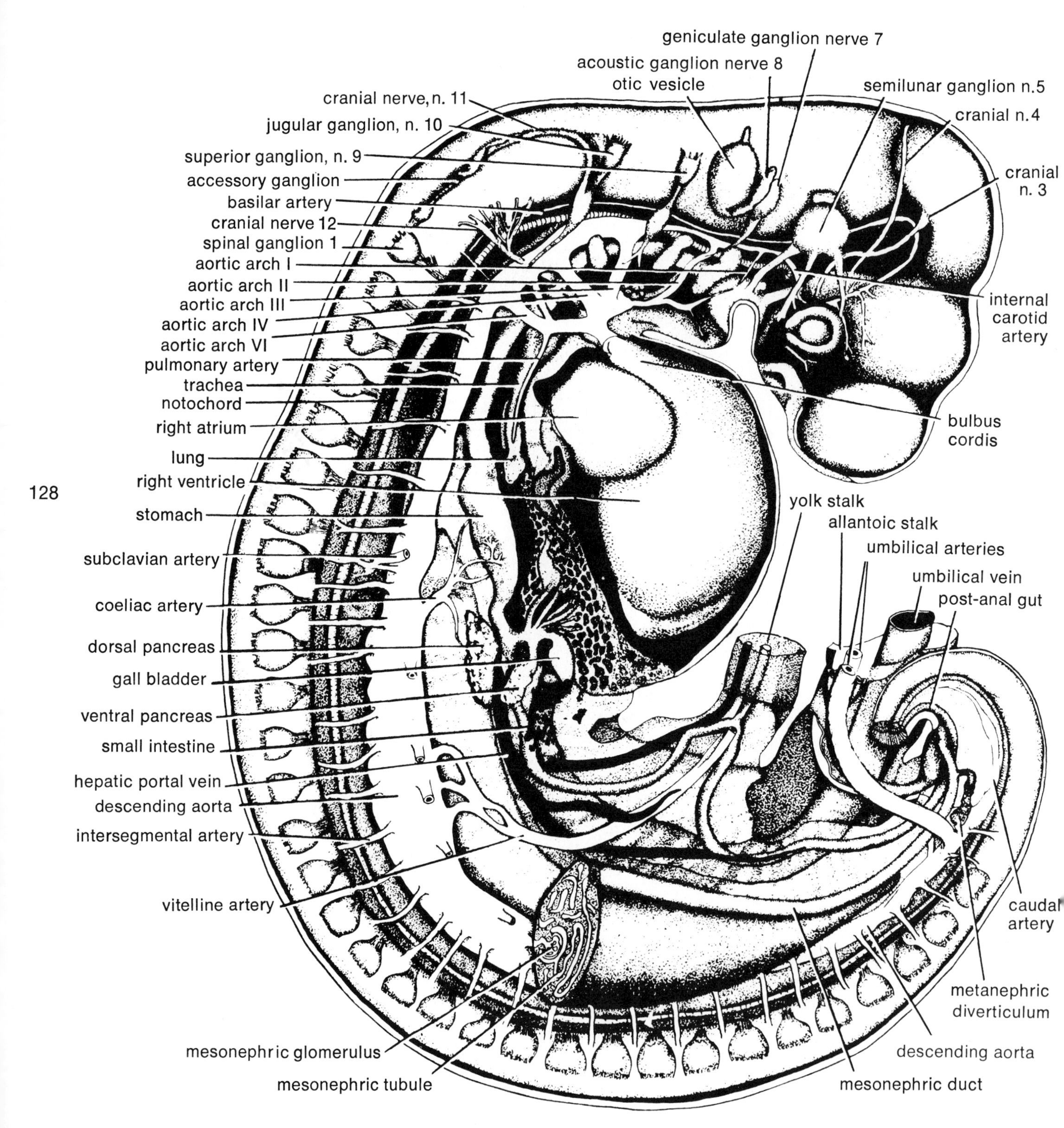

Figure 149 Reconstruction of 10-mm pig embryo (From *Comparative Embryology of the Vertebrates* by Olin E. Nelsen. Copyright 1953 by the Blakiston Company, Inc. Used with permission of McGraw-Hill Book Company.)

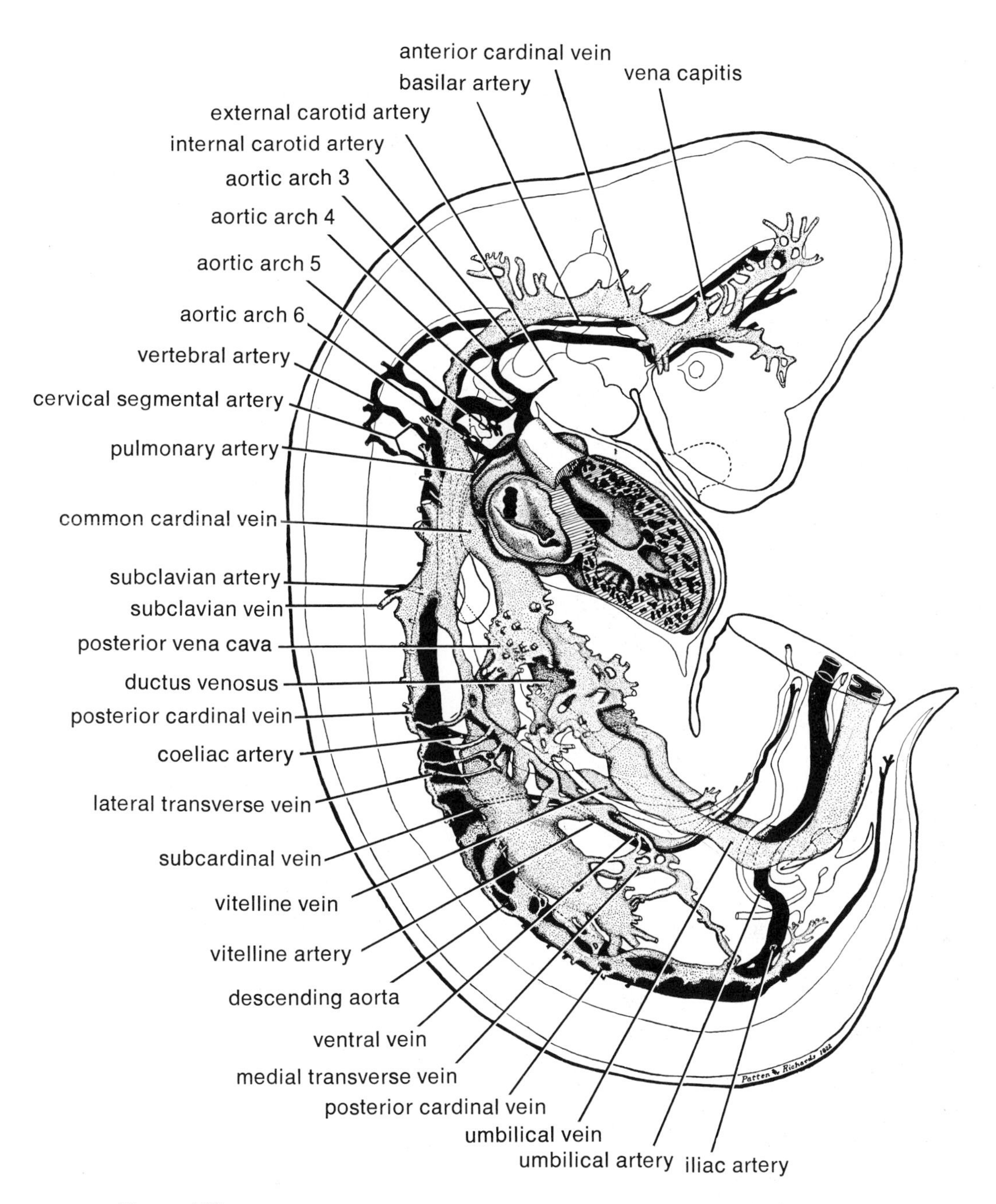

Figure 150

Reconstruction of the circulatory system of a 9.4-mm pig embryo (mag. 14×) (From *Embryology of the Pig*, 3rd ed., *by* Bradley M. Patten. Copyright 1959 by B. M. Patten. Used with permission of the McGraw-Hill Book Co.)

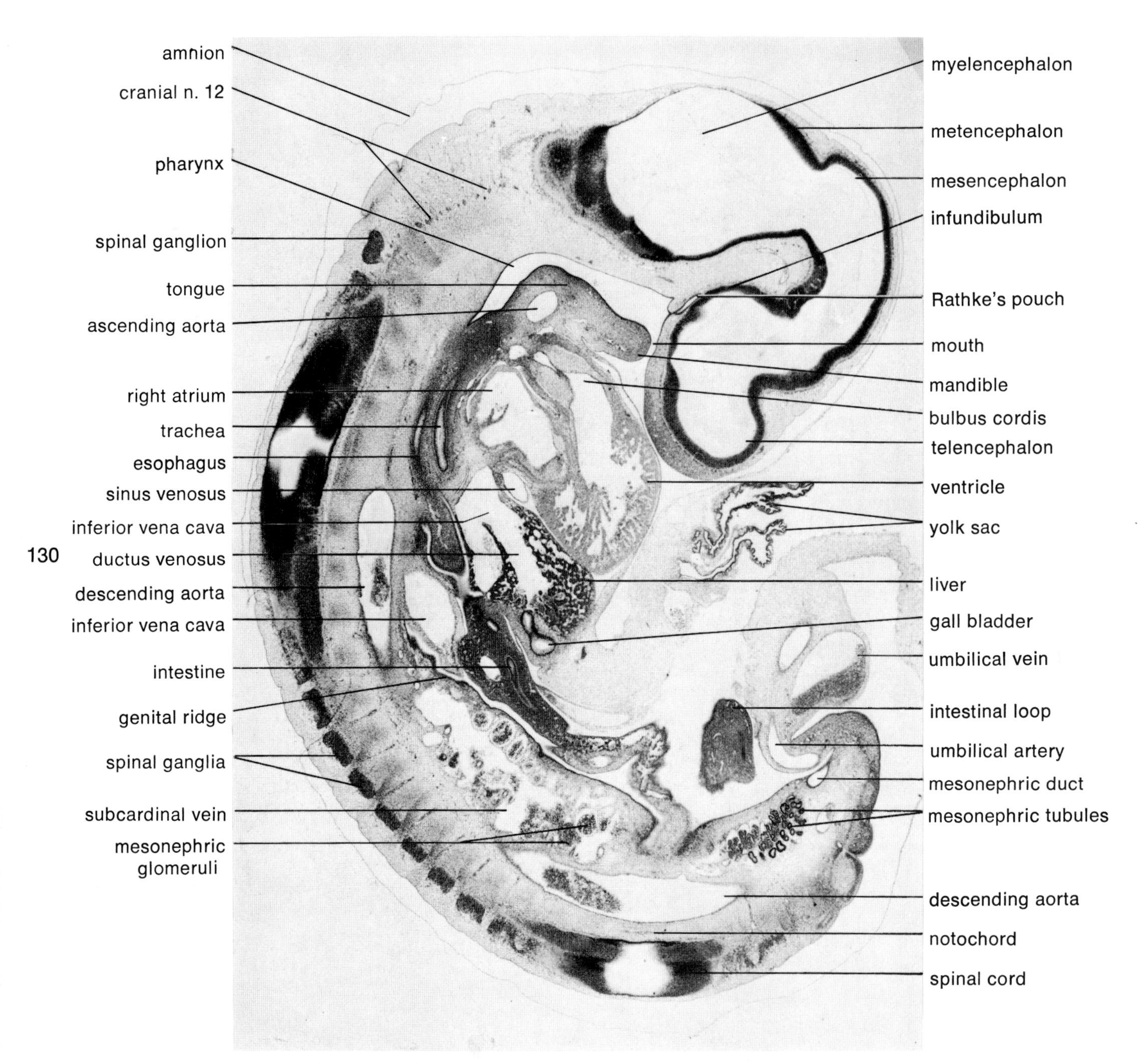

Figure 151 10-mm pig embryo, sagittal section (mag. 16X)

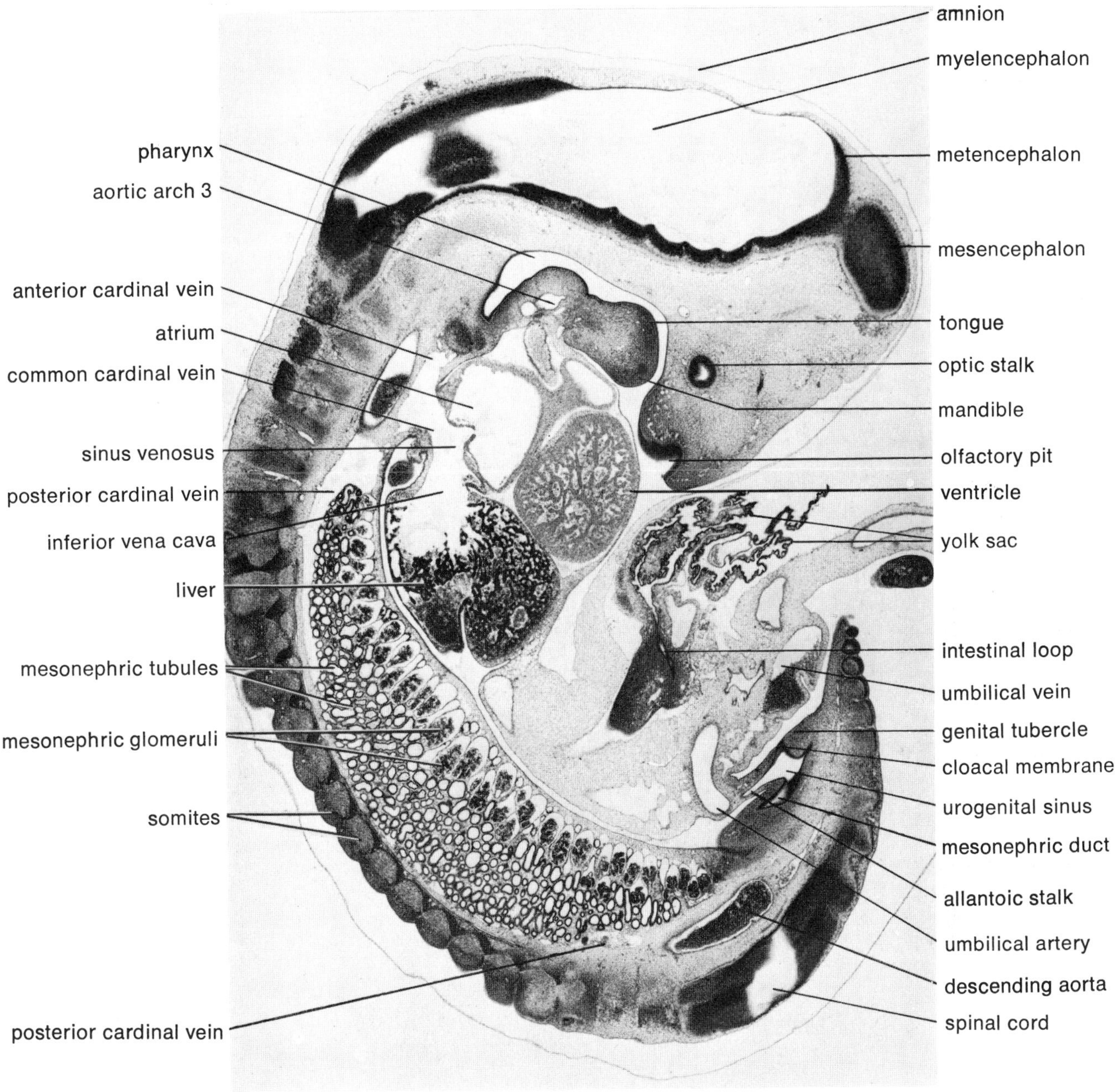

amnion
myelencephalon
pharynx
metencephalon
aortic arch 3
mesencephalon
anterior cardinal vein
tongue
atrium
optic stalk
common cardinal vein
mandible
sinus venosus
olfactory pit
posterior cardinal vein
ventricle
inferior vena cava
yolk sac
liver

intestinal loop
mesonephric tubules
umbilical vein
genital tubercle
mesonephric glomeruli
cloacal membrane
urogenital sinus
somites
mesonephric duct
allantoic stalk
umbilical artery
descending aorta
posterior cardinal vein
spinal cord

Figure 152 10-mm pig embryo, parasagittal section (mag. 16X)

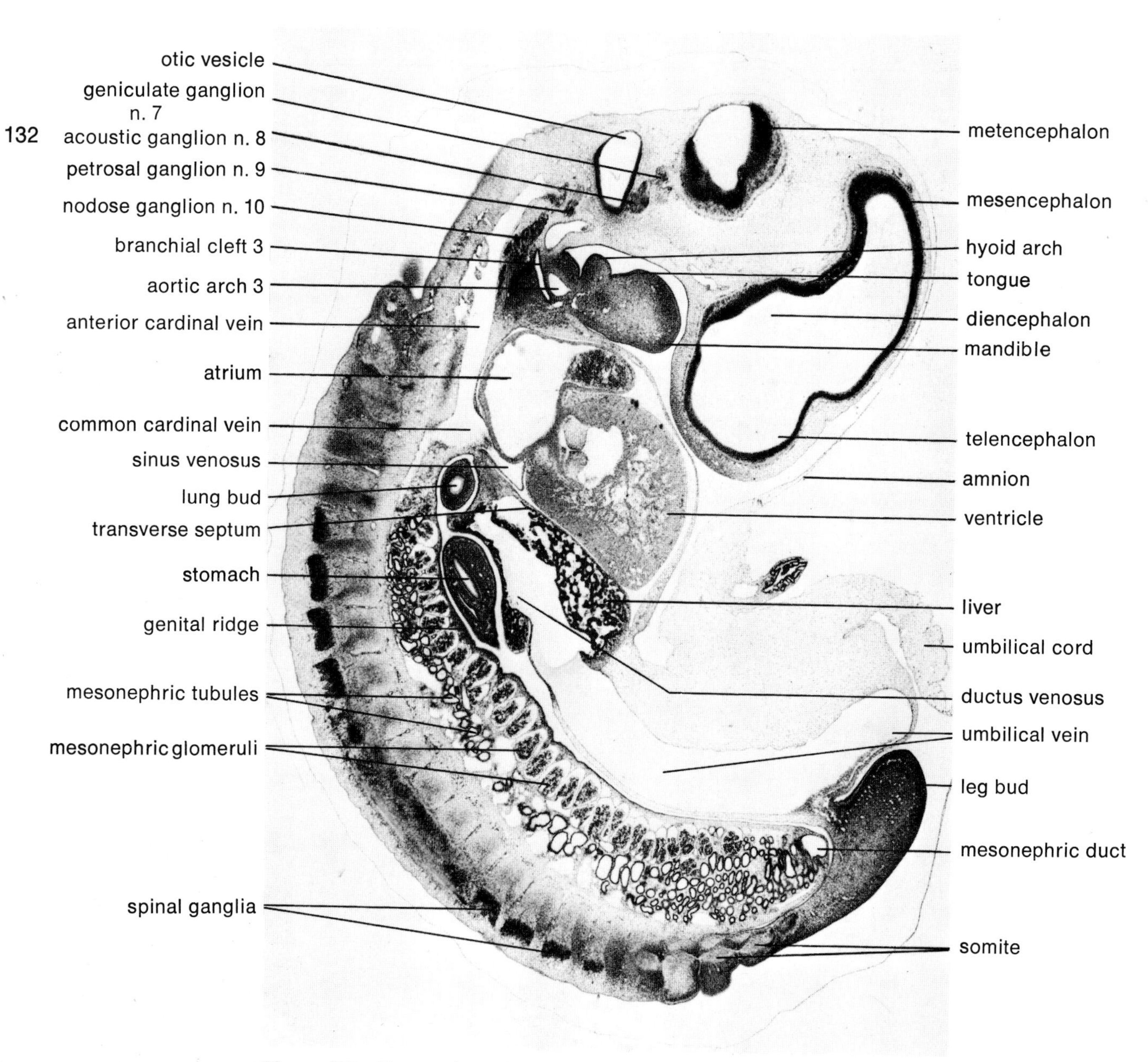

Figure 153 10-mm pig embryo, parasagittal section (mag. 16X)

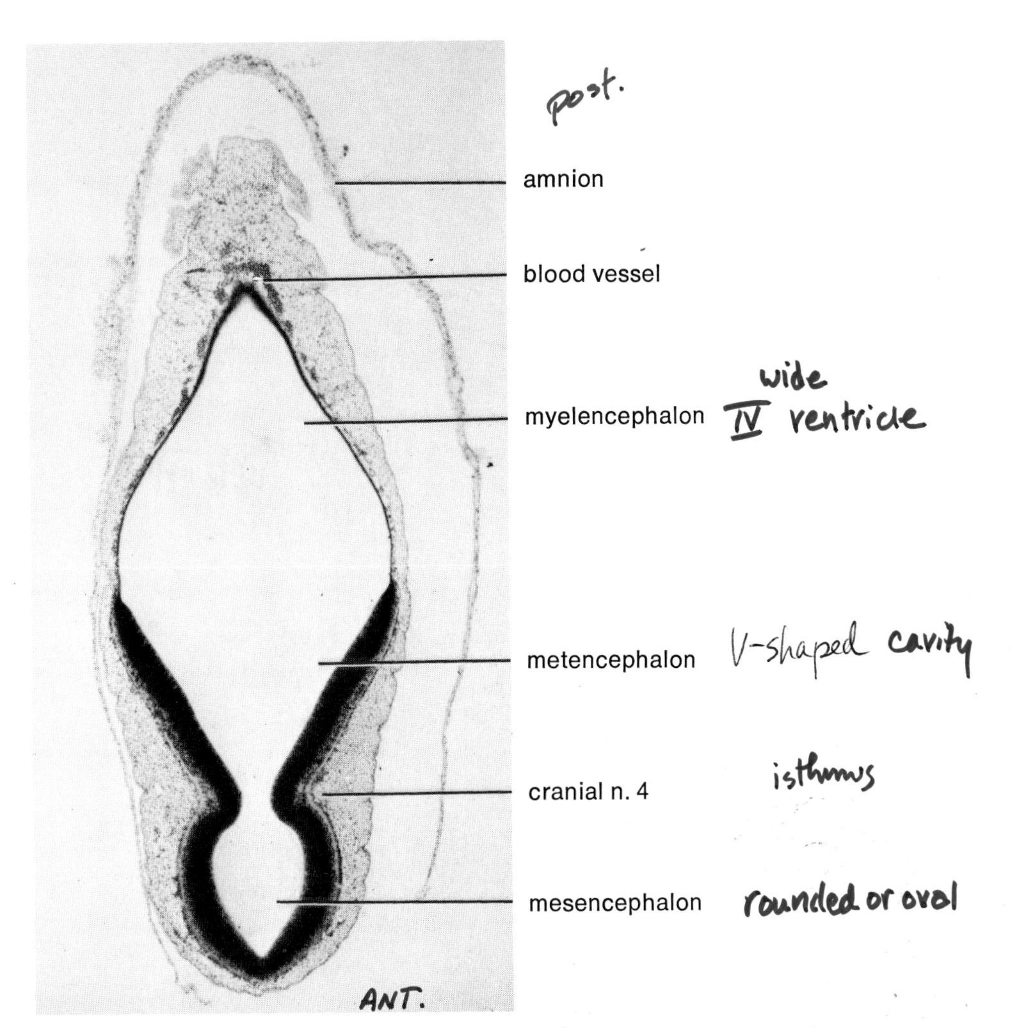

Figure 154 10-mm pig embryo, transverse section through cranial nerve 4 (mag. 30×)

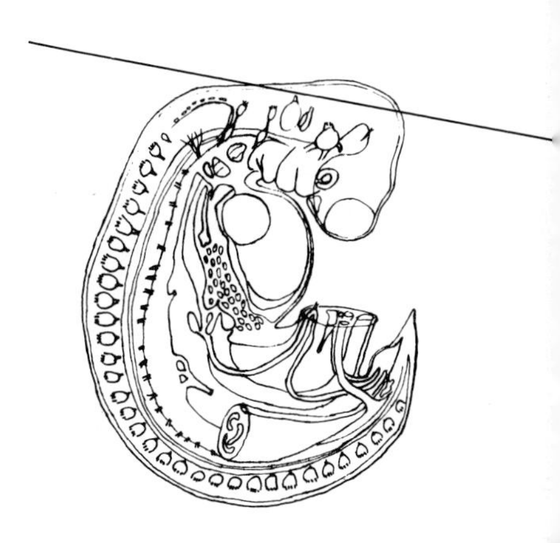

amnion

myelencephalon

ependymal layer

marginal layer

mantle layer

accessory ganglia n. 10

otic vesicle

endolymphatic duct

myelencephalon

metencephalon

cranial n. 3

mesencephalon

Figure 155 10-mm pig embryo, transverse section through accessory cranial ganglia (mag. 30X)

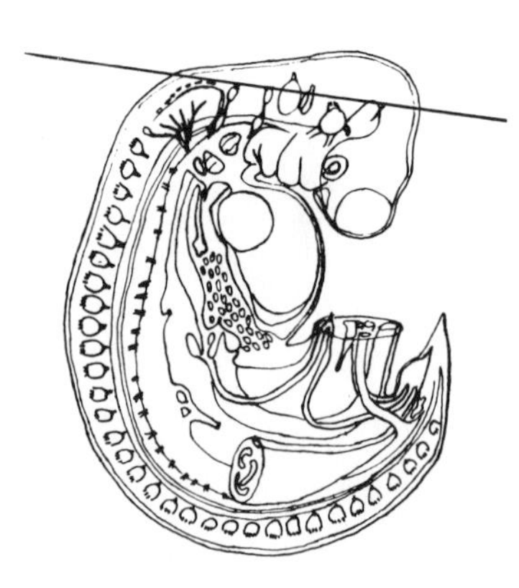

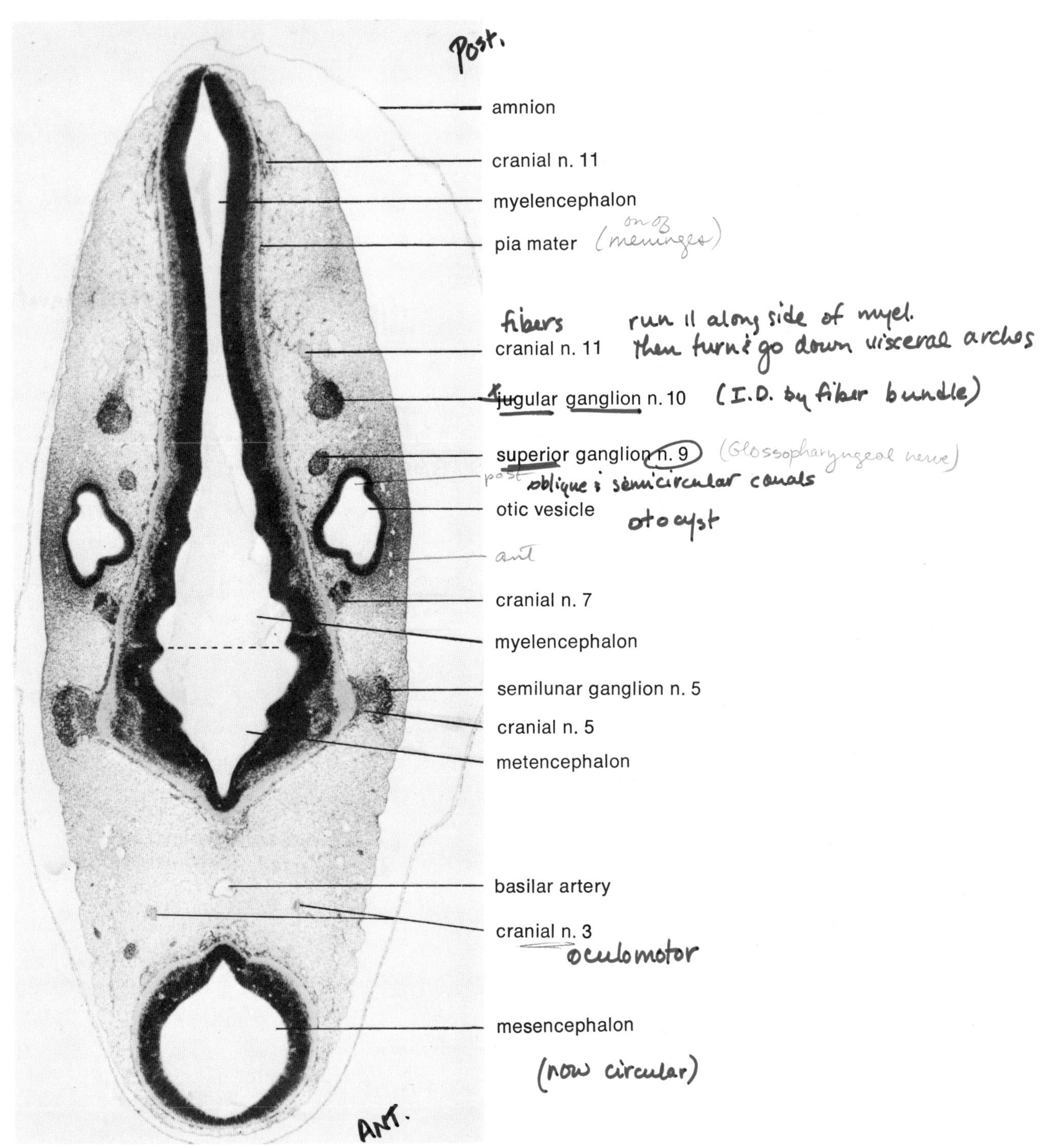

Figure 156 10-mm pig embryo, transverse section through jugular and superior cranial ganglia (mag. 30X)

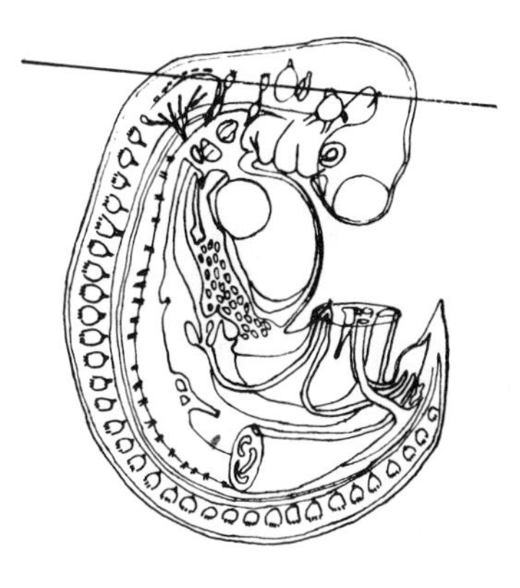

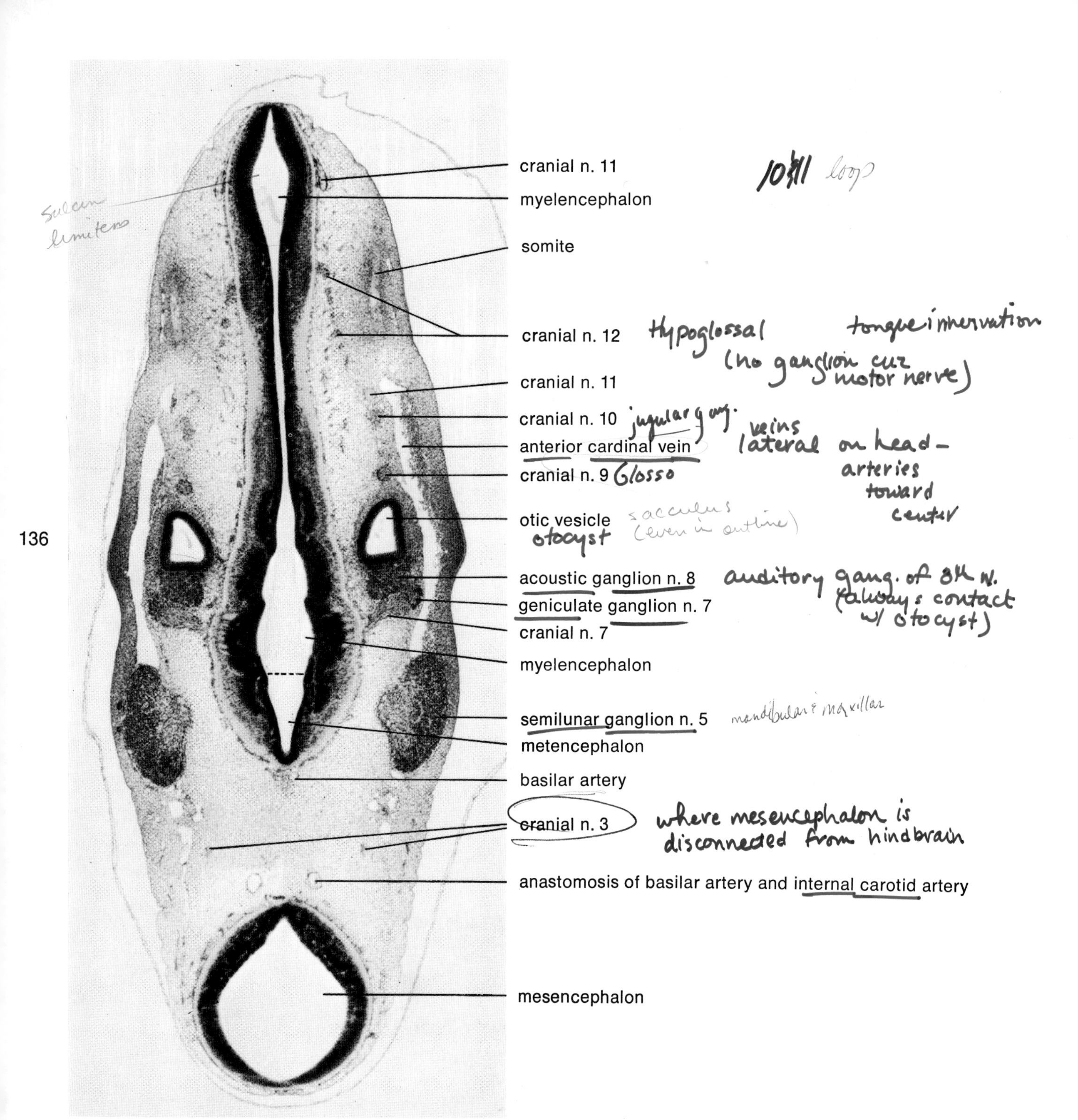

Figure 157 10-mm pig embryo, transverse section through semilunar and geniculate cranial ganglia (mag. 30X)

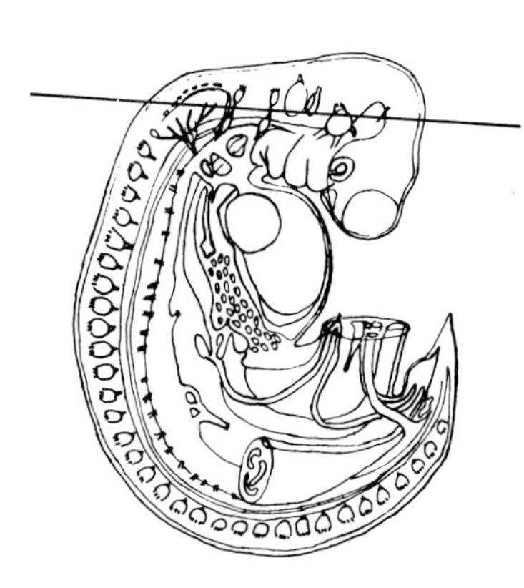

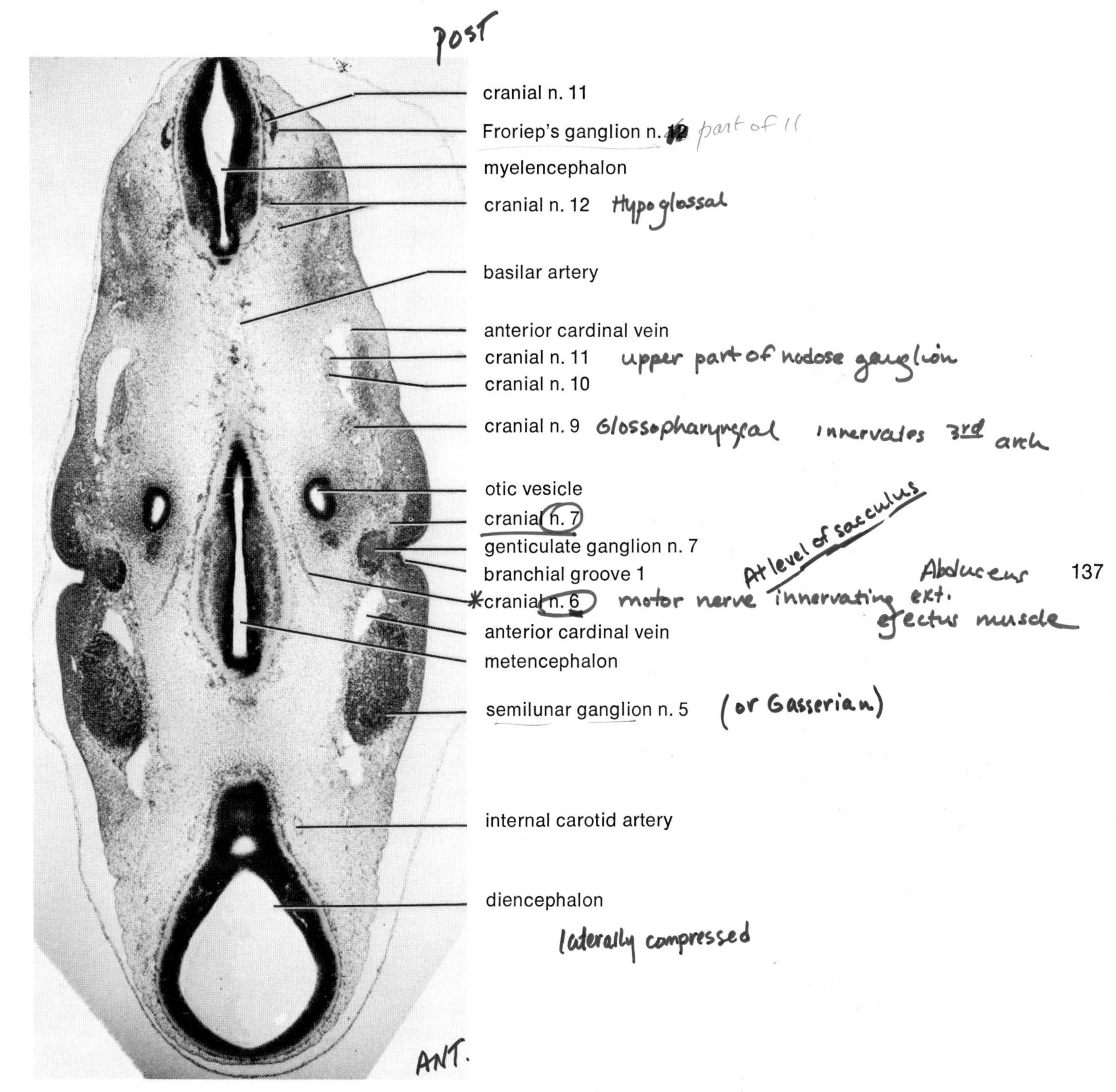

Figure 158 10-mm pig embryo, transverse section through cranial nerve 6 (mag. 30 X)

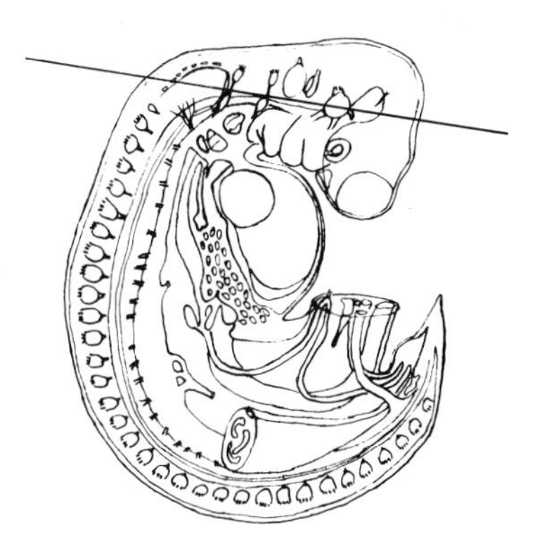

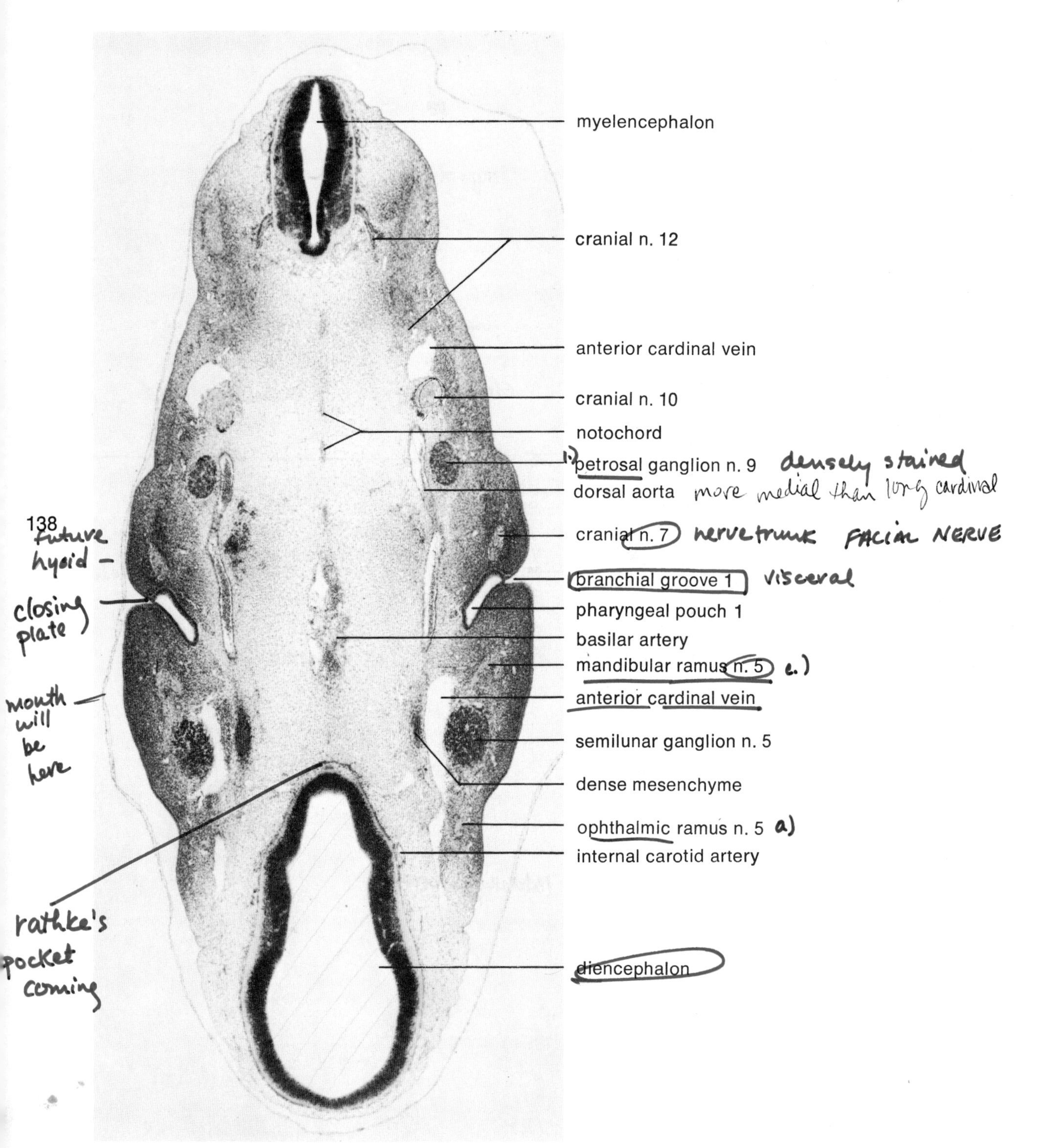

Figure 159 10-mm pig embryo, transverse section through pharyngeal pouch 1 (mag. 30X)

cutting thru upper part of mandibular arch.

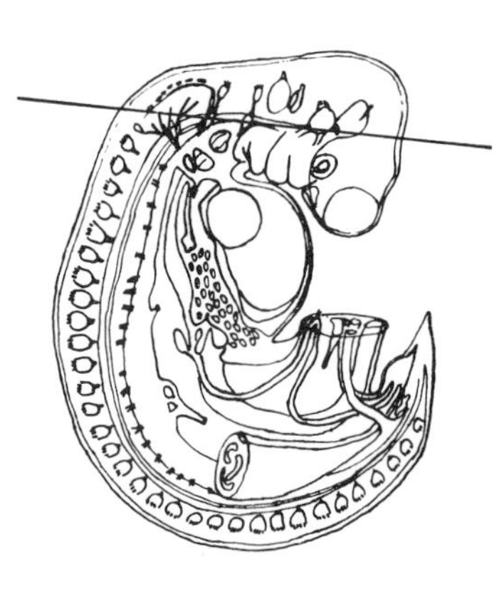

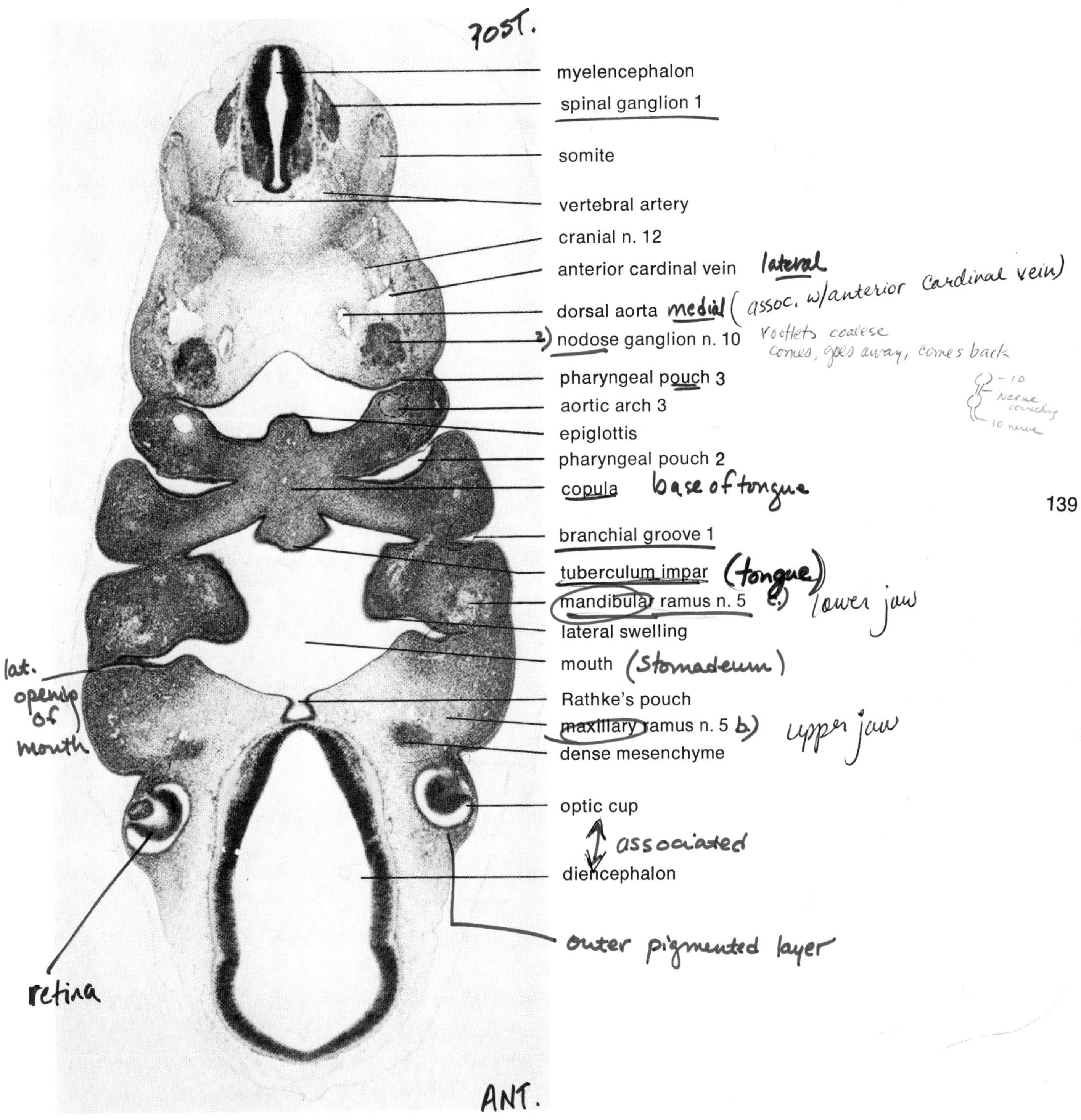

Figure 160 10-mm pig embryo, transverse section through Rathke's pouch (mag. 30X)

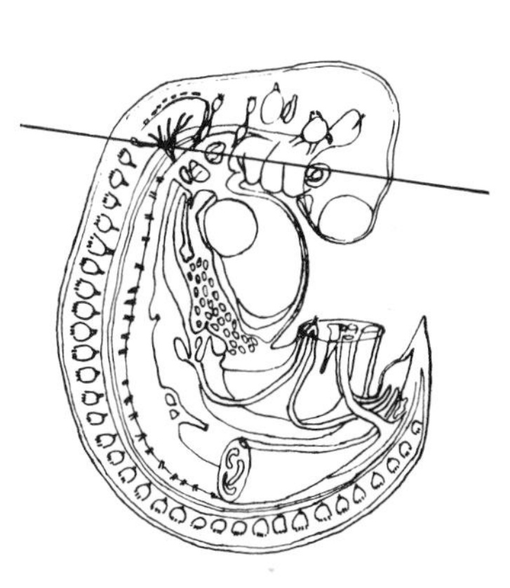

hyaloid artery
temp. supply eye

Tubucular Impar
(forms tongue)

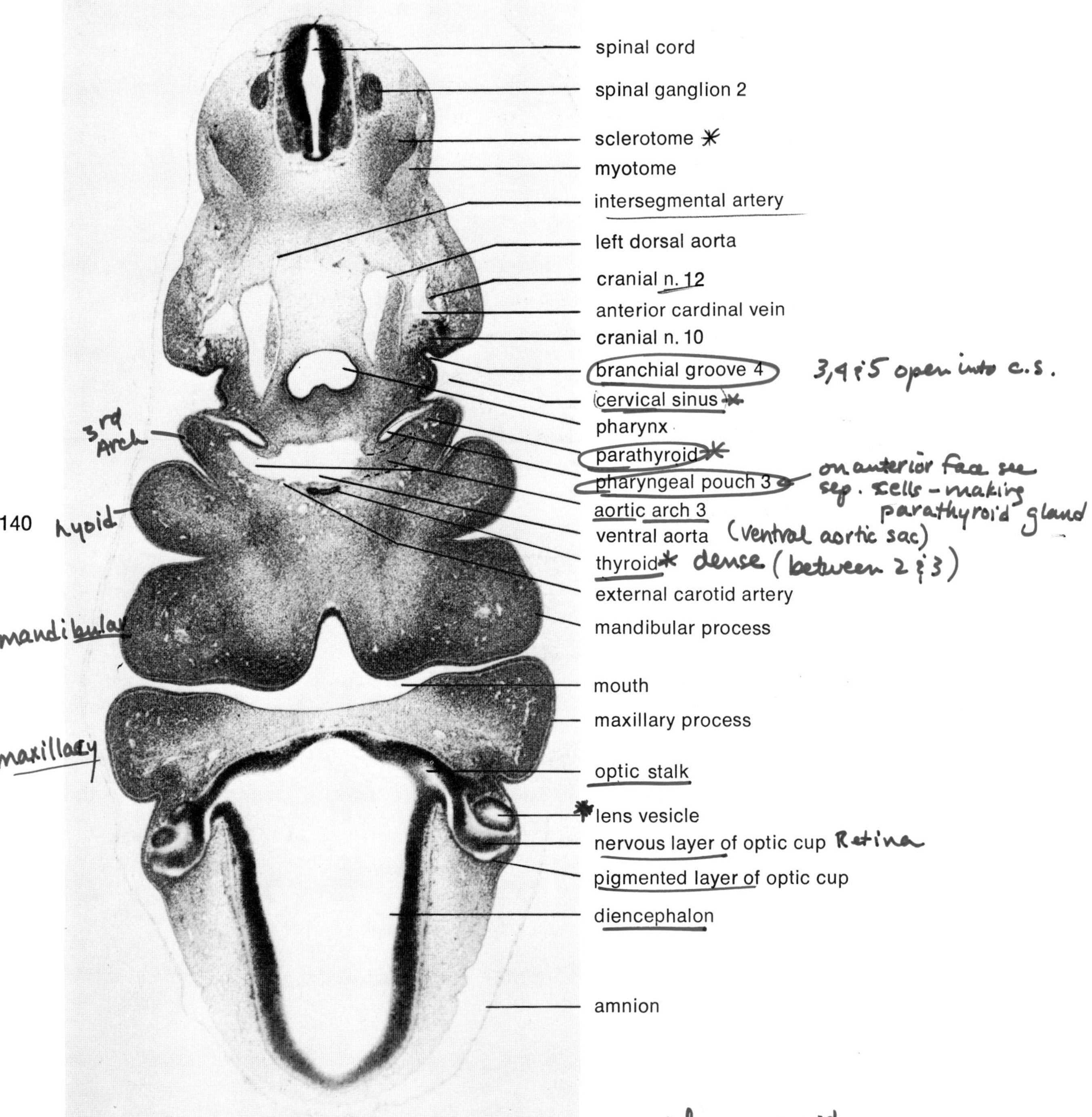

Figure 161 10-mm pig embryo, transverse section through thyroid (mag. 30X)

when pull 3rd pouch away = thymus gland
2 Moves medially & meet.

Choroid fissure - along lateral side (incomplete on top side)

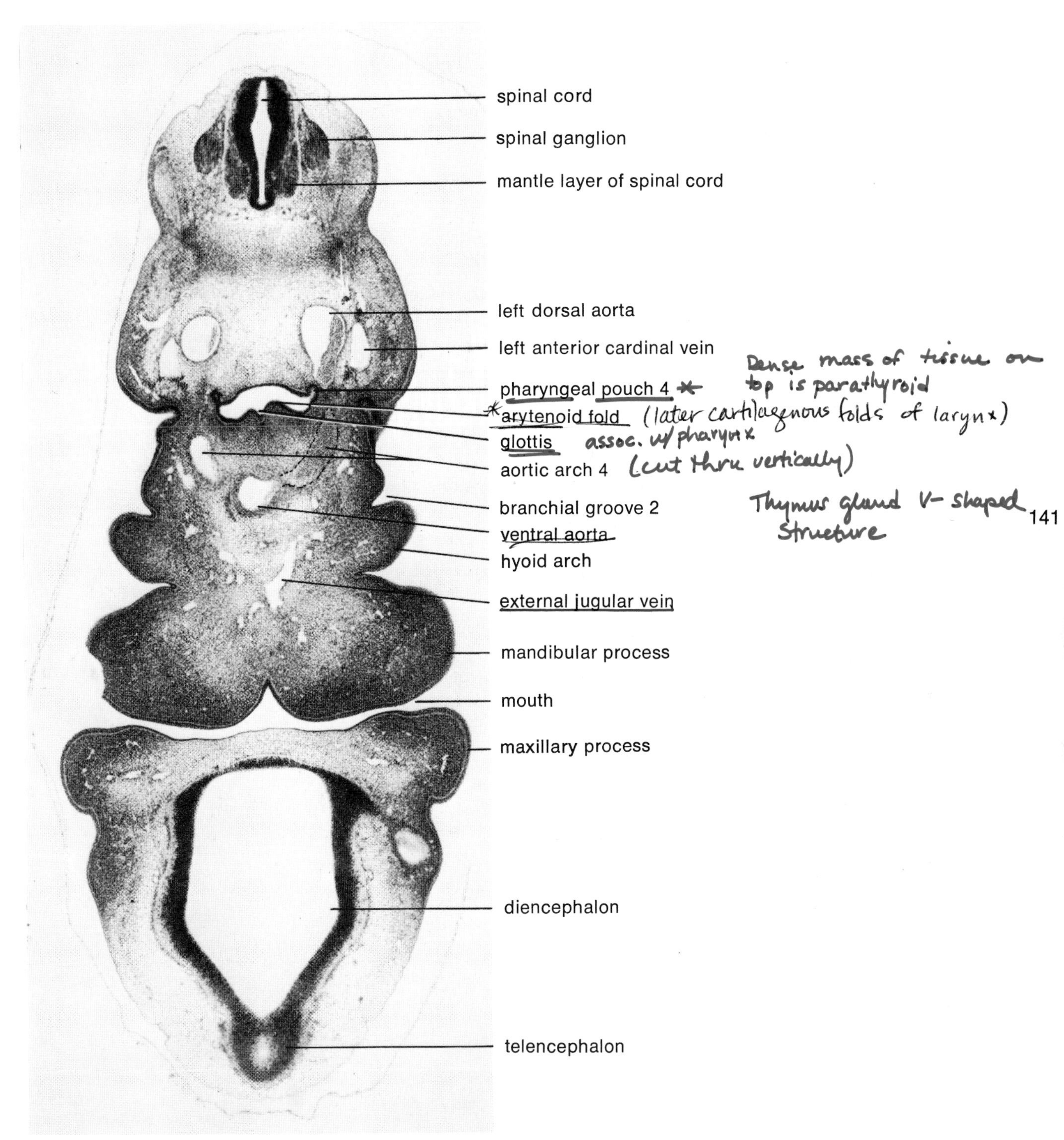

Figure 162 10-mm pig embryo, transverse section through fourth aortic arch (mag. 30X)

Systemic truck (ventral aorta)

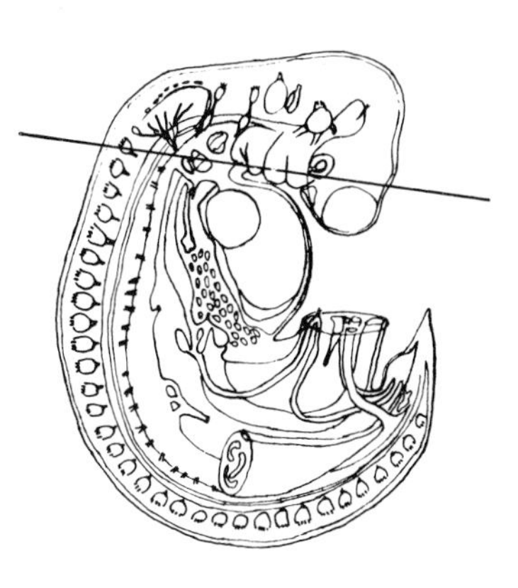

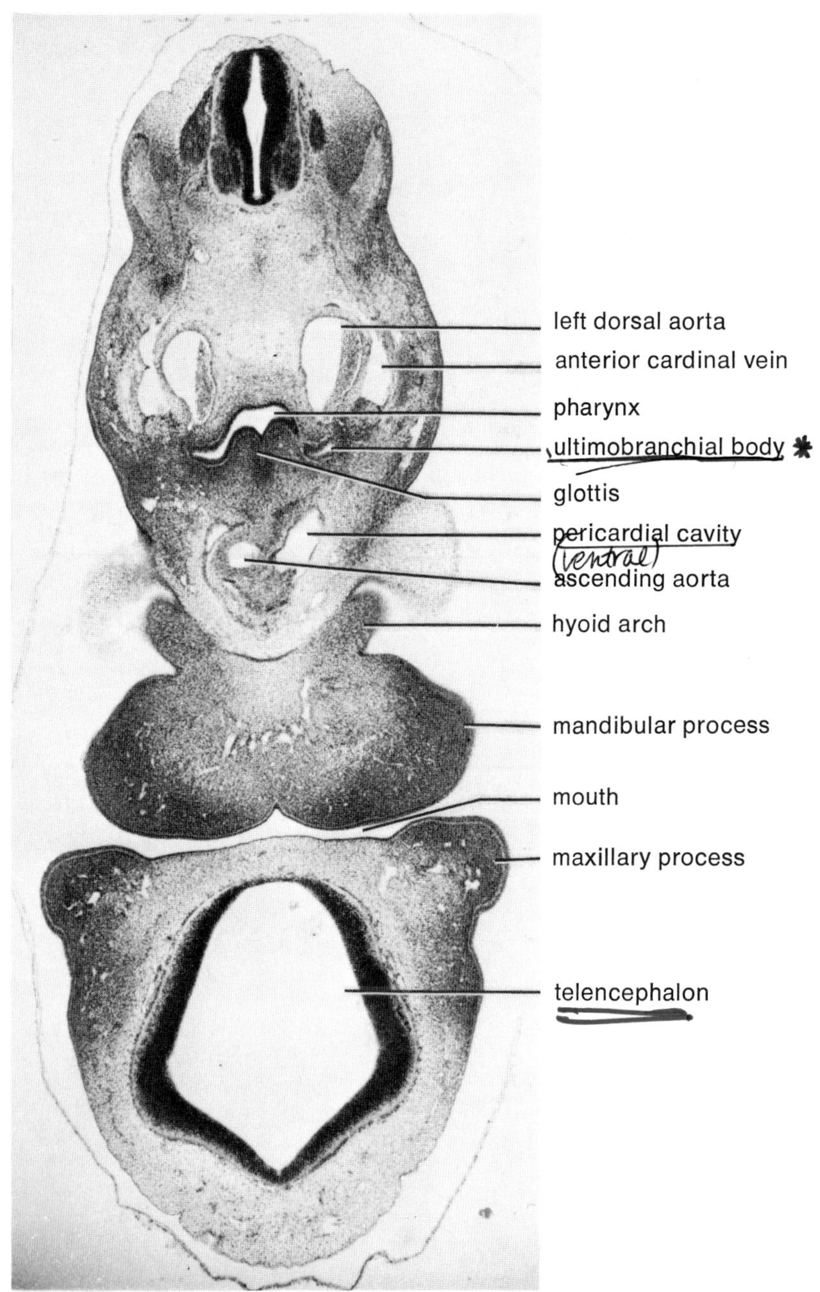

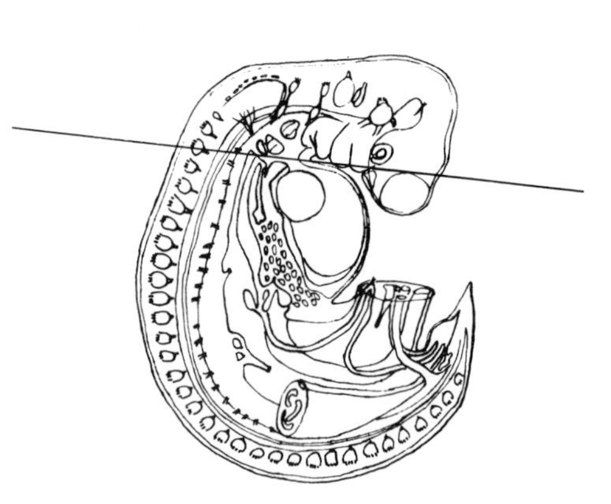

Figure 163 10-mm pig embryo, transverse section through ultimobranchial body (mag. 30 X)

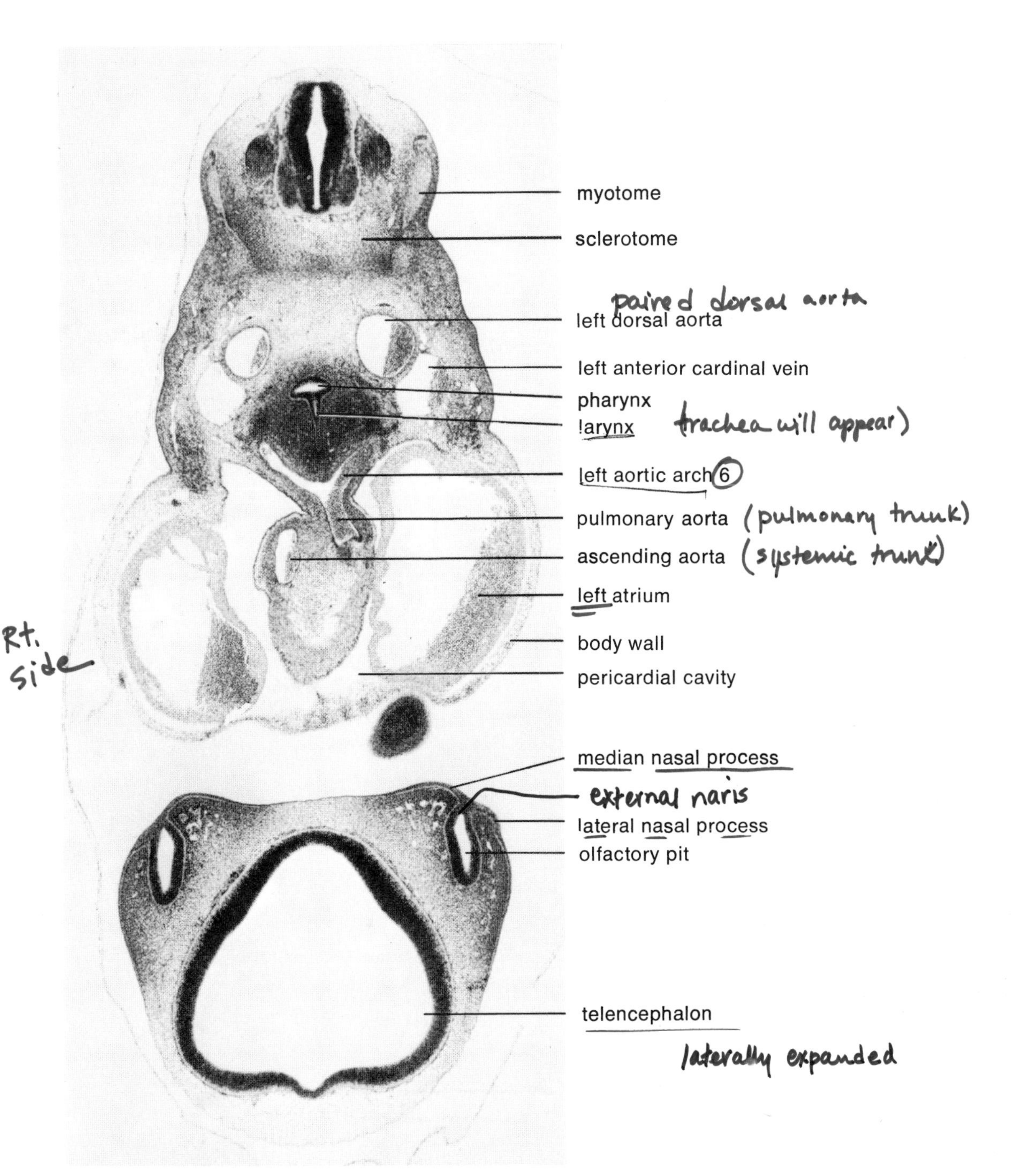

Figure 164 10-mm pig embryo, transverse section through pulmonary aorta (mag. 30X)

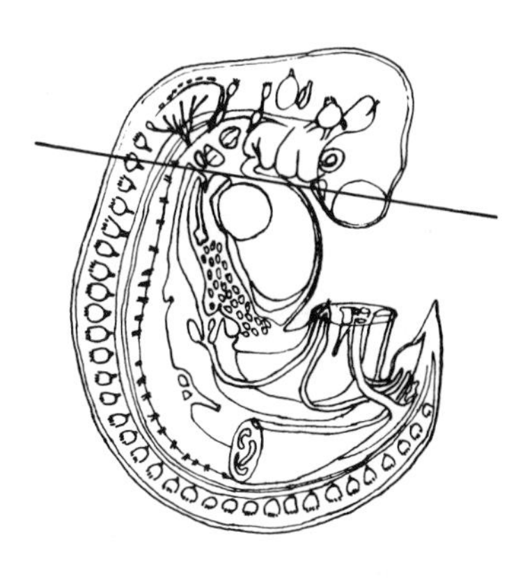

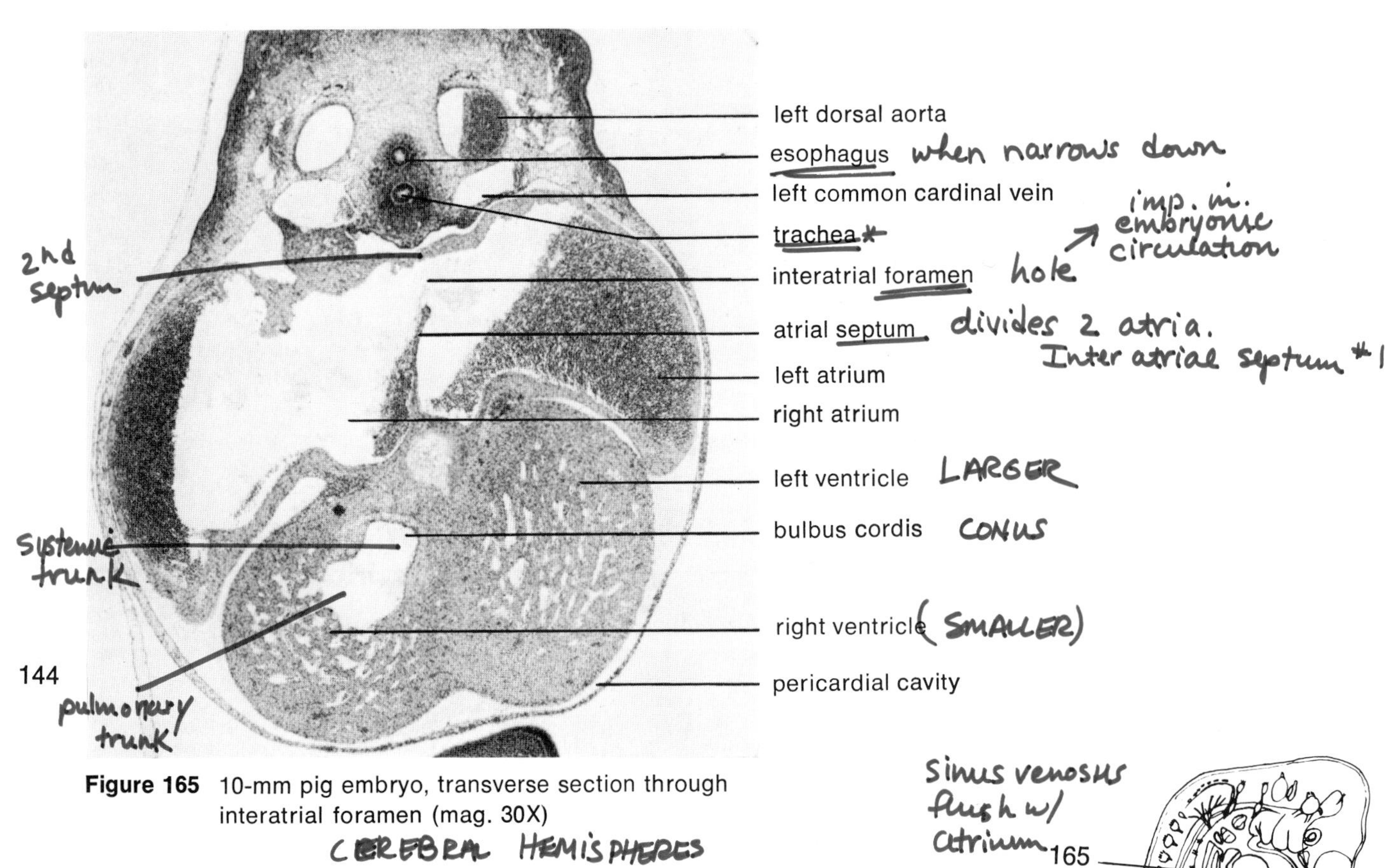

Figure 165 10-mm pig embryo, transverse section through interatrial foramen (mag. 30X)

CEREBRAL HEMISPHERES

Figure 166 10-mm pig embryo, transverse section through interventricular foramen (mag. 30X)

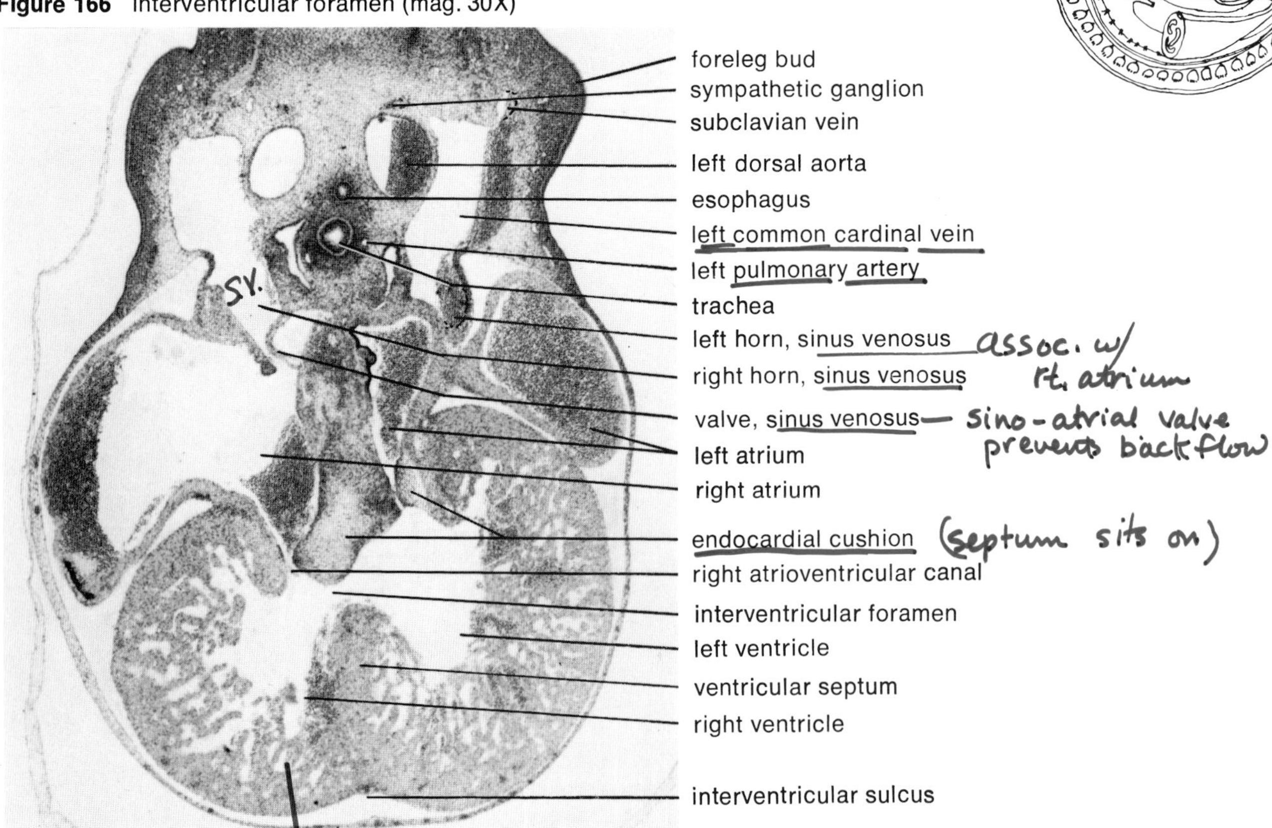

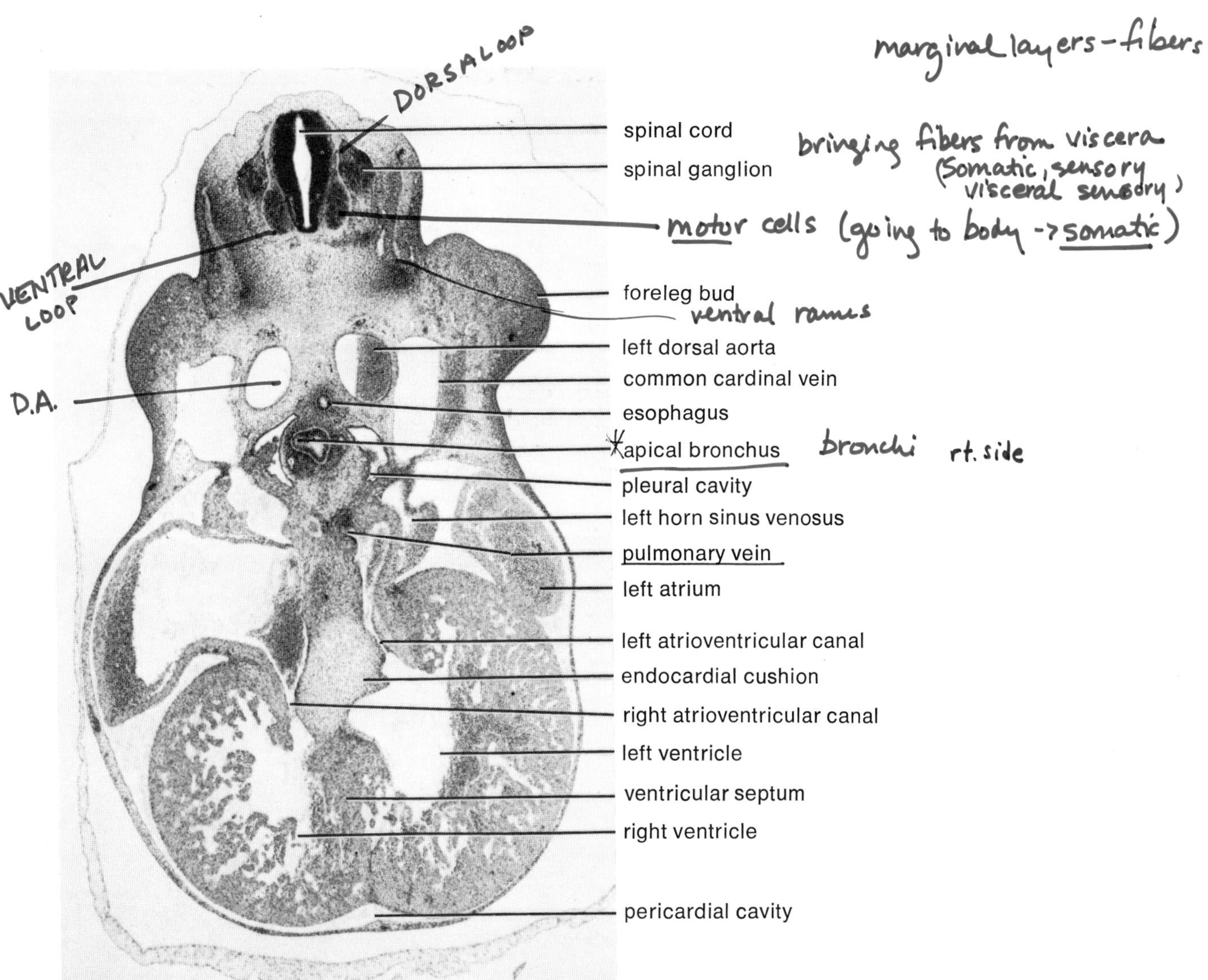

Figure 167 10-mm pig embryo, transverse section through apical bronchus (mag. 30 X)

Bicuspid mitral

Tricuspid

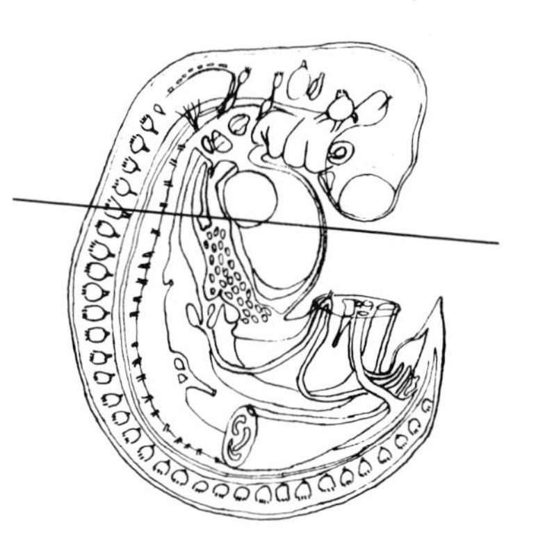

146

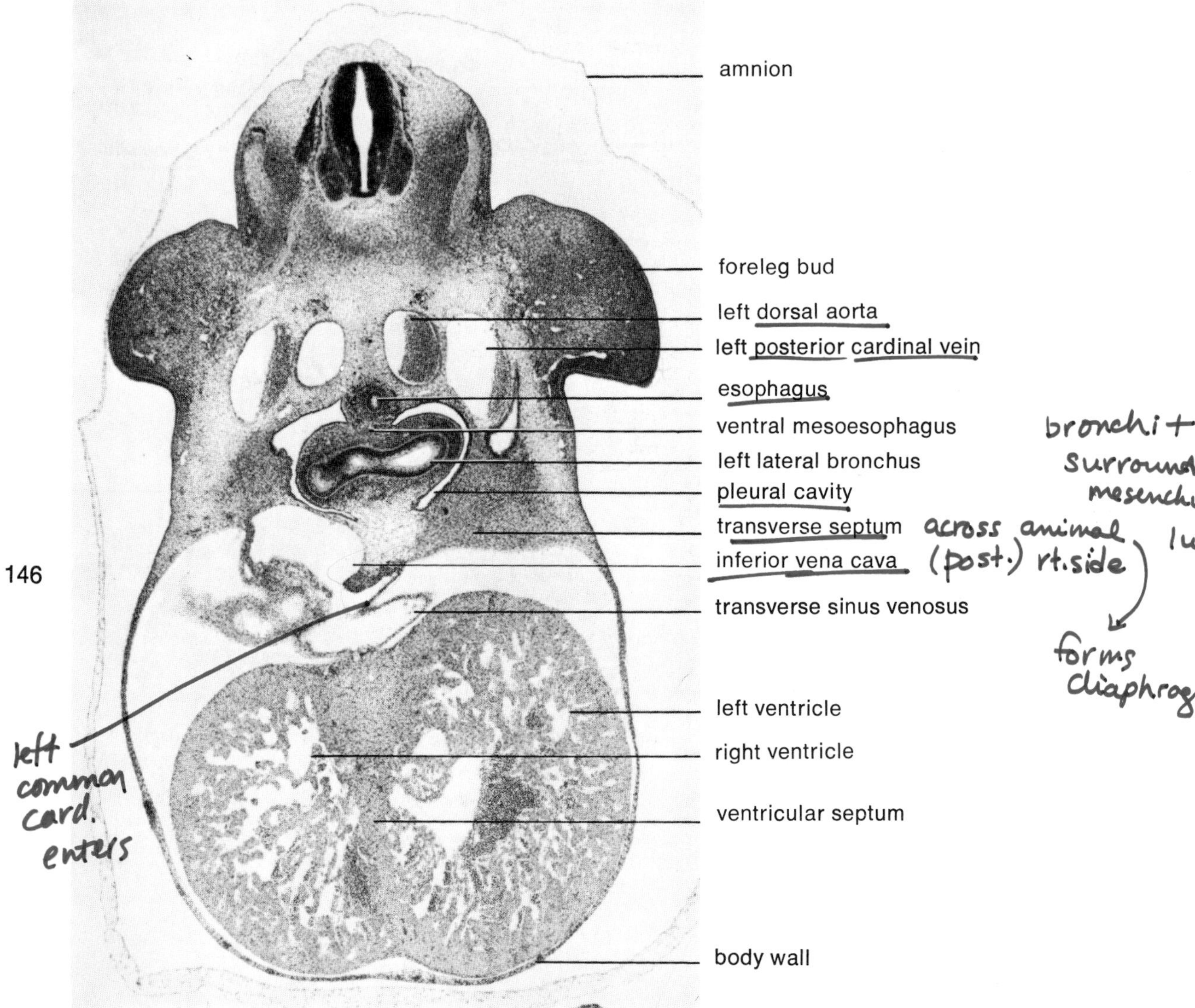

Figure 168 10-mm pig embryo, transverse section through lateral lung buds (mag. 30X)

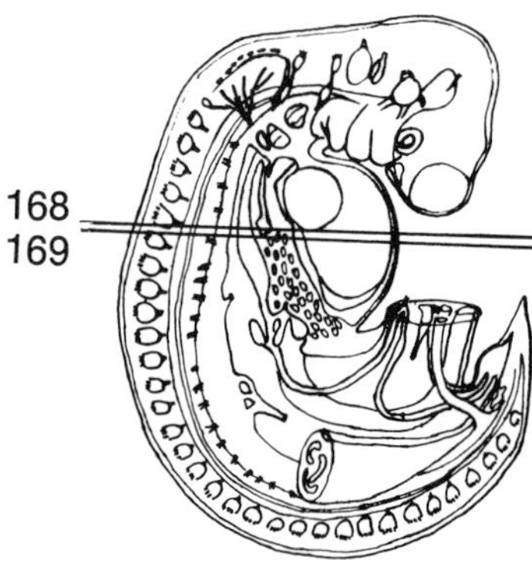

Figure 169 10-mm pig embryo, transverse section through caudal lung buds (mag. 30X)

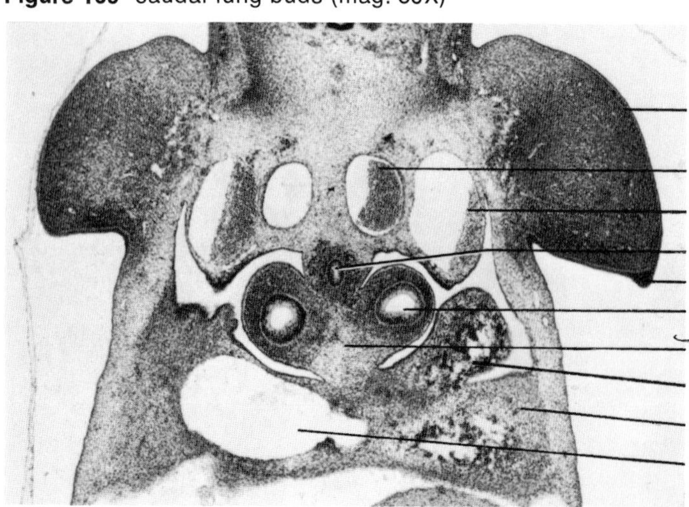

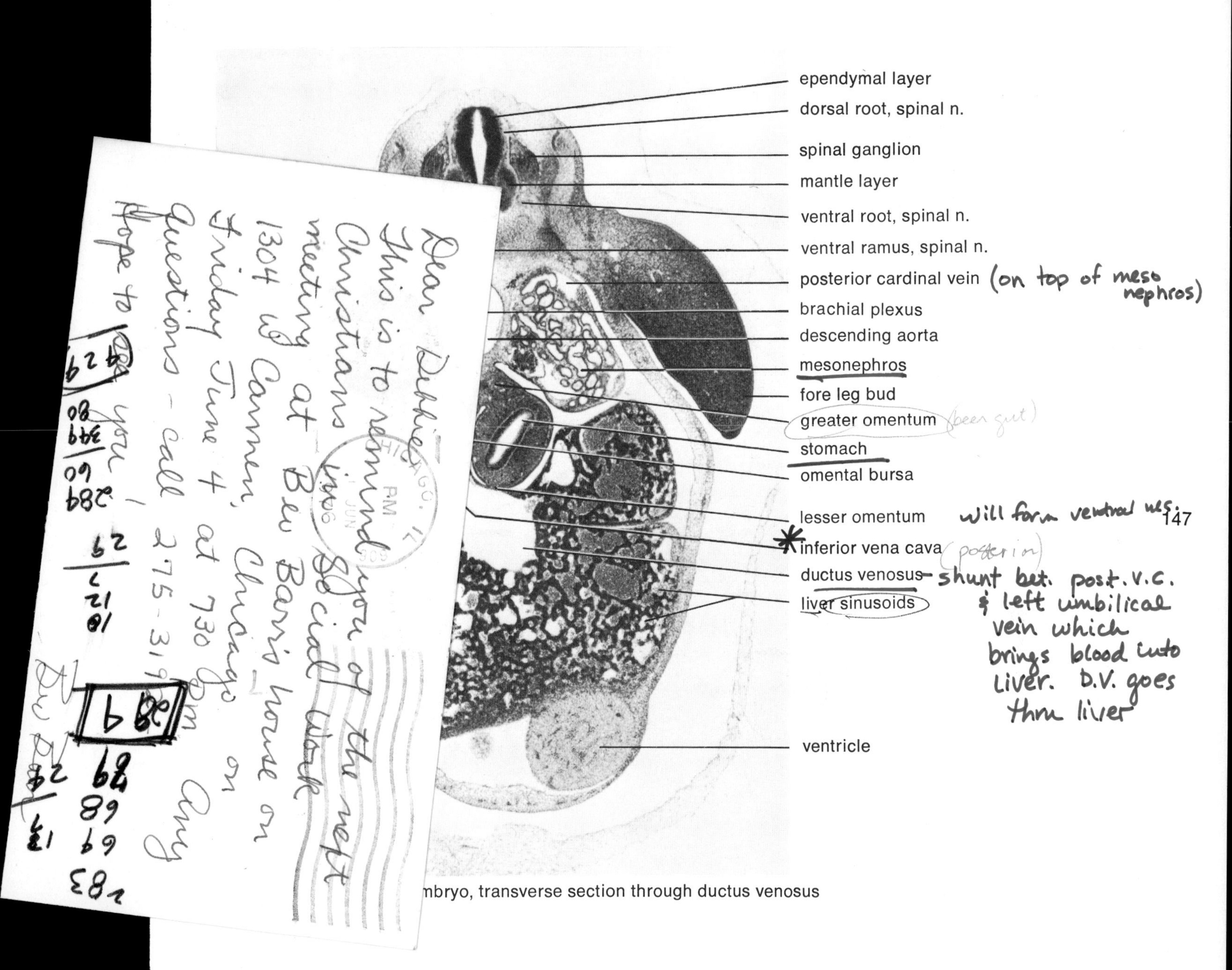

mbryo, transverse section through ductus venosus

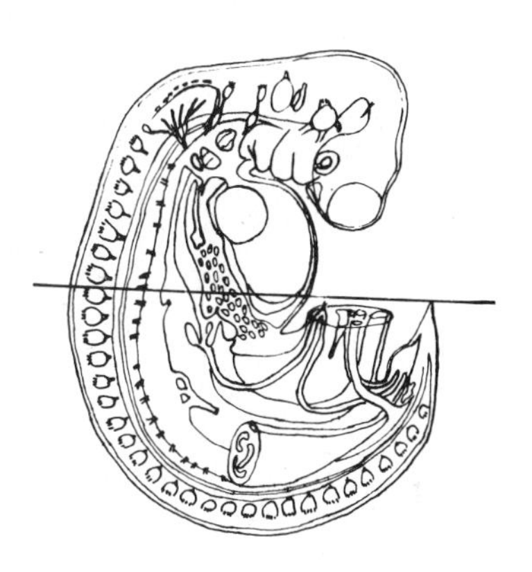

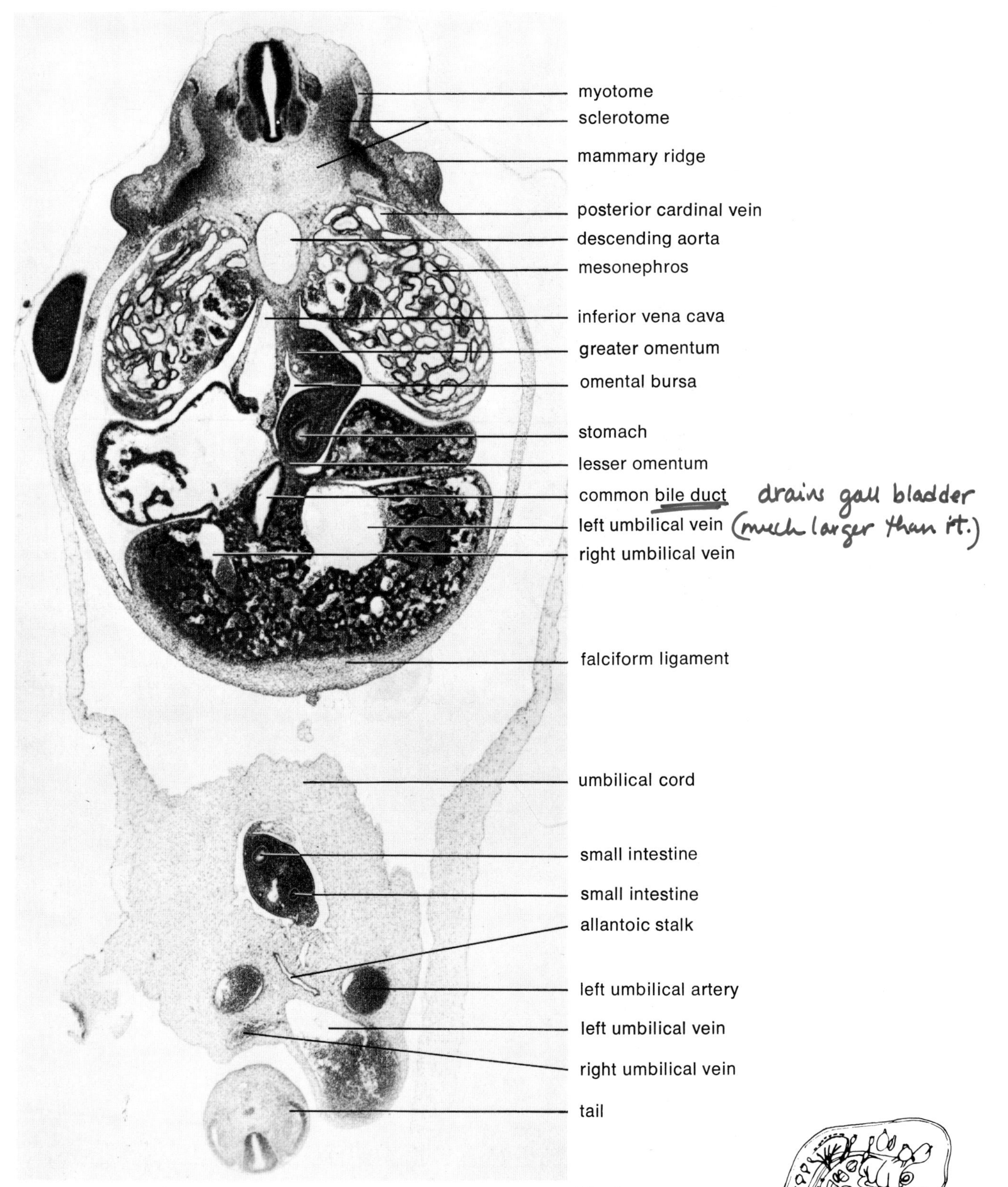

Figure 171 10-mm pig embryo, transverse section through common bile duct (mag. 30X)

drains gall bladder (much larger than it.)

hepatic portal
from rt. & left vitelline vein.
Drains from int. directly into liver.

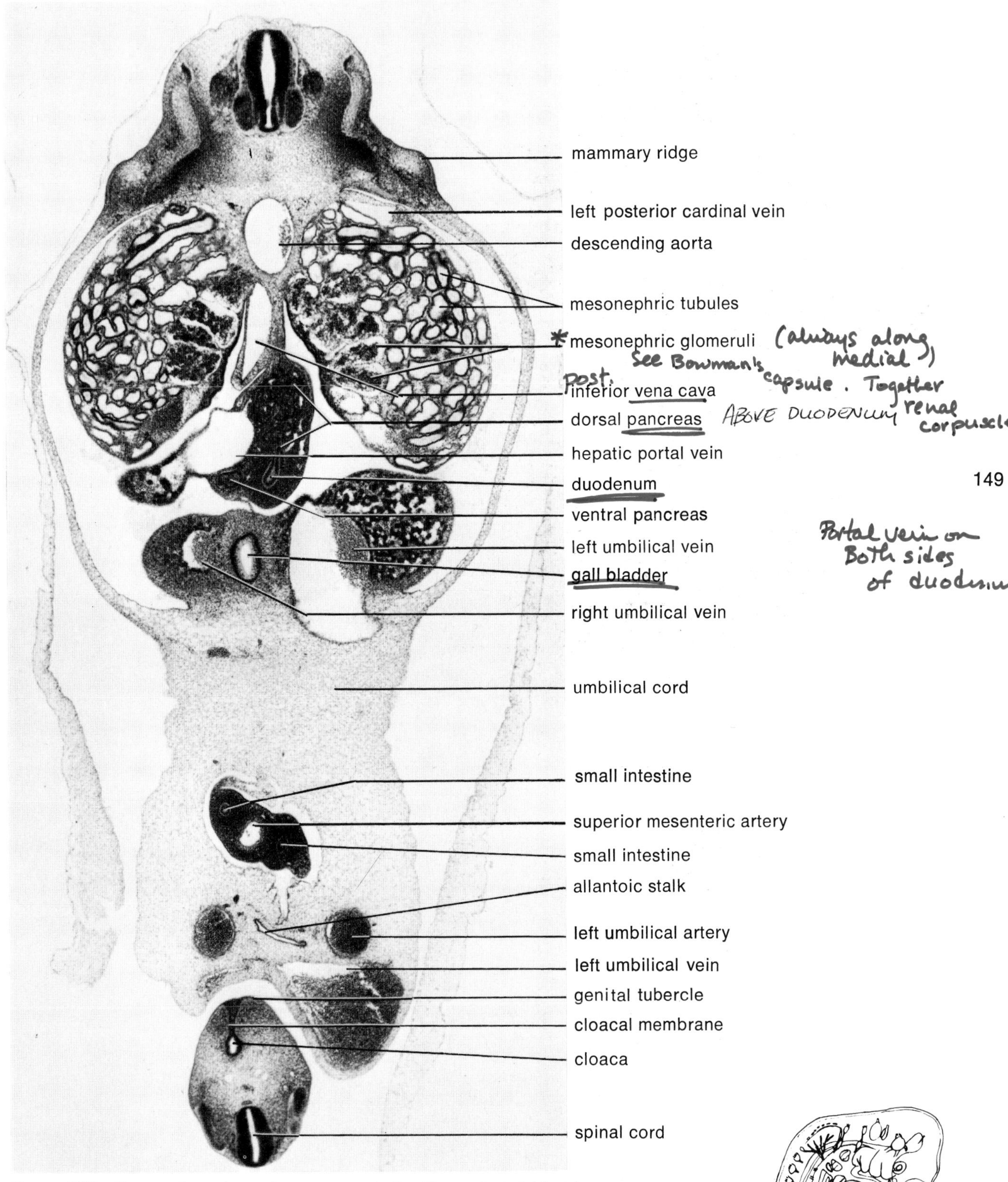

Figure 172 10-mm pig embryo, transverse section through gall bladder (mag. 30X)

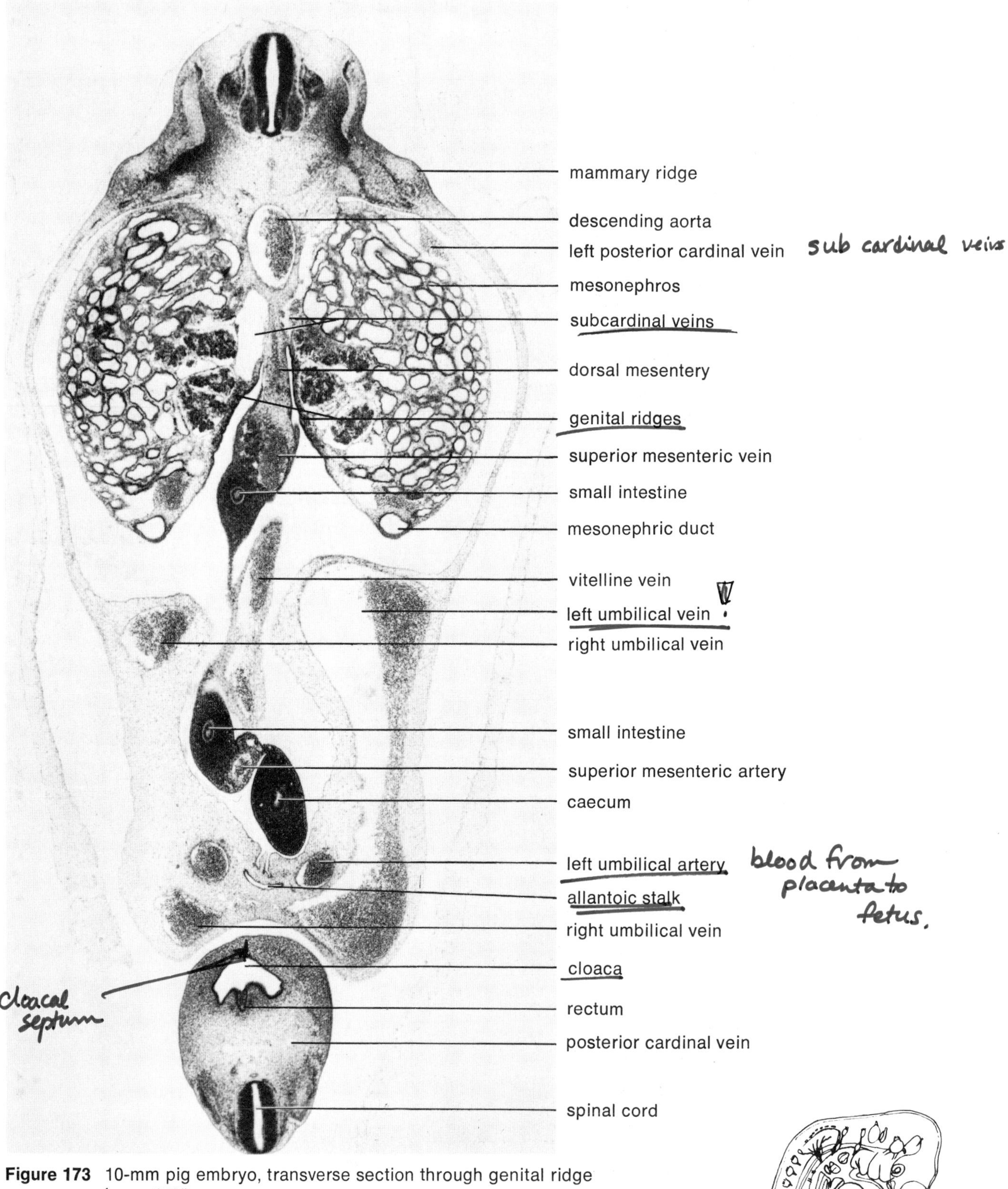

Figure 173 10-mm pig embryo, transverse section through genital ridge (mag. 30X)

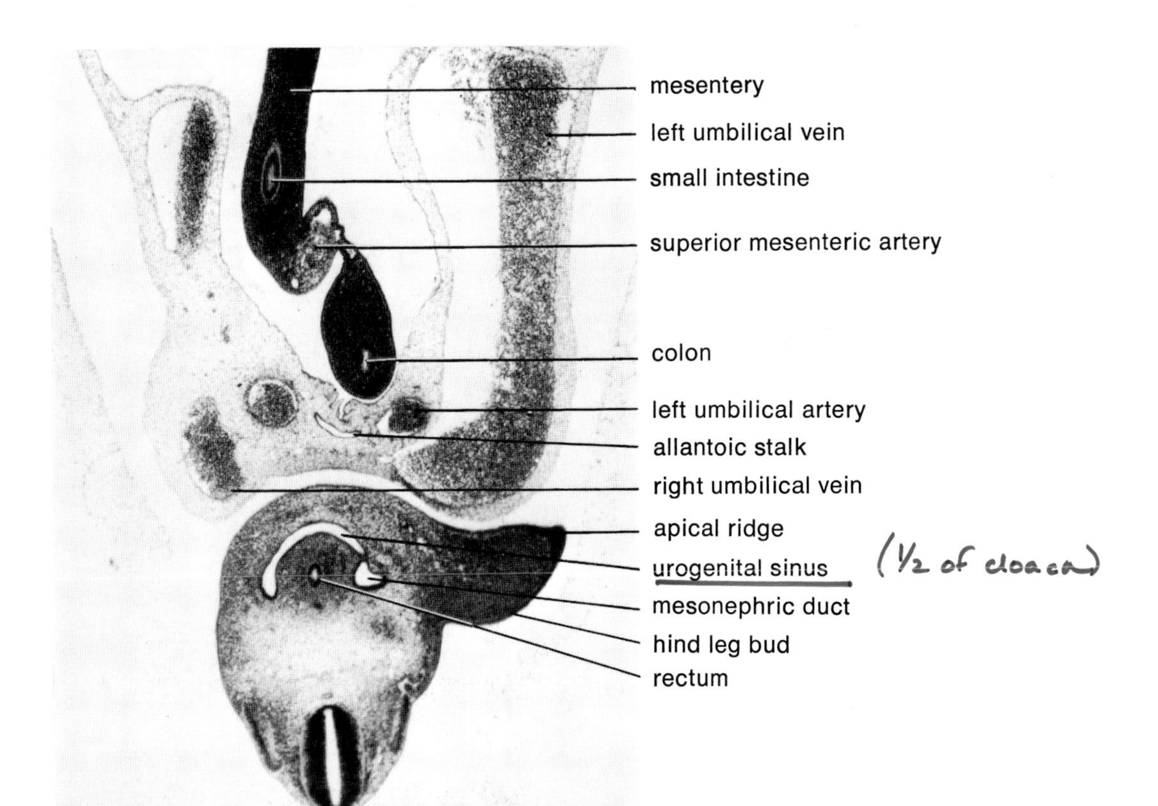

Figure 174 10-mm pig embryo, transverse section through urogenital sinus (mag. 30X)

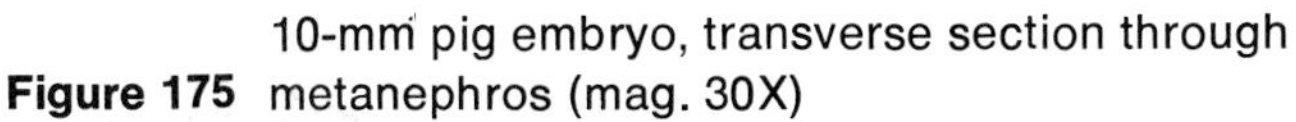

Figure 175 10-mm pig embryo, transverse section through metanephros (mag. 30X)

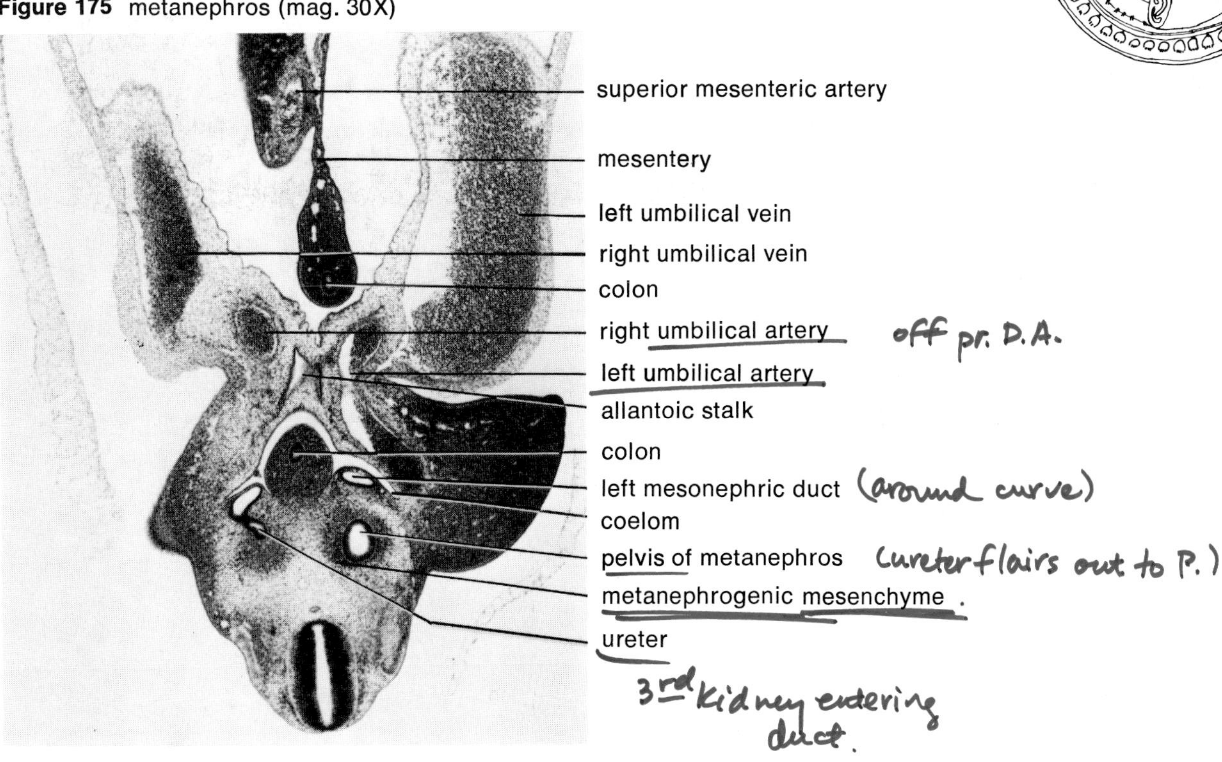

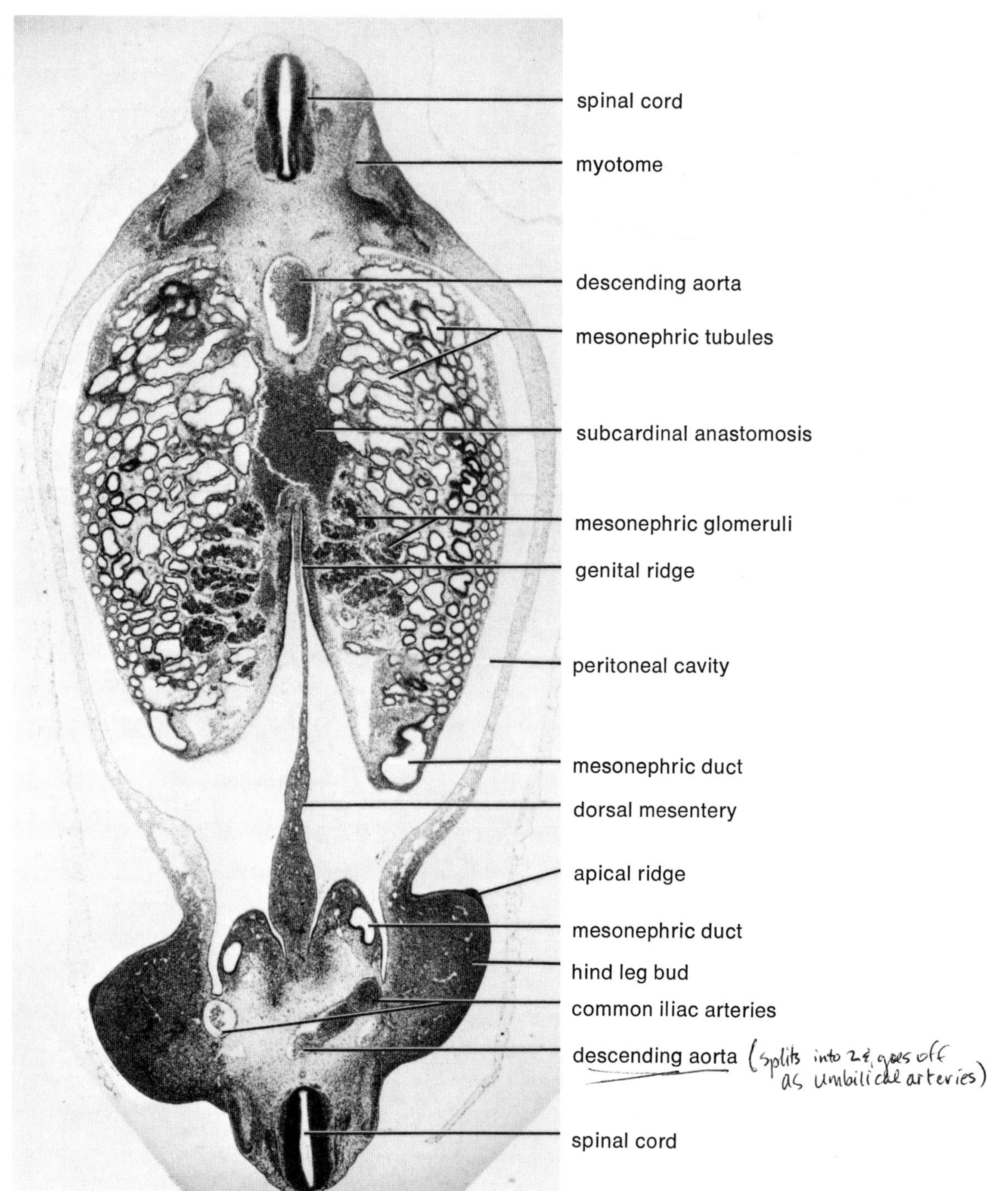

Figure 176 10-mm pig embryo, transverse section through common iliac artery (mag. 30X)

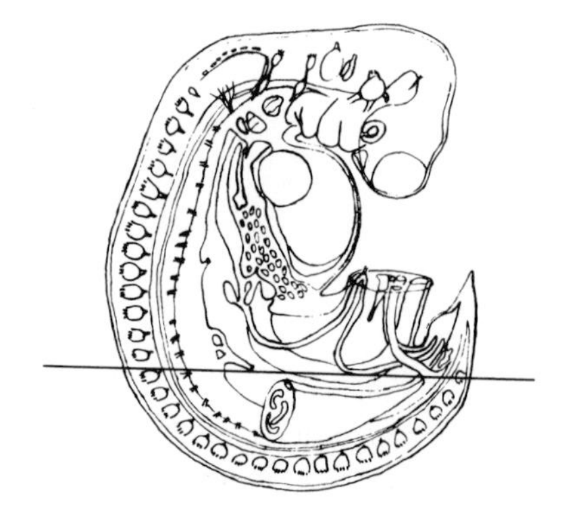

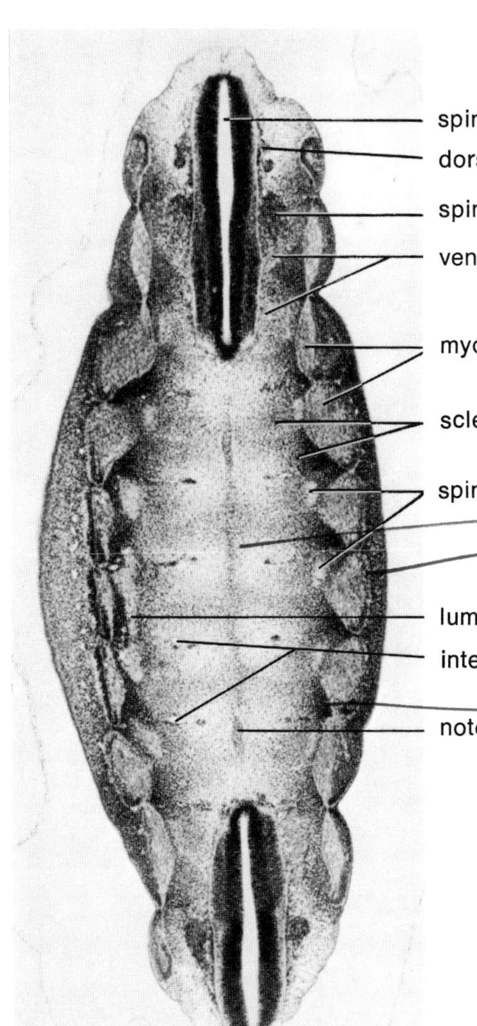

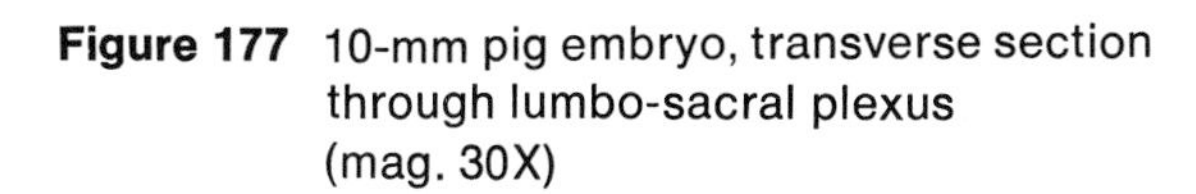

Figure 177 10-mm pig embryo, transverse section through lumbo-sacral plexus (mag. 30X)

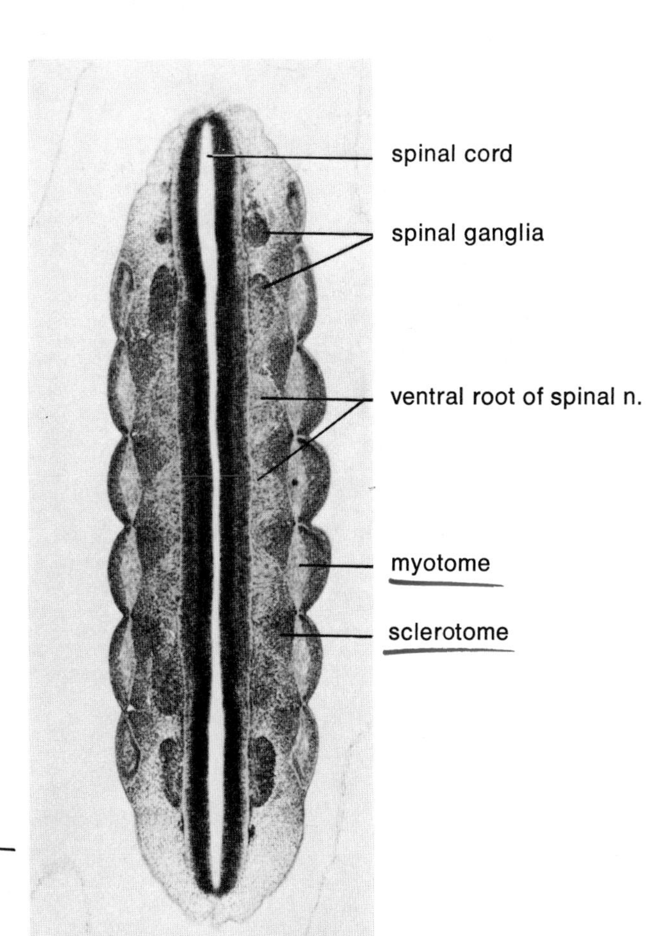

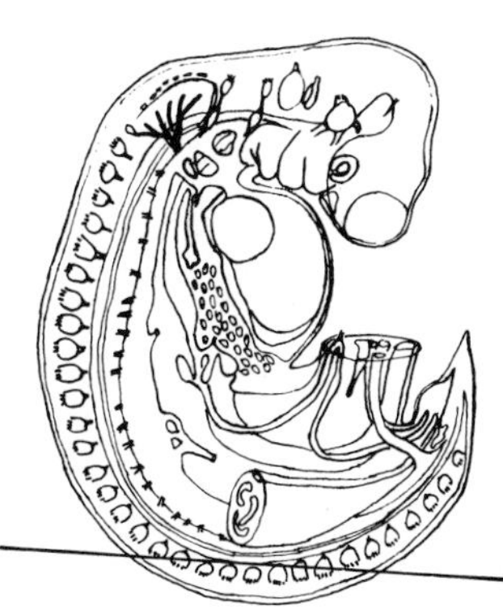

Figure 178

10-mm pig embryo, transverse section through spinal nerves (mag. 30X)

16. Human Placenta

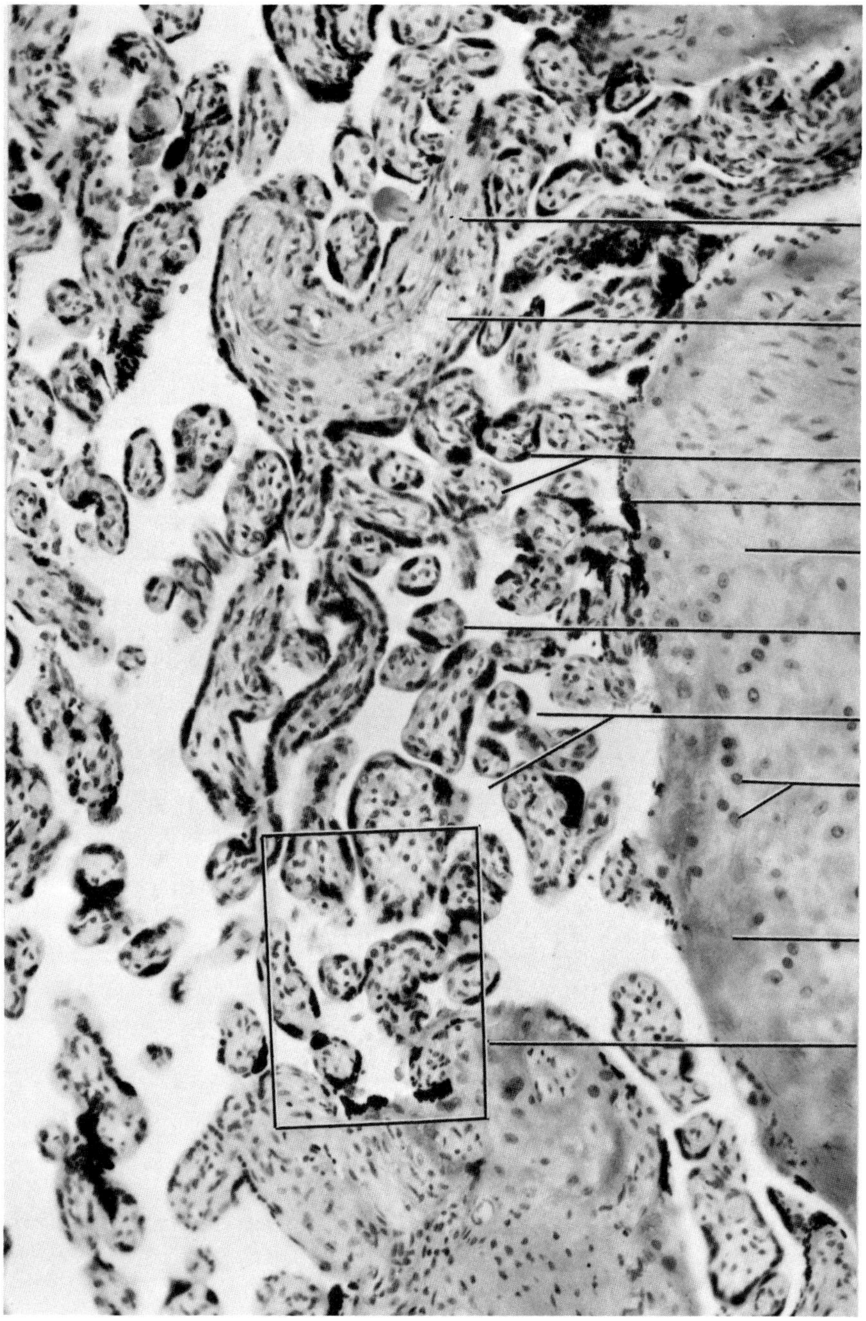

stem placental villus
fetal artery
placental villi
peripheral syntrophoblast
basal plate
villus syntrophoblast
intervillus space
decidual cells
fibrinoid
area in fig. 180

Figure 179 Human placenta, section (mag. 150 X)

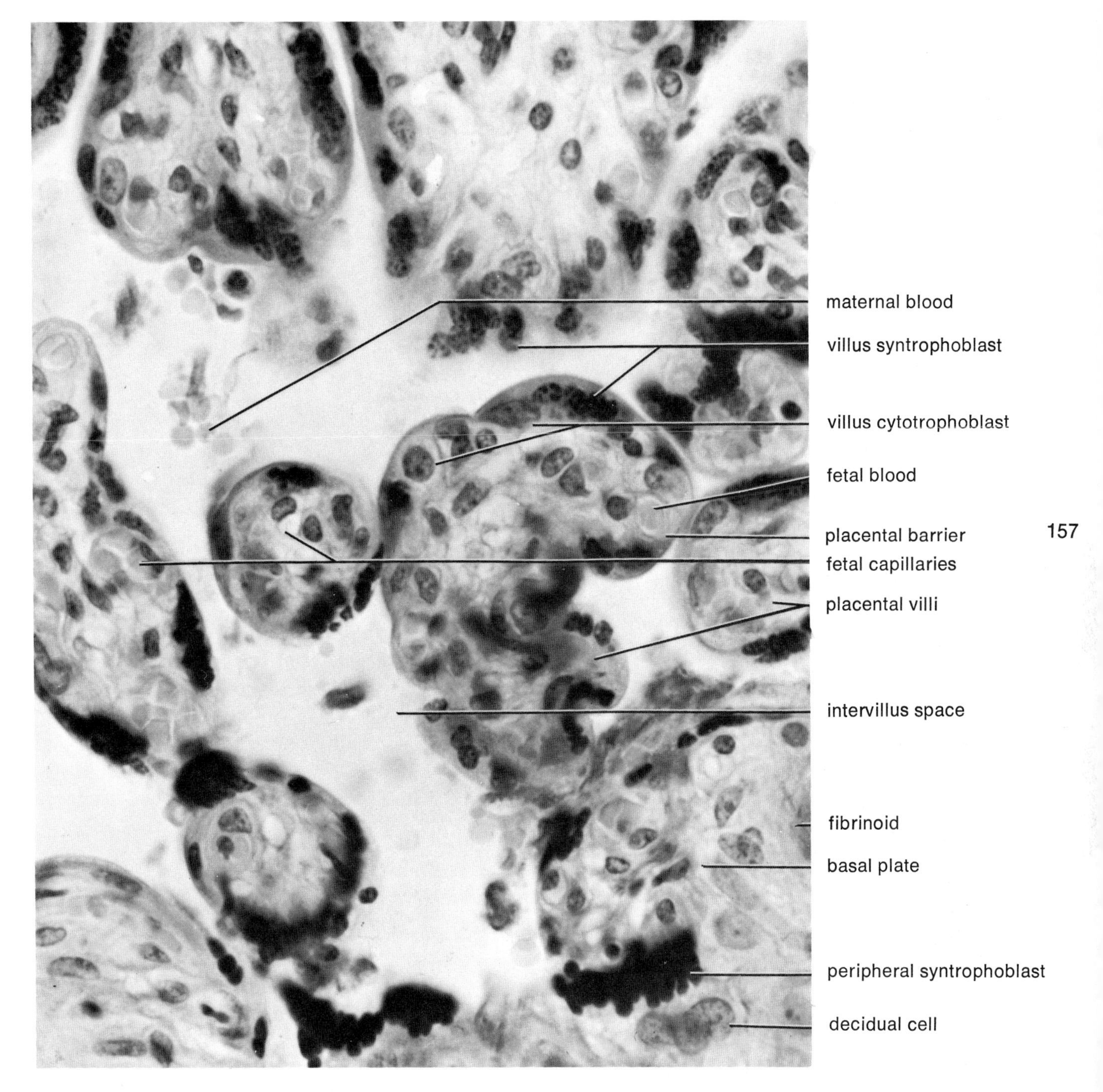

Figure 180 Human placenta, section (mag. 725X)

Glossary and Synopsis of Development

a

accessory cells fig. 20
Ovarian nurse cells of sea urchins containing yolk granules, lipid droplets, and glycogen which are probably transferred to the oocytes.

accessory ganglia of cranial n. 10 figs. 149, 155
A chain of small ganglia along the myelencephalon which contributes sensory fibers to the 10th and 12th cranial nerves.

acoustic ganglion figs. 149, 153, 157
see ganglion of cranial nerve 8.

adhesive glands figs. 51, 61, 62, 70
Paired ectodermal thickenings on the ventral side of the head of anuran tadpoles; secrete adhesive mucus; used for attachment.

allantoic stalk figs. 146, 149, 152, 171
A canal in the umbilical cord connecting the cloaca or, later, the urogenital sinus with the allantois; carries urine to the allantois in many amniotes.

allantoic vein fig. 126
see umbilical vein

allantois figs. 122, 127, 129, 141, 143, 145, 146
An extraembryonic membrane of amniotes; grows out of the hindgut to fuse with the chorion and form the chorioallantois; vascularized by the umbilical vessels; functions as the main embryonic respiratory organ of birds, stores embryonic urine, forms the albumen sac; in mammals, contributes to the placenta.

amnion figs. 108, 109, 110, 119, 127, 128, 130, 139, 142, 146, 151, 154, 155
also see amniotic fold
An extraembryonic membrane of amniotes which encloses the embryo and amniotic fluid; usually arises from folds of somatopleure.

amniotic cavity figs. 108, 127, 130
also see amnion
The lumen of the amnion; contains amniotic fluid and the embryo.

amniotic fold figs. 96, 97, 104, 105, 119, 120, 141
also see amnion; chorion
Folds of somatopleure surrounding the embryo, arising first anterior to the head, then extending along the sides and encircling the tail; the folds are drawn over the embryo, enclosing it with two membranes, an inner amnion, and an outer chorion.

anal arms figs. 32–34
Slender, paired extensions of the ventral body wall of the pluteus; supported by skeletal rods and bear ciliated bands; function to stabilize and propel the larva.

anal pit *see* proctodeum

anastomosis of basilar artery and internal carotid artery fig. 157
A ring-like linkage of basilar and internal carotid arteries encircling the infundibulum; forms the circle of Willis.

animal plate figs. 29, 30
A thickening of ectoderm near the animal pole of sea urchin embryos; bears long cilia of the apical tuft.

animal pole figs. 27, 28, 43
The end of the embryonic axis centered in the most active region; opposite the vegetal pole.

anterior cardinal veins figs. 72, 95, 106, 107, 110, 111, 115, 125, 126, 128, 130, 131, 133, 144, 146, 150, 152, 153, 157, 158, 159, 160, 163, 164
The primitive, paired veins of the head; drain with the posterior cardinals into the common cardinal veins; form the cerebral veins, dural sinuses, internal jugulars, and superior vena cava. *Syn.:* precardinal veins.

anterior intestinal portal figs. 85, 88, 92, 93, 94, 100, 104, 105, 106, 118, 122, 123, 137
also see midgut
In amniotes, the opening from midgut into foregut; moves posteriorly, lengthening the foregut, and meets the posterior intestinal portal at the level of the small intestine to form the yolk stalk.

anterior liver diverticulum figs. 117, 135
also see liver; posterior liver diverticulum
One of two outgrowths of the duodenum of birds which grow, branch, and anastomose to form the liver. *Syn.:* cranial liver bud; dorsal liver bud.

anterior neuropore figs. 85, 87, 94
A temporary anterior opening into the

neural tube; obliterated by complete closure of the prosencephalon.

anterior vitelline veins fig. 107
also see vitelline veins
Paired veins draining the anterior part of the yolk sac and the sinus terminalis; flow into the right and left vitelline veins.

antrum *see* follicular cavity

anus figs. 71, 72
The posterior opening of the digestive tract; derives from the blastopore in the sea urchin and from the proctodeum in vertebrates.

aorta *see* descending aorta

aortic arch 1 figs. 72, 95, 106, 107, 111, 112, 125, 149
The anterior member of a series of six paired arterial connections between the dorsal and ventral aortae; lies within the mandibular arch; degenerates at an early stage.

aortic arch 2 figs. 106, 107, 113, 125, 128, 129, 131, 132, 144, 146, 149
The second of a series of six paired arterial connections between the dorsal and ventral aortae; lies within the hyoid arch; degenerates at an early stage.

aortic arch 3 figs. 72, 106, 107, 123, 125, 128, 129, 131, 132, 133, 144, 146, 149, 150, 152, 153, 160, 161
The third of a series of six paired arterial connections between the dorsal and ventral aortae; lies within the third branchial arch; persists as part of the internal carotid arteries and, in mammals, also as part of the common carotids.

aortic arch 4 figs. 125, 144, 146, 149, 150, 162
The fourth of a series of six paired arterial connections between the dorsal and ventral aortae; lies within the fourth branchial arch. The right fourth aortic arch forms the arch of the aorta in birds, while the left fourth arch forms the arch of the aorta of mammals; in amphibians both right and left persist as aortae. *Syn.:* systemic arch.

aortic arch 5 fig. 150
The fifth of a series of six paired arterial connections between the dorsal and ventral aortae; vestigial and inconstant; atrophies completely.

aortic arch 6 figs. 144, 149, 150, 164
The sixth of a series of paired arterial connections between the dorsal and ventral aortae; forms part of the pulmonary arteries of tetrapods and, in late embryonic stages, the ductus arteriosus.

aortic sac *see* ventral aorta

apical bronchus fig. 167
also see lateral bronchus; stem bronchus
An unpaired evagination of the trachea or the right primary bronchus which, together with the two lateral bronchi and the stem bronchi, constitute the secondary bronchi of pig and human embryos; subsequently forms the upper lobe of the right lung.

apical ridge figs. 138, 141, 169, 174, 176
An ectodermal thickening on the margin of the limb buds of amniotes.

archenteron figs. 29–31, 44, 45, 46, 48, 49
The primitive gut formed by the gastrulation movements; at least part of its wall is endoderm; opens through a blastopore to the exterior. *Syn.:* gastrocoel.

archenteron roof fig. 45
The dorsal wall of the archenteron; in amphibians, forms the notochord and dorsal mesoderm.

archinephric duct *see* mesonephric duct

area opaca figs. 76, 77, 78, 81, 82, 83, 84
The transluscent peripheral zone of the chick blastoderm which is attached to the yolk underneath; surrounds the area pellucida.

area opaca vasculosa figs. 80, 85
also see blood island
The inner region of the area opaca; contains thickenings of splanchnic mesoderm, the blood islands, which differentiate into red blood cells and vitelline blood vessels.

area opaca vitellina fig. 85
The outer region of the area opaca; extends outward from the area vasculosa and is temporarily free of blood and blood vessels.

area pellucida figs. 76, 77, 78, 80, 85
The relatively transparent central region of the chick blastoderm; underlaid by the fluid-filled subgerminal cavity. The primitive streak and embryonic axis form within it.

artifact figs. 37, 43, 75
A change in the structure of a microscopic preparation caused by the processing technique; often appears as spaces or cracks from shrinkage or as granules and fibers precipitated by fixing fluids and solvents.

arytenoid folds fig. 162
Paired ridges lateral to the glottis; derived from the 4th and 5th branchial arches; contribute to the wall of the larynx.

ascending aorta figs. 151, 163, 164
The part of the aorta which extends anteriorly from the heart; forms from the longitudinal division of the bulbus cordis and from the ventral aorta.

atretic follicle figs. 4, 5
A degenerating ovarian follicle in mammals. Atresia may appear at any stage of follicular growth, destroying the oocyte, follicle cells, and theca; in some species (cat) atresia involves dispersal of some follicle cells, which then form interstitial cells.

atrial septum fig. 165
also see interatrial foramen
A partition arising from the wall of the primitive atrium near its sagittal plane in tetrapods; eventually divides it into right and left atria. An interatrial foramen through the septum prevents complete separation of the atria until its closure after hatching in birds or after birth in mammals.

atrioventricular canal figs. 166, 167
The passage within the heart connecting the atrium with the ventricle; encircled and constricted by the endocardial cushion; divides during the partitioning of the heart.

atrium figs. 92, 94, 104, 105, 106, 107, 108, 115, 122, 123, 125, 126, 127, 129, 134, 143, 146, 148
also see left atrium; right atrium
The heart chamber which in the embryo receives blood from the sinus venosus and delivers it to the ventricle; in tetrapods, becomes partitioned into right and left atria and incorporates the sinus venosus.

auditory nerve *see* ganglion of cranial nerve 8

auditory placode *see* otic placode

auditory vesicle *see* otic vesicle

axial filament fig. 13
also see centrioles; sperm tails
A central fiber in cilia and the tail or flagellum of sperm; composed of a ring of nine double filaments and a central pair of filaments; appears to arise from the centrioles near the head of the sperm; functions in the tail movements.

basal plate figs. 179, 180
also see peripheral syntrophoblast; decidual cells; fibrinoid
A sheet of chorionic tissue in contact with the maternal decidua; composed of peripheral syntrophoblast, peripheral cytotrophoblast, fibrinoid; forms part of the afterbirth.

basement membrane figs. 2, 74
A fibrous sheet beneath many kinds of epithelia; supports the epithelia.

basilar artery figs. 144, 149, 150, 156–159
A median vessel beneath the hindbrain; connects the vertebral arteries with the internal carotids.

bivalent fig. 15
A pair of homologous chromosomes in synapsis; found within the primary spermatocyte or primary oocyte during the 1st maturation division. *Syn.:* tetrad.

blastocoel figs. 27–31, 42–45, 46, 49, 53, 87
The cavity of the blastula; a closed space which is invaded early by mesenchyme and in some species is greatly reduced or obliterated by the enlarging archenteron. *Syn.:* segmentation cavity.

blastopore figs. 30, 44, 45, 47, 48, 49, 51
The opening of the archenteron to the exterior; marks the point of origin of the archenteron and the caudal end of the embryo.

b

blastula figs. 27, 28, 42, 43
The final stage of cleavage—typically, a hollow sphere—with the embryonic cells surrounding a cavity, the blastocoel.

blood island figs. 80, 85
A mass of splanchnic mesodermal cells in the gut wall of amphibians or the yolk sac of amniotes; the first blood-forming tissue of the embryo, forms red blood cells and vitelline blood vessels.

body wall figs. 104, 105, 108, 127, 135, 146, 164, 168
The outer layer of the body, enclosing the body cavities and viscera; derives from epidermis, dermatome, myotome, and somatic mesoderm of the embryo.

brachial plexus figs. 148, 170
An interconnection of cervical and thoracic spinal nerves from which branch the nerves of the forelimb.

branchial arch 1 *see* mandibular arch

branchial arch 2 *see* hyoid arch

branchial arch 3 fig. 146
also see aortic arch 3
The third of a series of paired bars in the wall of the pharynx, within which are formed the third aortic arch and a cartilage bar of the visceral skeleton; in aquatic vertebrates, forms and supports gills; in mammals, contributes to the epiglottis and in tetrapods to the hyoid bone.

branchial arch 4 fig. 132
also see aortic arch 4
The fourth of a series of paired bars in the wall of the pharynx, within which are formed the fourth aortic arch and a cartilage bar of the visceral skeleton; in aquatic vertebrates, forms and supports gills; in tetrapods, contributes to the larynx, and in mammals, to the epiglottis.

branchial cleft 1 figs. 104, 105, 112, 122, 123, 126, 128, 129, 143, 147, 148
also see branchial groove 1; pharyngeal pouch 1
A slit-like perforation in the wall of the pharynx between the mandibular and hyoid arches; formed from first pharyngeal pouch and first branchial groove; forms the eustachian tube, middle ear cavity, and external ear canal; in lower vertebrates (frog), contributes to the thymus. *Syn.:* gill slit 1; hyomandibular cleft; visceral cleft 1.

branchial cleft 2 figs. 61, 104, 105, 113, 122, 123, 126, 128, 129, 143
also see branchial groove 2; pharyngeal pouch 2
A slit-like perforation in the wall of the pharynx between the hyoid and third branchial arches; formed from the second pharyngeal pouch and second branchial groove; subsequently closes and obliterates, except contributes to the thymus in lower vertebrates (frog). *Syn.:* gill slit 2; visceral cleft 2.

branchial cleft 3 figs. 61, 122, 126, 128, 129, 143, 153
also see pharyngeal pouch 3
A slit-like perforation in the wall of the pharynx between the third and fourth branchial arches; formed from the third pharyngeal pouch and third branchial groove; subsequently closes, parts of its wall contributing to the thymus and parathyroids. *Syn.:* gill slit 3; visceral cleft 3.

branchial cleft 4 figs. 126, 129, 143
also see branchial groove 4; pharyngeal pouch 4
A thin region in the wall of the pharynx between the fourth and fifth branchial arches; formed from the fourth pharyngeal pouch and fourth branchial groove; breaks through only in lower vertebrates (frog); contributes to the parathyroids. *Syn.:* gill slit 4; visceral cleft 4.

branchial groove 1 figs. 111, 132, 133, 158–60
also see branchial cleft 1
An ectodermal invagination meeting the first pharyngeal pouch to form the first branchial cleft; forms the external ear canal in amniotes. *Syn.:* visceral groove 1.

branchial groove 2 figs. 133, 162
also see branchial cleft 2
An ectodermal invagination meeting the second pharyngeal pouch to form the second branchial cleft; obliterates in tetrapods. *Syn.:* visceral groove 2.

branchial groove 4 fig. 161
also see branchial cleft 4

An ectodermal invagination meeting the fourth pharyngeal pouch to form the fourth branchial cleft; obliterates in tetrapods. *Syn.:* visceral groove 4.

branchial pouch *see* pharyngeal pouch

bulbus cordis figs. 72, 92, 94, 98, 104–108, 114, 115, 122, 123, 125–128, 134, 143, 145, 148, 151, 165
The heart chamber, originally most anterior in position, connecting the ventricle with the ventral aorta; in tetrapods, partitioned longitudinally to form the ascending aorta and pulmonary aorta. *Syn.:* bulbus arteriosus; conus arteriosus.

caecum figs. 118, 173
A pocket in the large intestine of mammals near its connection to the small intestine; appears as an enlargement on the intestinal loop early in the development of the intestine.

caudal artery figs. 72, 149
The extension of the aorta into the tail.

caudal flexure figs. 126, 143
The ventral bend in the tail and posterior trunk; together with the flexures in the head and anterior trunk, forms the embryo into a compact C-configuration.

caudal liver bud *see* posterior liver diverticulum

caudal vein fig. 72
The principal vein of the tail; drains into the posterior cardinals.

cell center *see* centrosome

centrioles fig. 13
also see sperm tails; axial filament
A pair of minute cytoplasmic granules, usually near the nucleus, surrounded by a zone of gelated cytoplasm; self-replicating; composed of a ring of nine sets of ultramicroscopic tubules; associated with the formation of spindle fibers and axial filaments of cilia and flagellae.

centrosome figs. 16, 17
A cytoplasmic organelle composed of two minute granules, the centrioles, and surrounding fibrous protein; forms the mitotic spindle in dividing cells and in sperm also the axial filament. *Syn.:* cell center.

cerebral hemispheres figs. 143, 146, 148
also see telencephalon
Paired dorso-lateral bulges of the telencephalon; form the cerebrum.

cervical flexure fig. 126
also see caudal flexure; cranial flexure; dorsal flexure
One of several ventral bends in the body axis giving the amniote embryo a compact C-configuration; forms in region of the hindbrain and anterior trunk.

cervical segmental artery figs. 144, 150
One of a series of small branches of the aorta arising between the cervical somites; contributes to the subclavian and vertebral arteries.

cervical sinus figs. 147, 161
A depressed region in the sides of the neck bearing the third and fourth branchial grooves in its floor; subsequently closes and obliterates.

chitinous layer figs. 15–19
A thick, clear layer of the egg shell of *Ascaris;* formed from the egg after separation of the fertilization membrane; composed of chitin and protein.

chorda *see* notochord

chorion figs. 109, 110, 111, 114, 115, 116, 118, 119, 128, 130, 134, 137, 139, 140, 141, 142
also see amniotic fold
The outermost extraembryonic membrane of amniotes; arises from somatopleure and is usually drawn over the embryo by the amniotic folds; later fuses with the allantois to form the chorioallantois, which is vascularized by the umbilical vessels and functions as the main embryonic respiratory organ of birds; the chorioallantois contributes to the placenta of mammals. *Syn.:* serosa.

choroid fissure figs. 106, 113, 126
A groove in the ventral wall of the optic cup and optic stalk; after mesenchyme and blood vessels invade the fissure, its lips fuse.

chromosomes figs. 16, 37
Threads of chromatin in the cell nucleus or in the mitotic and meiotic division figures; contain the genes in linear order; composed of DNA, RNA, and protein.

C

cleavage division, 1st fig. 18
The first mitotic division of the egg after fertiization; forms the two-cell stage.

cleavage division, 2nd fig. 19
The second mitotic division after fertilization; forms the four-cell stage.

cleavage furrow fig. 41
A.constriction in the cytoplasm which divides the egg or blastomere; forms during the telophase of the cleavage divisions.

cleavage stage figs. 18, 19, 23–28, 41–43
The period of development beginning with the first mitotic division of the egg and ending with the blastula; a period of rapid mitoses during which no growth occurs, the cells becoming smaller, as they become more numerous.

cloaca figs. 72, 122, 128, 143, 146, 172, 173
also see urogenital sinus
The posterior chamber of the vertebrate digestive tract; receives the allantoic stalk, urinary ducts, and reproductive ducts; partitioned in mammals to form the rectum, urinary bladder, and urogenital sinus.

cloacal membrane figs. 142, 152, 172
A double-layered membrane formed where the ventral wall of the cloaca fuses with ventral ectoderm; ruptures to open the anus and, in mammals, the urogenital sinus as well.

coeliac artery figs. 144, 149, 150
A ventral branch of the aorta supplying the anterior digestive system; derives from vitelline arteries.

coelom figs. 20, 50, 57, 68, 88, 89, 95, 98, 102, 103, 121, 127–129
also see coelomic vesicles; embryonic coelom; extraembryonic coelom; pericardial coelom
A cavity within the mesoderm which forms the body cavities of the adult; in vertebrates, arises as a cleft in the lateral plate mesoderm, which is thereby divided into somatic and splanchnic layers.

coelomic vesicles fig. 32
Two pouches arising from the inner end of the archenteron of sea urchins; divide to form coelom and water vascular cavities.

collecting tubules fig. 36
In the frog testis, conducting vessels carrying mature sperm from the seminiferous tubules to the vasa efferentia.

colon figs. 174–175
also see caecum
Part of the large intestine; arises from hindgut.

common bile duct fig. 171
A vessel connecting the cystic and hepatic ducts to the duodenum; arises from the stem of the liver diverticulum. *Syn.:* ductus choledochus.

common cardinal vein figs. 72, 95, 106, 116, 125, 126, 128, 134, 144, 146, 150, 152, 153, 165, 166, 167
The trunk of the anterior and posterior cardinal veins; connects to the sinus venosus; contributes to the anterior vena cava of the adult. *Syn.:* duct of Cuvier.

common iliac arteries figs. 144, 150, 176
also see umbilical arteries
The large terminal branches of the aorta; arise from the proximal segments of the umbilical arteries and a pair of dorsal intersegmental arteries; persist as the common trunks of the external and internal iliac arteries.

connective tissue figs. 4, 73, 75
One of the main kinds of tissue; characterized by much intercellular material, including fibers.

conus arteriosus *see* bulbus cordis

copula fig. 160
also see tongue
A median elevation on the floor of the mouth arising from the hyoid arch and contributing to the root of the tongue.

corpus luteum (*pl.* **corpora lutea**) fig. 6
A mass of endocrine gland tissue in the ovary of viviparous vertebrates; forms from an ovulated follicle and persists intc pregnancy.

cortex fig. 4
The outer zone of an organ or egg; contains the follicles and corpora lutea in the ovary.

cranial flexure fig. 126
also see caudal flexure; cervical flexure; dorsal flexure

One of several ventral bends in the body axis giving the amniote embryo a compact C-configuration; forms in the midbrain; the only permanent flexure.

cranial nerve 3 figs. 132, 133, 149, 155–157, 158
A pair of motor nerves arising from the floor of the mesencephalon; innervate some extrinsic and all inner eye muscles. *Syn.:* oculomotor nerve.

cranial nerve 4 figs. 149, 154
A pair of motor nerves arising from the mesencephalon and emerging from the roof of the brain at the isthmus; innervate the superior oblique ocular muscles. *Syn.:* trochlear nerve.

cranial nerve 5 figs. 131, 156
also see ganglion of cranial nerve 5; mandibular ramus of cranial nerve 5; maxillary ramus of cranial nerve 5; ophthalmic ramus of cranial nerve 5; root of cranial nerve 5
A pair of mixed nerves arising from the sides of the myelencephalon and semilunar ganglia; three divisions—the ophthalmic, maxillary, and mandibular rami—innervate the mandibular arch region. *Syn.:* trigeminal nerve.

cranial nerve 6 fig. 158
A pair of motor nerves emerging from the floor of the myelencephalon; innervate the external rectus eye muscles.

cranial nerve 7 figs. 131, 149, 156–159
also see ganglion of cranial nerve 7
A pair of mixed nerves arising from the myelencephalon at the anterior margin of the otic vesicle and the geniculate ganglion; innervate the hyoid arch. *Syn.:* facial nerve.

cranial nerve 9 figs. 157, 158
also see ganglion of cranial nerve 9; superior ganglion; petrosal ganglion
A pair of mixed nerves arising from the myelencephalon at the caudal margin of the otic vesicles; they bear the superior and petrosal ganglia and innervate the third branchial arch. *Syn.:* glossopharyngeal nerve.

cranial nerve 10 figs. 149, 157–161
also see ganglion of cranial nerve 10; jugular ganglion; accessory ganglia of cranial nerve 10; nodose ganglion
A pair of mixed nerves arising from the myelencephalon and bearing the jugular and nodose ganglia; innervate branchial arches 4, 5, and 6, and extend parasympathetic fibers to the viscera; in aquatic vertebrates (frog tadpole), innervate the lateral line. *Syn.:* vagus nerve.

cranial nerve 11 figs. 156–158
A pair of motor nerves arising from the myelencephalon and spinal cord and innervating muscles of the pharynx and shoulder; part of the vagus nerve in aquatic vertebrates. *Syn.:* accessory nerve.

cranial nerve 12 figs. 149, 151, 157–161
A pair of nerves arising from many rootlets on the ventral wall of the myelencephalon; innervate the tongue muscles; evolved from cervical spinal nerves of aquatic vertebrates. *Syn.:* hypoglossal nerve.

cumulus oophorus fig. 5
also see stratum granulosa
A thickening in the stratum granulosa containing the mammalian oocyte.

decidual cells figs. 179, 180
Enlarged connective tissue cells of the maternal decidua; distributed beneath the basal plate; contain abundant stores of glycogen and lipid.

dense mesenchyme figs. 30, 160
A condensation of mesenchyme cells preparatory to the formation of cartilage in these locations.

dermatome figs. 118, 134
also see somites
The outermost division of the somite; lies under and in contact with epidermis forming the dermis or connective tissue of the skin.

descending aorta figs. 107, 108, 115, 116, 122, 123, 125, 127, 131, 132, 138, 143, 145, 150, 151, 170, 173, 176
also see dorsal aortae
The principal artery of the trunk; a median vessel formed by the fusion of the paired dorsal aortae; extends from the subclavian to the common iliac arteries. *Syn.:* aorta.

d

diencephalon figs. 104, 105, 108, 110–113, 122, 123, 126–129, 134, 135, 143, 145, 148, 153, 158, 159, 162
The posterior division of the prosencephalon; a deep, laterally compressed region to which the optic stalks, infundibulum, and epiphysis attach; its cavity is the third ventricle of the brain; it forms the epithalamus, thalamus, and hypothalamus; its roof forms the choroid plexus.

differentiating spermatid figs. 2, 7, 13
also see spermatid
The spermatid during its transformation into a sperm, a process called spermiogenesis; during differentiation the spermatid is embedded in a pocket within the Sertoli cell. *Syn.:* immature sperm.

diplotene stage fig. 11
also see pachytene stage
A stage of the first maturation division in spermatogenesis and oogenesis; follows the pachytene stage; chromosomes in synapsis separate except at points of crossing over (chiasmata); followed by diakinesis during which the chromosomes separate farther moving the chiasmata to the ends of chromosomes (terminalization); metaphase of the first maturation division follows diakinesis.

dormant primary follicles fig. 4
Small follicles just within the tunica albuginea of the mammalian ovary; consist of an immature oocyte and a thin layer of follicle cells; formed during fetal life. *Syn.:* primordial follicle.

dorsal aortae figs. 72, 92–95, 97, 98, 101, 106, 107, 111, 112, 114, 119, 120, 125, 126, 129, 131, 139, 140, 146, 149, 159–161, 163, 167, 169
also see descending aorta
The primitive paired, longitudinal arteries of the trunk which fuse together posterior to the pharynx to form the descending aorta; in pharyngeal region, contribute to the internal carotids, descending aorta, and, in mammals, the right subclavian artery.

dorsal fin fig. 69
A flat extension of the body wall along the dorsal midline of the trunk and tail; degenerates during metamorphosis in anurans.

dorsal flexure fig. 126
also see caudal flexure; cervical flexure; cranial flexure
One of several ventral bends in the body axis giving the amniote embryo a compact C-configuration; forms in the trunk region.

dorsal hindgut diverticulum fig. 51
also see hindgut
A dorsal extension of the hindgut connecting to the caudal neural tube to form the postanal gut and neurenteric canal which subsequently disappear.

dorsal lip figs. 44, 45, 46, 47, 48
The margin of the blastopore toward the animal pole and at the dorsal side of the embryo; the first blastopore lip to form; derives from the gray crescent area of amphibian eggs; forms the foregut roof, notochord, and dorsal mesoderm.

dorsal liver bud *see* anterior liver diverticulum

dorsal mesentery figs. 117, 118, 135–137, 173, 176
also see greater omentum
A double layer of splanchnic mesoderm suspending the gut from the dorsal body wall and extending from the esophagus to the cloaca; provides a path and support for nerves and vessels of the gut; forms the mediastinum, greater omentum, and mesenteries of the intestine; contributes to the diaphragm.

dorsal mesocardium figs. 99, 114, 116, 134
The dorsal mesentery of the heart; derives from the ventral mesentery of the foregut; soon ruptures and disappears.

dorsal mesoderm fig. 53
also see head mesenchyme; prechordal plate; segmental mesoderm; somites
The thickened axial and paraxial mesoderm; closely associated with the neural plate and neural tube; extends from the prechordal plate and head mesenchyme through the somites and segmental mesoderm of the trunk.

dorsal pancreas figs. 136, 149, 172
also see ventral pancreas
A dorsal evagination of the duodenum; together with a ventral evagination (two in frogs and birds), forms the rudiments of the pancreas, which fuse to form one glandular mass.

dorsal root ganglia *see* spinal ganglia

dorsal root of spinal nerve figs. 170, 177
also see spinal ganglia
The dorsal division of a spinal nerve connecting the trunk of the nerve to the alar plate of the spinal cord; composed of sensory nerve fibers; bears the dorsal root ganglion. *Syn.:* sensory root.

duct of Cuvier *see* common cardinal vein

ductus choledochus *see* common bile duct

ductus venosus figs. 126, 135, 145, 150, 151, 153, 170
A vein within the liver of amniotes carrying blood from the vitelline and left umbilical veins to the sinus venosus; derives from vitelline veins; obliterates after hatching in birds or after birth in mammals.

duodenum figs. 136, 172
The first segment of the small intestine; arises from foregut; forms the diverticula of the liver and pancreas; later connects with the common bile duct and pancreatic ducts.

ear placode *see* otic placode

ear vesicle *see* otic vesicle

ectoderm figs. 46, 48, 87, 90, 91, 95, 103, 106
also see epithelial layer of ectoderm; head ectoderm; nervous layer of ectoderm
The outermost of the three primary germ layers; develops into epidermis, skin glands, hair, feathers, nails, scales, nervous system, lining of the nose, inner ear, retina and lens of the eye, pituitary gland, mouth, pigment cells, anus.

egg pronucleus fig. 20
also see mature ova
The haploid nucleus of the mature egg formed by the completion of the second maturation division.

embryonic coelom figs. 100, 101, 104, 105, 108, 116, 117, 119, 120, 138, 141, 145, 146
also see coelom; extraembryonic coelom
The division of the coelom within the head and body folds; forms the pericardial, pleural, and peritoneal cavities of the adult.

embryonic shield fig. 76
A thickening of the blastoderm during the pre-streak stage of chick embryos; marks the posterior end of the future embryonic axis.

endocardial cushion figs. 166, 167
A ring of connective tissue which encircles and then divides the atrioventricular canal; forms the atrioventricular valves.

endocardium figs. 116, 135
The lining of the heart; arises from splanchnic mesoderm; fuses with the epimyocardium to form the heart wall.

endoderm figs. 46, 48, 49, 52–54, 56, 87, 88, 90, 91, 103, 138
also see hypoblast; yolk endoderm
The innermost of the three primary germ layers, inward movement of which is part of gastrulation; forms the lining of the digestive and respiratory tracts, the pancreas, liver, thyroid, parathyroids, thymus and primordial germ cells (except in urodeles), the bladder, and urethra.

endolymphatic duct fig. 155
The stalk of the otic vesicle; except in elasmobranchs, soon loses its connection with the body surface; forms part of the inner ear.

ependymal layer figs. 155, 170
The inner layer of primitive neuroepithelial cells of the neural tube; by proliferation of cells, supplies neuroblasts for the mantle and marginal layers and subsequently forms the ependyma of the spinal cord and brain.

epiblast figs. 77, 78, 84
The upper or outer layer of the blastoderm of birds; in the area pellucida the original caudal half of the epiblast migrates through the primitive streak to form mesoderm; the anterior half of the epiblast becomes ectoderm.

epibranchial placodes fig. 67
Ectodermal thickenings dorsal to the branchial clefts; contribute cells to the cranial ganglia.

epidermis figs. 49, 51, 56, 57, 63
The outer, epithelial layer of the skin; derives from the ectoderm.

epiglottis fig. 160
An elevation on the floor of the pharynx anterior to the glottis in mammals; composed of cartilage in the adult; derives from branchial arches 3 and 4.

e f

epimere *see* somites

epimyocardium figs. 94, 116, 135
The outer layer of the heart, including the heart muscle; forms from splanchnic mesoderm; fuses with the endocardium to form the heart wall.

epiphysis figs. 61, 62, 70, 72, 106, 122, 123, 126, 127, 143
An evagination from the roof of the diencephalon; forms the pineal gland.

epithelial layer of ectoderm fig. 45
The darkly pigmented surface layer of the ectoderm in early amphibian embryos.

esophagus figs. 32–34, 134, 145, 151, 165–169
Part of the digestive tract that connects the pharynx (or mouth in the sea urchin) with the stomach; lengthens markedly in amniotes during development; arises from foregut in vertebrates and from archenteron in the sea urchin; forms the crop of birds.

exocoel *see* extraembryonic coelom

external carotid artery figs. 72, 144, 150, 161
An artery of the head; arises as an outgrowth of the ventral aorta near the base of aortic arch 3; initially, supplies the mandibular and hyoid arches.

external jugular vein figs. 72, 162
A vein of the head; arises as a branch of the anterior cardinal; initially, drains the mandibular and hyoid arches.

external layer figs. 18, 19
The outermost layer of the egg shell of *Ascaris;* adherent to the fertilization membrane; formed from uterine secretion.

extraembryonic coelom figs. 96, 97, 101, 110, 119, 120, 130, 134, 139, 141
also see coelom
The division of the coelom outside the head and body folds; lies between the chorion and amnion and between chorion and yolk sac; in chick and pig, receives the expanding allantois. *Syn.:* exocoel.

eye cup *see* optic cup

eye vesicle *see* optic vesicle

facial nerve *see* cranial nerve 7

falciform ligament fig. 171
The ventral ligament of the liver; attaches the liver to the ventral body wall; derives from the ventral mesentery.

fertilization membrane figs. 15–17, 22–25, 44, 45
A membrane separated from the surface of the egg after fertilization in many aquatic species; derives from the vitelline membrane, often thickened by material from the cortical granules; contributes to the egg shell in *Ascaris.*

fetal artery fig. 179
also see stem placental villus
In the placenta, a branch of the umbilical artery found in the placental plate and in the villi; supplies the capillaries of the villi with fetal blood.

fetal blood fig. 180
also see fetal capillaries; fetal artery
In the human placenta, the blood within the umbilical and placental vessels; separated from maternal blood by the placental barrier except during labor when tears in the barrier may permit some mixing.

fetal capillaries fig. 180
also see placental barrier
In the placenta, the anastomosing capillary bed of the placental villi; contains fetal blood supplied by the umbilical arteries and drained by the umbilical vein; forms part of the placental barrier.

fibrinoid figs. 179, 180
also see basal plate
An intercellular matrix of the basal plate in which peripheral cytotrophoblast cells are frequently embedded; gives a positive periodic acid-Schiff reaction for polysaccharide.

follicle cells fig. 37
also see stratum granulosa
The epithelial cells enclosing the oocyte; probably regulate the transfer of materials to the oocyte; form the stratum granulosa of bird and mammalian follicles.

follicular cavity figs. 4, 5
The space within the Graafian follicle; filled with a viscous follicular fluid. *Syn.:* antrum.

forebrain *see* prosencephalon

foregut figs. 51, 52, 53, 67, 85, 87, 88, 92, 93, 117, 118
also see pharynx
The part of the gut extending into the head from midgut; in amniotes, formed by the head fold as it passes under the head and trunk; eventually forms the pharynx, respiratory tract excepting the nasal passages, esophagus, stomach, duodenum, liver, and pancreas.

foreleg bud figs. 147, 148, 166, 168–170
also see hindleg bud; leg bud
The rudiment of the foreleg; arises as a thickening of somatic mesoderm in the body wall, an ectodermal thickening, the apical ridge forms on its margin; homologous to the wing bud of birds and arm bud of humans.

Froriep's ganglion fig. 158
In some mammals, the most caudal of the accessory cranial ganglia; may contribute to the 12th cranial nerve.

gall bladder figs. 149, 151, 172
A sac-like vessel connected by the cystic duct to the common bile duct; arises from a caudal extension of the liver diverticulum.

ganglion of cranial nerve 5 figs. 105, 110, 123, 126, 128–132, 143, 148
also see semilunar ganglion
Ganglion arising from anterior neural crest and from the epibranchial placode above the first branchial cleft, also from the dorsolateral placodes in lower vertebrates; supplies sensory fibers to the 5th cranial nerve. *Syn.:* Gasserian ganglion; semilunar ganglion; trigeminal ganglion.

ganglion of cranial nerve 7 figs. 104, 105, 111, 122, 123, 126, 143, 148, 153, 158
also see geniculate ganglion
Ganglion arising from preotic neural crest and from the epibranchial placode above the first branchial cleft, also from the dorsolateral placodes in lower vertebrates; forms beside the ganglion of the 8th cranial nerve and supplies sensory fibers to the 7th cranial nerve. *Syn.:* geniculate ganglion.

ganglion of cranial nerve 8 figs. 104, 105, 111, 122, 123, 126, 128, 130, 143, 148, 153
also see acoustic ganglion
Ganglion formed by aggregating cells detached from the otic placode and from the otic vesicle; lies between the geniculate ganglion and the otic vesicle; later divides into the spiral and vestibular ganglia of the inner ear; supplies the fibers of the 8th cranial nerve. *Syn.:* acoustic ganglion.

ganglion of cranial nerve 9 figs. 104, 105, 112, 113, 122, 123, 126, 128, 130, 143, 153
also see superior ganglion; petrosal ganglion
Ganglion formed from postotic neural crest and cells from the epibranchial placode above the 2nd branchial cleft; also from the dorsolateral placodes in lower vertebrates; divides subsequently into the superior and petrosal ganglia and supplies sensory fibers to the 9th cranial nerve.

ganglion of cranial nerve 10 figs. 123, 126, 131, 153
also see accessory ganglia; jugular ganglion; nodose ganglion
Ganglion formed from postotic neural crest and cells from the epibranchial placode above the 3rd branchial cleft, also from the dorsolateral placodes in lower vertebrates; divides subsequently into the jugular and nodose ganglia and supplies sensory fibers to the 10th cranial nerve.

Gasserian ganglion *see* ganglion of cranial nerve 5

gastrocoel *see* archenteron

gastrula figs. 29–31, 44, 45
The embryonic stage during which the primitive gut or archenteron is formed; the period following the blastula stage and during which extensive cell migrations form the primary germ layers.

geniculate ganglion figs. 149, 157, 158
see ganglion of cranial nerve 7

genital ridge figs. 138, 151, 153, 173, 176
A thickening of splanchnic mesoderm (germinal epithelium) and of the underlying mesenchyme on the medial edge of the mesonephros; in the early stages, contains large primordial germ cells; forms the testis or ovary (except that in

g h

female birds the right genital ridge fails to develop). *Syn.:* germinal ridge.

genital tubercle figs. 148, 152, 172
An elevation on the ventral body surface of mammals anterior to the cloacal membrane; enlarges into the phallus and eventually forms the penis of males and clitoris of females.

germinal epithelium fig. 4
The epithelial covering of the adult ovary and the embryonic gonad; derived from splanchnic mesoderm and primordial germ cells; probably forms the germ cells of the gonad.

germinal vesicle figs. 14, 20, 21, 37, 75
The much enlarged nucleus of the oocyte; during the prophase of the 1st maturation division, its membrane breaks, releasing most of its contents into the cytoplasm.

germ wall figs. 77, 78
The outer circular zone surrounding the cellular blastoderm; a region of nuclear proliferation and cell organization which contributes cells to the margin of the epiblast and hypoblast.

gill plate fig. 50
A thickened region of the ectoderm lateral to the anterior neural folds in amphibian embryos; the gill arches and clefts posterior to the first, otic vesicles, and neural crest differentiate from it.

gill pouch *see* pharyngeal pouch

gill slit *see* branchial cleft

glossopharyngeal nerve *see* cranial nerve 9

glottis figs. 162, 163
The opening from the pharynx into the trachea of early embryos or into the larynx of later embryos; in mammals, acquires lateral borders (the arytenoid folds) from the 4th and 5th branchial arches.

Graafian follicle figs. 4–6
The ovarian follicle of mammals containing a follicular cavity; derive from a primary follicle and either atrophy (atresia) or ovulate to form a corpus luteum. *Syn.:* vesicular follicle.

greater omentum figs. 170, 171
A sac-like membrane attached to the greater curvature of the stomach of birds and mammals; contains a cavity, the omental bursa; derives from the dorsal mesentery of the stomach.

head ectoderm figs. 52, 85, 87–89, 97
also see epibranchial placodes; lens placode; olfactory placodes; otic placode
The epithelial outer covering of the head; mostly forms epidermis but some placodes (thickenings) arise which contribute to the sense organs and cranial ganglia.

head fold figs. 80, 88, 98, 99
A downward bend of membranes around the head which mark the boundaries of the embryonic area; undercuts the head, separating it from the extraembryonic area; forms by invagination, the ventral surface of the head and the foregut and is posteriorly continuous with the body folds.

head mesenchyme figs. 48, 52, 55, 56, 65, 81, 85, 88, 89, 97, 100, 109, 110, 130
also see dense mesenchyme
A loose tissue surrounding the brain and foregut, mostly of mesodermal origin; derives from the paraxial mesoderm anterior to the somites, the prechordal plate, and the neural crest; forms the following head structures: blood vessels, skull, head muscles, and connective tissue.

head organizer fig. 49
Inductor of head parts; consists of foregut roof, prechordal plate and anterior notochord; the first area to pass over the dorsal lip of the blastopore during gastrulation in amphibians. *Syn.:* head inductor.

head plexus figs. 107, 125
A dense capillary network surrounding the brain anterior to the myelencephalon; supplied by the internal carotid arteries and drained by the anterior cardinal veins; some head vessels develop later from the plexus.

head process *see* notochordal process

heart figs. 61, 62, 66, 70, 71, 95, 147
also see atrium; bulbus cordis; sinus venosus; ventricle
In early stages, a tubular organ divided by constrictions into sinus venosus, atrium, ventricle, and bulbus cordis; its wall is formed of an inner endocardial layer and an outer epimyocardium; arises from

paired heart tubes derived from splanchnic mesoderm which fuse beneath the foregut; in air-breathing vertebrates, becomes more or less completely divided longitudinally in the later stages to provide a separate pulmonary circulation.

Hensen's node *see* primitive knot

hepatic portal vein figs. 149, 172
The vessel which carries blood from the superior mesenteric and splenic veins to the ductus venosus and sinusoids of the liver; derives from parts of the right and left vitelline veins.

hindbrain *see* rhombencephalon

hindgut figs. 51, 58, 59, 60, 62, 69, 121, 127, 140, 141
also see cloaca
The posterior part of the embryonic gut; extends from the midgut to the tail in amniotes; formed as the tail fold passes under the posterior trunk region; forms successively tail gut, cloaca, colon, and posterior small intestine; in amphibians, forms the rectum.

hindleg bud figs. 126, 147, 148, 174, 176
also see foreleg bud; leg bud.
The rudiment of the hindleg; arises as a thickening of somatic mesoderm in the body wall; an ectodermal thickening, the apical ridge, forms on its margin; homologous to the leg bud of birds and humans.

horny teeth fig. 72
Larval teeth arising from cornified epithelium of the mouth; lost during metamorphosis and replaced by permanent teeth of the adult.

hyoid arch figs. 60, 112, 113, 122, 133, 143, 146, 147, 153, 162, 163
also see aortic arch 2
The second branchial arch; arises as a thickening of the pharyngeal wall between the first and second branchial clefts; contains the second aortic arch in the early stages, forms the stapes (columella), styloid process, stylohyoid ligament, part of the hyoid bone, root of the tongue, and the facial muscles. *Syn.:* branchial arch 2.

hyomandibular cleft *see* branchial cleft 1

hypoblast figs. 77, 78, 81–84
The inner or lower layer of the blastoderm of birds; lies between the epiblast and subgerminal cavity or between epiblast and yolk; formed by inward migration and aggregation of large yolky cells; mostly forms endoderm.

hypochordal rod *see* subnotochordal rod

hypoglossal nerve *see* cranial nerve 12

hypomere *see* lateral plate mesoderm

hypophysis figs. 51, 55, 59, 62, 64, 71, 72
also see infundibulum; Rathke's pouch
An endocrine gland beneath the hypothalamus; derives from the infundibulum and Rathke's pouch or, in amphibians, from infundibulum and a solid ingrowth from stomodeum. *Syn.:* pituitary.

immature sperm figs. 12, 13, 73
also see differentiating spermatid
The sperm during its final stage of differentiation; some further elongation of the head with its dense nucleus and of the tail will occur.

inferior ganglion *see* petrosal ganglion

inferior vena cava figs. 72, 144, 150–152, 168–172
The principal systemic vein of the trunk; derives from several primitive paired veins, including the right vitelline, right subcardinal, right supracardinal, and right posterior cardinal; originally drains into the sinus venosus but is carried into the right atrium as the sinus venosus merges with the atrium.

infundibulum figs. 59, 60, 62, 65, 71, 72, 92–96, 104–106, 108, 111, 123, 126, 127, 133, 143, 145, 148, 151
also see hypophysis
A ventral evagination of the prosencephalon; becomes located in the floor of the diencephalon and later in the hypothalamus; subsequently evaginates the neural (posterior) lobe of the hypophysis.

interatrial foramen fig. 165
also see atrial septum
An opening in the atrial septum allowing blood to pass from the right to the left side of the heart in tetrapods; closes soon after breathing begins, completing the longitudinal division of the heart.

i j l

intermediate mesoderm *see* nephrotome

internal carotid arteries figs. 107, 125, 126, 131-133, 144, 149, 150, 158, 159
The main arterial supply to the brain; arise as anterior extensions of the dorsal aortae; later acquire additions from the dorsal aortae and 3rd aortic arches.

intersegmental arteries figs. 125, 150, 161, 177
Originally, small paired branches of the dorsal aortae arising between the somites; contribute to the following arteries: vertebrals, subclavians, intercostals, and lumbars.

intersomitic grooves fig. 85
also see somites
The spaces separating somites; the first to form lies between the 1st and 2nd somites; the others form within the segmental mesoderm in an anterio-posterior sequence; subsequently obliterated by fusion of adjacent somites.

interstitial cells figs. 2, 4, 36
In the testis, clusters of endocrine gland cells (Leydig cells) between the seminiferous tubules which probably secrete testosterone; in the ovary, clusters of gland cells scattered in the cortex which probably secrete estradiol; derive from atretic follicles.

interventricular foramen fig. 166
also see ventricular septum
An opening in the ventricular septum allowing blood to cross between the right and left ventricles in tetrapods; closes during division of the bulbus cordis and the atrioventricular canal.

interventricular sulcus fig. 166
A longitudinal groove on the surface of the primitive ventricle marking the plane of its impending division into right and left ventricles.

intervillus space figs. 179, 180
also see placental villi
In the placenta, region filled with maternal blood and in which the placental villi are suspended; maternal blood supply to the intervillus space is by way of the spiral arteries of the basal plate; veins of the basal plate return the blood to the maternal circulation.

intestinal loop figs. 148, 151, 152
also see caecum; colon; small intestine
A ventral extension of the intestine into the umbilical cord of mammals bearing an enlargement, the caecum; coiling soon retracts it into the peritoneal cavity.

intestine figs. 32–34, 72, 151
also see caecum; colon; duodenum; intestinal loop; rectum
Segment of gut following the stomach; derives from both foregut and hindgut in amniotes, from the midgut as well in amphibians, and from archenteron in the sea urchin; the intestinal lining develops from gut endoderm; muscle, connective tissue, blood vessels, and serosa develop from splanchnic mesoderm.

jugular ganglion figs. 149, 156
also see ganglion of cranial nerve 10
A large ganglion of the 10th cranial nerve dorsal to the nodose ganglion; contributes sensory fibers to the 10th nerve.

labia fig. 72
The lips bordering the frog larval mouth.

laryngotracheal groove figs, 108, 115, 127, 133
A trough in the floor of the posterior pharynx from which arise the lung buds; also contributes to the larynx and trachea.

larynx fig. 164
The voice box; derives from upper trachea, the floor of the pharynx, and branchial arches 4 and 5.

lateral body fold figs. 119, 120, 137, 138
A depressed fold in the somatopleure beside the embryonic trunk; together with the head and tail folds, to which it connects, forms the boundary between the embryonic and extraembryonic regions.

lateral bronchus fig. 168
A secondary branch of the respiratory tract lateral and anterior to the stem bronchus.

lateral nasal process fig. 164
An elevation on the embryonic face lateral to the olfactory pit; fuses with the maxillary and median nasal processes and forms the sides of the nose.

lateral plate mesoderm figs. 49, 53, 54, 57, 60, 67, 68, 90, 91, 94, 106
also see ventral mesoderm; somatic mesoderm; splanchnic mesoderm
The mesodermal layer lateral and ventral to the nephrotome; split by the coelom into somatic and splanchnic mesoderm. *Syn.:* hypomere.

lateral swellings fig. 160
also see tongue
Paired elevations on the mandibular process within the mouth; fuse with tuberculum impar to form the body of the tongue.

lateral transverse vein fig. 150
In the pig embryo, part of a plexus of small veins draining the mesonephros.

left atrium figs. 164–167
also see atrial septum; atrium
The left division of the primitive atrium in tetrapods; separated from the right atrium by the atrial septum; receives blood through the interatrial foramen and the pulmonary veins before breathing begins; after closure of interatrial foramen, supplied by increased pulmonary flow. *Syn.:* left auricle.

left auricle *see* left atrium

left horn of sinus venosus figs. 166, 167
also see sinus venosus
The part receiving the left common cardinal, left vitelline, and left umbilical veins; conducts the flow to the transverse sinus venosus whence it passes into the right atrium through the sinoatrial opening; as the venous return is directed to the right horn of the sinus venosus, the left horn is reduced, forming finally the oblique vein of the left atrium.

left ventricle figs. 165–168
also see ventricle
Heart chamber formed from the partitioning of the primitive ventricle by the ventricular septum; receives blood from the left atrium and delivers it under high pressure to the ascending aorta.

leg bud figs. 122, 127, 128, 140, 143–146, 153
also see foreleg bud; hindleg bud
The rudiment of the leg; arises as a thickening of somatic mesoderm of the body wall, later bearing an ectodermal thickening, the apical ridge.

lens placode figs. 41, 61, 64, 96
also see lens vesicle
A thickening of head ectoderm overlying the optic vesicle; invaginates to form the lens vesicle and subsequently the eye lens.

lens vesicle figs. 104, 105, 106, 112, 122, 123, 126, 134, 161
also see lens placode
An ectodermal sac within the optic cup; derives from lens placode, forms the eye lens.

leptotene stage figs. 8, 9
also see prochromosome stage
An early stage of the first maturation division in spermatogenesis and oogenesis; chromosomes have the form of separate long thin threads except that the X-chromosome may be a dense contracted body; followed by the zygotene stage; during which synapsis or pairing of homologous chromosomes occurs.

lesser omentum figs. 170, 171
A membrane which attaches the lesser curvature of the stomach to the liver; derives from the ventral mesentery of the stomach.

liver figs. 51, 72, 126, 127, 143, 145, 147, 148, 151–153, 169
also see liver diverticulum; liver sinusoid; anterior liver diverticulum; posterior liver diverticulum
The largest of the digestive glands; important in fetal life for blood homeostasis and blood formation; arises as a ventral diverticulum of the foregut in amphibians and mammals, and as two buds on the duodenum in birds; the buds branch and anastomose around the ductus venosus.

liver diverticulum figs. 48, 53, 57, 59, 61, 62, 67, 70, 71, 72
also see anterior liver diverticulum; liver; posterior liver diverticulum
The rudiment of the liver, gall bladder, and common bile duct; arises as a ventral evagination of the foregut in amphibians and mammals, and as two buds on the duodenum of birds.

liver sinusoids fig. 170
The smallest blood vessels of the liver; differ from capillaries in that their walls contain phagocytes; derive originally from the ductus venosus and vitelline veins.

lumbosacral plexus fig. 177
An anastomosis of spinal nerves in the posterior trunk region which supplies the nerves of the hind limb.

lungs fig. 149
also see laryngotracheal groove; lateral bronchus; lung bud; stem bronchus
Arise as a ventral diverticulum of the pharynx which branches repeatedly to form the endodermal lining of the trachea, bronchi, and lungs; the mesoesophagus and lining of the pleural cavities form the mesodermal parts—muscle, connective tissue, and pleura; the pulmonary arteries arise from the 6th aortic arch and invade the lung; the pulmonary veins grow in from the atrium.

lung buds figs. 72, 126, 129, 134, 153
also see laryngotracheal groove; lateral bronchus; lungs; stem bronchus
The paired rudiments of the lungs and bronchi; arise from the laryngotracheal groove of the pharynx. *Syn.:* primary bronchi.

lymph sinus fig. 37
A large, lymph-filled space; drains into the veins; the cavity of the hollow amphibian ovary.

macromeres figs. 26, 42, 43, 47
The largest blastomeres or cells formed during cleavage; form endoderm and in the sea urchin ventral ectoderm also.

mammary ridge figs. 171–173
The rudiment of the mammary gland; initially appears as an ectodermal thickening extending longitudinally between the bases of the limb buds; at the site of the mammary glands, the mammary ridges form tubular ingrowths, the milk ducts; renewed development of the mammary glands at puberty in females is a response to the ovarian hormones; development completed during pregnancy.

mandible figs. 143, 145, 148, 151–153
also see mandibular arch; mandibular process
The lower jaw; formed by the fusion of the mandibular processes of the first branchial arch.

mandibular arch figs. 64, 104, 105, 112, 131
also see aortic arch 1; mandibular process; maxillary process
The most anterior branchial arch, composed of a mandibular process forming the posterior border of the stomodeum, and a maxillary process anterior to the stomodeum; in the early stages it contains the first aortic arch; its posterior boundary is the 1st branchial cleft. *Syn:* branchial arch 1.

mandibular process figs. 122, 123, 127–129, 132, 133, 146, 147, 161–162
also see mandible; mandibular arch
The posterior division of the mandibular arch; forms the mandible, Meckel's cartilage, body of tongue, malleus, salivary glands, jaw muscles.

mandibular ramus of cranial nerve 5 figs. 159–160
also see cranial nerve 5
The posterior division of the fifth cranial nerve; innervates the mandible and jaw muscles.

mantle layer figs. 155, 162, 170
The middle layer of the developing neural tube; contains neuroblasts from the ependymal layer and their nerve fibers; forms the gray matter.

marginal layer fig. 155
The outer layer of the developing neural tube; contains neuroblasts from the inner layers and nerve fibers; forms the white matter.

marginal zone figs. 43, 44, 46, 47
The part of the animal hemisphere nearest the vegetal hemisphere; is turned in during gastrulation to form mesoderm and foregut endoderm.

maternal blood fig. 180
also see intervillus space
In the living placenta, maternal blood fills the intervillus spaces; mostly drained

out in microscopic preparations of placenta; separated from fetal blood by the placental barrier.

maturation division I figs. 2, 7, 10, 15, 36, 74
The first of two specialized cell divisions during the formation of sperm and eggs; the prophase is long and includes synapsis, chromosome replication, and crossing over; reduces the chromosome number to the haploid state; forms the secondary spermatocytes of males and secondary oocyte and a polar body in females. *Syn.:* meiosis I, reduction division I.

maturation division II figs. 7, 10, 16
The second of two specialized cell divisions during the formation of sperm and eggs; begins immediately after the first maturation division but in many eggs is not completed until after fertilization; no chromosome replication occurs; forms the spermatids of males and the second polar body and mature egg in females. *Syn.:* reduction division II, meiosis II.

mature ova fig. 20
also see oocyte; primary oocyte; egg pronucleus
The female germ cells after completion of the maturation divisions; derived from oocytes; in sea urchins, found in the lumen of the ovary where they may be fertilized.

maxillary process figs. 122, 123, 128, 129, 132, 133, 146, 148, 161, 162, 163
also see mandibular arch
The anterior division of the mandibular arch; forms the cheek, lateral part of upper jaw, palate, incus.

maxillary ramus of cranial nerve 5 fig. 160
also see cranial nerve 5
The middle division of the 5th cranial nerve; innervates the upper jaw and face.

medial transverse vein fig. 150
In the pig embryo, part of a plexus of small veins draining the mesonephros.

median nasal processes fig. 164
Elevations on the embryonic face medial to the olfactory pit; the two median nasal processes fuse to form median part of the upper jaw.

medulla fig. 6
The inner or deep division of an organ; in the ovary, a region of connective tissue and blood vessels lacking follicles.

medullary groove *see* neural groove

medullary plate *see* neural plate

meiosis *see* maturation division

mesencephalon figs. 59, 61, 62, 64, 70, 71, 92, 93, 97, 104, 105, 106, 108, 109, 122, 123, 127–133, 143–148, 151–157
The middle primary vesicle of the brain; forms visual and auditory centers (corpora quadrigemina in mammals) and motor centers for movements of the head; its cavity narrows to form the cerebral aqueduct; bears the 3rd and 4th cranial nerves. *Syn.:* midbrain.

mesentery figs. 174, 175
A supporting membrane attached to organs within the coelom; carries the vascular and nerve supply of organs; formed by fusion of two layers of splanchnic mesoderm.

mesoderm figs. 46, 48, 49, 51, 56, 72, 80, 82–84, 103
also see primary mesenchyme; secondary mesenchyme, coelomic vesicles; head mesenchyme; lateral plate mesoderm, dorsal mesoderm; segmental mesoderm; somite; nephrotome: ventral mesoderm
The middle primary germ layer; formed mostly by the gastrulation movements; forms dermis, muscle, skeleton, blood vessels, blood (excepting possibly lymphocytes), connective tissues, kidneys, ureters, gonads (excepting possibly germ cells), reproductive tracts, peritoneum.

mesomeres fig. 26
also see nephrotome
The cells of intermediate size that compose the animal hemisphere of sea urchin embryos during cleavage; form the ectoderm of the gastrula and larva. Also the nephrotome of vertebrates.

mesonephric duct figs. 119, 120, 138, 140–142, 146, 149, 151, 153, 173–176
The excretory duct of the mesonephros; formed initially as the pronephric duct by the caudal growth of the pronephric

buds to the cloaca; contributes to the metanephros of amniotes by forming one of its rudiments, the ureteric bud; mostly degenerates in female amniotes but in males forms the ductus epididymis, ductus deferens, and seminal vesicles; forms the duct of the adult kidney (opisthonephros) of amphibians. *Syn.:* archinephric duct; Wolffian duct.

mesonephric glomeruli figs. 149–153, 172, 176
Tufts of capillaries within Bowman's capsules which together form the renal corpuscles of the mesonephros; similar structures form in the metanephros; glomeruli of the pronephros are associated with the coelom rather than with Bowman's capsule.

mesonephric ridge fig. 138
also see mesonephros
A bulge or thickening extending into the dorsal part of the embryonic coelom at midtrunk levels; formed by the growing mesonephros.

mesonephric tubules figs. 105, 119, 138, 149–153, 172, 176
The kidney tubules of the mesonephros; possess glomeruli and a well-formed, coiled tubular structure; excretory during the embryonic period of amniotes; most degenerate but some form the efferent ductules of male amniotes.

mesonephros figs. 122, 127, 139, 145, 147, 148, 170, 171, 173
also see mesonephric tubules; mesonephric glomeruli; mesonephric ridge; mesonephric duct
The second or middle kidney of amniotes; contains well-formed tubules with glomeruli that produce urine during the embryonic period; the arrangement of the tubules is not segmental; the pronephric duct is appropriated as the mesonephric duct; mostly degenerates in the adult amniote except that in males the caudal parts form the male reproductive tract (efferent ductules, ductus epididymis, ductus deferens, seminal vesicles). *Syn.:* Wolffian body.

mesovarium fig. 6
The mesentery of the ovary; provides support for the ovary.

metanephric diverticulum fig. 149
also see pelvis of metanephros; ureter
A rudiment of the metanephros; arises as an outgrowth of the mesonephric duct near its junction with the cloaca; the stalk becomes the ureter and the expanded distal end, the pelvis and collectings ducts of the metanephros. *Syn.:* ureteric bud.

metanephric duct *see* ureter

metanephrogenic mesenchyme fig. 175
A strand of dense mesenchyme surrounding the pelvis of the metanephros; continuous anteriorly with the mesonephrogenic tissue; derives from the caudal nephrotomes; forms the metanephric tubules. *Syn.:* metanephrogenous tissue.

metanephrogenous tissue
see metanephrogenic mesenchyme

metencephalon figs. 104, 105, 106, 108, 109, 122, 123, 126–130, 143–148, 151–158
The anterior division of the rhombencephalon; its roof expands greatly to form the cerebellum while the pons forms in its floor; nerve centers (nuclei) for several cranial nerves develop within, including those of the 5th, 6th, 7th, and 8th; its cavity becomes the 4th ventricle of the brain.

micromeres: figs. 26, 27, 42, 43
The smallest cells of the cleavage stages; lie near the vegetal pole in the sea urchin and migrate into the blastocoel to form the mesenchyme; compose the entire animal hemisphere of amphibian embryos, forming ectoderm and mesoderm after gastrulation.

midbrain *see* mesencephalon

midgut figs. 54, 57, 59, 60, 62, 68, 89, 101, 119, 129
also see anterior intestinal portal
In amphibians the middle part of the gut with a small lumen and thick yolky floor; derives from archenteron and will form the small intestine; in amniotes, the middle part of the gut whose floor is

the cavity of the yolk sac (yolk-filled in reptiles and birds); it is steadily diminished by the lengthening of the foregut and hindgut to a mere yolk stalk attached to the small intestine.

mitotic figure fig. 42
also see cleavage division
The mitotic apparatus, consisting of chromosomes, spindle fibers, and centrioles; appears during each cleavage division; often indistinct in preparations used for class work.

motor root *see* ventral root of spinal nerve

mouth figs. 32–34, 143, 145, 151, 160–163
also see oral evagination; stomodeum
The anterior opening of the digestive tract; derived partly from an ectodermal invagination on the ventral side of the head, the stomodeum; the endodermal rudiment arises from the anterior wall of the foregut and is for a time separated from the stomodeum by the pharyngeal membrane; rupture of the membrane opens the mouth.

myelencephalon figs. 104, 105, 106, 108–114, 122, 123, 126, 127, 129, 130, 143–148, 151, 152, 154–160
The posterior division of the rhombencephalon and the most posterior part of the brain, a transition region from brain to spinal cord; contains nerve centers (nuclei) of cranial nerves 9 to 12; cranial nerves 9 and 10 are attached to its sides and in amniotes, nerves 11 and 12 as well; its cavity becomes part of the 4th ventricle of the brain and its roof, the choroid plexus.

myotome figs. 118, 134, 161, 164, 171, 176, 177, 178
also see somites
The division of the somite which forms skeletal muscle of the body wall; lies at first in the dorsal region of the somite then migrates ventrally under the dermatome; its cells elongate longitudinally forming a muscle segment; may also contribute to limb musculature; myotomes are comparatively large in amphibian embryos.

nasal cavity fig. 72
also see olfactory pits
A canal extending from the nostril to the mouth; formed from the olfactory pits as they deepen and break through the roof of the mouth; forms the olfactory organ. *Syn.:* nasal passage.

nasal pit *see* olfactory pits

nasal placode *see* olfactory placodes

nephrotome figs. 57, 101, 120
also see pronephros
Stalk-like connection between the somite and lateral plate; forms segmental buds in the anterior region which hollow out to form the pronephric tubules and ducts; at posterior levels, forms mesenchyme which develops into tubules of the mesonephros, metanephros, and the gonads. *Syn.:* mesomere, intermediate mesoderm.

nervous layer of ectoderm figs. 45, 48
The inner layer of ectoderm, covered by the epithelial layer; thickens in the regions of the neural plate and ectodermal placodes.

nervous layer of optic cup figs. 134, 161
also see optic cup
The inner layer of the optic cup; arises from the lateral wall of the optic vesicle; forms the nervous or sensory layer of the retina and the optic nerve fibers.

neural crest figs. 52, 53, 56, 65, 88, 96, 100
also see ganglion of cranial nerve 5, etc.; spinal ganglia
An ectodermal mesenchyme arising from the neural folds; aggregates in many locations to form cranial ganglia, spinal ganglia, autonomic ganglia and adrenal medulla; attaches to epidermis to form pigment cells and to the neural tube to form the pia mater; forms the neurolemma sheath cells of nerves; some migrate into the branchial arches to form the visceral skeleton.

neural folds figs. 48, 51–54, 85, 89
The elevated edges of the neural plate; opposite neural folds are brought together by bending and closing of the neural groove; fuse and bud off streams of neural crest cells.

n o

neural groove figs. 47, 48, 51–54, 87, 89, 92, 103
also see neural plate; neural folds; neural tube
A trough formed by the bending or rolling up of the neural plate; closes to form neural tube. *Syn.:* medullary groove.

neural plate figs. 80–83, 91
also see neural folds; neural groove; neural tube
A thickening of the dorsal ectoderm caused by its transverse contraction; the placode or rudiment of the nervous system. *Syn.:* medullary plate.

neural tube figs. 51, 85, 88, 94, 95, 106, 126
also see mesencephalon; neural groove; neural plate; prosencephalon; rhombencephalon; spinal cord
The rudiment of the central nervous system; formed by bending and closing of the neural plate; the anterior large end forms the brain, the posterior part, the spinal cord; in amphibians the extreme posterior end contributes to the tail somites.

neurenteric canal figs. 51, 58
A temporary connection between the caudal end of the neural groove or neural tube and the archenteron or yolk sac cavity; occurs in many embryos, including frog and man.

neurocoel fig. 77
The lumen or cavity of the neural tube. *Syn.:* neural canal.

neuromere figs. 106, 130
A minor segment of the brain formed by transverse constrictions; particularly prominent in the myelencephalon; later vanishes.

nodose ganglion fig. 160
also see ganglion of cranial nerve 10
A large ganglion of the 10th cranial nerve lying ventral to the jugular ganglion; contributes sensory fibers to the 10th nerve.

notochord figs. 48, 49, 51, 53, 54, 56–59, 62, 65–72, 88, 91, 94, 95, 97, 102, 105, 106, 108, 111, 112, 123, 126, 127, 131, 132, 139, 145, 151, 159, 177
also see notochordal process
The initial axial skeleton of vertebrate embryos; arises in amphibians from cells turned in over the dorsal lip of the blastopore, and in amniotes from cells extending anteriorly from the primitive knot; shows marked elongation and in amphibians extends the embryonic axis; lies under the neural tube from the mesencephalon to the end of the spinal cord; its gelatinous cells acquire a tough sheath forming a flexible skeletal rod; in higher vertebrates it is small and mostly replaced by the vertebral column. *Syn.:* chorda.

notochordal process figs. 80, 81
also see notochord
In amniotes, a band of mesodermal cells extending anteriorly from the primitive knot; the rudiment of the notochord. *Syn.:* head process.

nuclear envelope *see* nuclear membrane

nuclear membrane figs. 21, 37
Encloses the substance of the nucleus; at the ultrastructural level, it is double-layered with "pores"; destroyed during prophase of mitosis and re-formed during telophase. *Syn.:* nuclear envelope.

nucleolus figs. 21, 37
One or more dense spherical granules in the nucleus of many cells; composed of ribosomal RNA and protein; disappears during mitosis to re-form in connection with a pair of nucleolar chromosomes; large in cells with high rates of protein synthesis.

nucleus figs. 22, 41, 78
also see germinal vesicle; nuclear membrane; nucleolus; pronucleus
A large body within the cell during interphase; contains the chromosomes and often one or more nucleoli; enclosed by a nuclear membrane; disappears during mitosis; site of synthesis of most of the DNA and RNA of the cell.

oculomotor nerve *see* cranial nerve 3

olfactory pits figs. 61, 63, 70, 72, 122, 123, 126, 135, 143, 146–148, 152, 164
also see olfactory placodes
Cavities on the lateral surfaces of the head anterior to the eyes; arise by invagination of the olfactory placodes;

deepen and break through the roof of the mouth in air breathers to form the nasal cavities; olfactory cells differentiate from their walls as do the olfactory nerves. *Syn.:* nasal pit.

olfactory placodes figs. 60, 114
also see olfactory pits
Paired ectodermal thickenings on the lateral surfaces of the head anterior to the eyes; invaginate to form the olfactory pits; the rudiments of the nasal passages. *Syn.:* nasal placode.

omental bursa figs. 170, 171
The cavity of the greater omentum; arises as an invagination into the dorsal mesentery of the stomach; connected with the peritoneal coelom; two bursae form in birds.

omphalomesenteric artery *see* vitelline arteries

omphalomesenteric veins *see* vitelline veins

oocyte figs. 4, 5, 14, 20, 21, 37, 75
also see primary oocyte
The immature egg, distinguishable from other ovarian cells by gigantic size and prominent nucleus (germinal vesicle); grows from the smaller oogonium; becomes a mature egg upon completion of its growth and two maturation divisions.

opercular opening fig. 72
A persistent excurrent opening at the posterior edge of the operculum on the left side of the larval frog body; closed during metamorphosis by proliferation of tissue which fills the opercular cavity. *Syn.:* spiracle.

ophthalmic ramus of cranial nerve 5 fig. 159
also see cranial nerve 5
The anterior division of the 5th cranial nerve; distributes sensory fibers to the anterior facial region.

optic cup figs. 61, 64, 70, 104, 105, 106, 112, 122, 123, 126, 143, 148, 160
also see nervous layer of optic cup; optic vesicle; pigmented layer of optic cup
A double-walled chamber formed by invagination of the optic vesicle; the lens vesicle lies in its "mouth"; remains connected to the diencephalon by the optic stalk; the outer wall forms the pigment epithelium of the retina; the inner wall forms the nervous layers of the retina and the optic nerve fibers which grow through the stalk to the brain; the rim of the cup contributes to the iris and ciliary body; the optic cup of mammals is small. *Syn.:* eye cup.

optic stalk figs. 64, 105, 113, 129, 134, 146, 152, 161
The narrow connection of the optic cup to the diencephalon; guides the growing optic nerves from the optic cup to the brain.

optic vesicle figs. 55, 60, 92, 95, 96, 123, 126
also see optic cup
A lateral evagination of the prosencephalon; by invagination of outer wall, forms the optic cup and subsequently the retina. *Syn.:* eye vesicle.

oral arms figs. 32–34
Slender, paired extensions of the dorsal body wall of pluteus larvae; supported by skeletal rods and bearing bands of cilia; function to stabilize and propel the larva and to collect food.

oral evagination figs. 62, 64, 71
The endodermal rudiment of the mouth; an anterior evagination of the pharynx toward the stomodeum in amphibians; contact with the stomodeum forms the pharyngeal membrane, which subsequently ruptures to open the mouth.

oral plate *see* pharyngeal membrane

otic placode figs. 93, 100
also see otic vesicle
A thickening of head ectoderm lateral to the myelencephalon; invaginates, forming the otic pit, and separates from the ectoderm as the otic vesicle; subsequently forms the inner ear and contributes cells to the ganglion of the 8th cranial nerve. *Syn.:* auditory placode; ear placode.

otic vesicle figs. 61, 66, 70, 72, 104, 105, 106, 111, 122, 123, 128, 130, 143, 148, 149, 153–158

o p

also see otic placode
A closed chamber formed by the invagination of the otic placode; separates from the head ectoderm and subsequently forms the inner ear. *Syn.:* auditory vesicle, ear vesicle.

ovarian lumen fig. 20
The cavity within the ovary of sea urchins into which eggs are ovulated; leads to the exterior through the genital pore.

ovarian wall fig. 20
Connective tissue capsule enclosing the germ cells and accessory cells in the ovary of sea urchins.

ovary figs. 4–6
The female gonad; organ where eggs differentiate; in vertebrates, also secretes the female sex hormones estradiol and progesterone.

pachytene stage figs. 8, 9, 10, 11
also see diplotene stage
A stage of the first maturation division in spermatogenesis and oogenesis; follows the zygotene stage during which synapsis of homologous chromosomes occurs; contains a haploid number of bivalents (tetrads) or double chromosomes in synapsis; bivalents shorten and thicken; followed by the diplotene stage.

pancreas fig. 72
also see dorsal pancreas; ventral pancreas
A digestive and endocrine gland arising as outgrowths of the duodenum and liver diverticulum.

parathyroids fig. 161
Masses of endocrine gland tissue usually derived from the 3rd and 4th pharyngeal pouches; migrate to the vicinity of the thyroid.

pelvis of metanephros fig. 175
also see metanephric diverticulum
The expanded distal end of the metanephric diverticulum; is surrounded by metanephrogenic mesenchyme; forms the pelvis, calyces, and collecting tubules of the metanephros.

pericardial cavity figs. 134, 163, 164, 165, 167
also see pericardial coelom
The body cavity around the heart; derives from the pericardial coelom as the latter becomes isolated from the pleuroperitoneal coelom.

pericardial coelom figs. 66, 71, 95, 99, 114–116
also see pericardial cavity
The large coelomic space around the heart; formed by a cleft in the lateral plate mesoderm of the head; part of the splanchnic mesodermal layer thus formed develops into heart; is cut off from the pleuroperitoneal coelom by the pleuropericardial membranes to form the pericardial cavity.

peripheral cytoplasm fig. 74
In the chicken oocyte, an outer or cortical layer of finely granular cytoplasm; as the oocyte matures it contributes to the blastodisc.

peripheral syntrophoblast figs. 179, 180
also see villus syntrophoblast; placental villi; basal plate
An extension of the villus syntrophoblast covering the fetal surface of the basal plate; like villus syntrophoblast it is in contact with maternal blood and underlaid by cytotrophoblast.

peritoneal cavity figs. 135, 136, 176
also see embryonic coelom
The body cavity of the abdomen; derives from the posterior region of the coelom; after the pericardial cavity is cut off from the pleuroperitoneal cavity, the latter is split into two pleural cavities and a peritoneal cavity by the pleuroperitoneal membranes.

perivitelline space figs. 16, 18, 19
The space between the fertilization membrane and the egg surface; between the zona pellucida and the egg in mammals; contains perivitelline fluid, the "culture medium" of the developing egg.

petrosal ganglion fig. 159
also see ganglion of cranial nerve 9
The more ventral of two ganglia of the 9th cranial nerve; contributes sensory fibers to the nerve. *Syn.:* inferior ganglion.

pharyngeal membrane figs. 87, 97, 108, 130
also see mouth; stomodeum
A double-layered membrane composed of the floor of the stomodeum and the anterior wall of the pharynx; rupture of the

membrane opens the mouth into the pharynx. *Syn.:* oral plate.

pharyngeal pouch 1 figs. 60, 65, 111, 131, 159
also see branchial cleft 1
Paired evaginations of the lateral pharyngeal wall posterior to the mandibular arch; form the endodermal part of the first branchial clefts; extend dorsally toward the otic vesicles to form the Eustachian tubes and tympanic cavities. *Syn.:* branchial pouch 1; gill pouch 1; visceral pouch 1.

pharyngeal pouch 2 figs. 56, 60, 131, 132, 160
also see branchial cleft 2
Paired evaginations of the pharyngeal wall posterior to the hyoid arch; form the endodermal part of the second branchial clefts; subsequently obliterates, except contributes to the thymus in lower vertebrates (frog). *Syn.:* branchial pouch 2; gill pouch 2; visceral pouch 2.

pharyngeal pouch 3 figs. 60, 66, 114, 131, 132, 160, 161
also see branchial cleft 3
Paired evaginations of the pharyngeal wall posterior to the third branchial arch; form the endodermal part of the third branchial clefts; contributes to the thymus and parathyroids. *Syn.:* branchial pouch 3; gill pouch 3; visceral pouch 3.

pharyngeal pouch 4 figs. 132, 133, 162
also see branchial cleft 4
Paired evaginations of the lateral pharyngeal wall posterior to the fourth branchial arch; contribute to the parathyroids.

pharynx figs. 51, 59–62, 65, 66, 70, 71, 95, 97–99, 104–106, 108, 111–115, 126, 127, 133, 145, 146, 148, 151, 152, 161, 163, 164
also see foregut
The region of the embryonic foregut bearing branchial clefts; large in fishes and amphibian tadpoles and forms gills; in air breathers extends posteriorly to the glottis; is much reduced and its pouches are transformed.

pia mater fig. 156
The inner layer of the meninges; a delicate membrane on the brain and spinal cord derived from head mesenchyme and neural crest.

pigmented cortex fig. 42
The surface coat of amphibian eggs and early embryos; a gelled layer containing much melanin in the animal hemisphere.

pigmented layer of optic cup figs. 134, 161
also see optic cup
The outer wall of the optic cup; formed from the medial half of the optic vesicle; forms the pigmented epithelium of the retina.

pituitary *see* hypophysis

placental barrier fig. 180
also see placental villi
The layers of the placenta interposed between maternal blood and fetal blood; includes in humans at least the chorionic endothelium of fetal capillaries and the syntrophoblast but in some areas also the cytotrophoblast and chorionic mesenchyme; controls the exchange between maternal and fetal bloods.

placental villi figs. 179, 180
also see stem placental villus; peripheral syntrophoblast; villus cytotrophoblast; fetal capillaries
A branching tree-like outgrowth of the chorion into the maternal blood of the intervillus spaces; covered by two epithelial layers, the outer syntrophoblast and inner cytotrophoblast; contains a mesenchyme connective tissue and fetal blood vessels; some villi attach to the maternal decidua; forms the placental barrier which interposes between the maternal and fetal bloods, controlling the exchange of substances between the two blood streams.

pleural cavities figs. 134, 167, 168
The body cavities surrounding the lungs; derive from anteriodorsal divisions of the pleuroperitoneal coelom, which become isolated from the pericardial and peritoneal cavities.

pluteus larva figs. 32–34
A bilateral, free-swimming larval stage of sea urchins, sand dollars, and brittle stars; possesses long ciliated arms.

polar body I figs. 16–19
A small cell separated from the primary oocyte by the first maturation division; may divide again but then degenerates.

p

polar body II fig. 17
A small cell separated from the secondary oocyte by the second maturation division; degenerates.

postanal gut fig. 149
The extension of the hindgut into the tail; gradually degenerates. *Syn.:* tail gut.

postcardinal veins *see* posterior cardinal veins

posterior cardinal veins figs. 72, 94, 106, 107, 108, 118, 119, 125, 126, 128, 134–139, 144, 145, 150, 152, 168–173
The primitive paired veins of the trunk; lie dorsal to the mesonephros and drain with the anterior cardinals into the common cardinals; as the mesonephroi grow, form the renal portal veins; mostly degenerate in amniotes with the mesonephros, but form the iliac veins. *Syn.:* postcardinal veins.

posterior intestinal portal figs. 106, 121, 122, 140
also see midgut
The opening from the midgut into the hindgut of amniotes; moves anteriorly, lengthening the hindgut; meets the anterior intestinal portal at the level of the small intestine to form the yolk stalk.

posterior liver diverticulum figs. 117, 136
also see anterior liver diverticulum; liver
One of two outgrowths of the duodenum of birds which grow, branch, and anastomose to form the liver. *Syn.:* caudal liver bud; ventral liver bud.

posterior vitelline vein fig. 107
also see vitelline veins
A branch of the left vitelline vein extending posteriorly to receive the sinus terminalis.

prechordal plate figs. 49, 51, 87
A mass of cells anterior to the notochord and between the foregut and prosencephalon; constitutes, with the notochord, the axial mesoderm; a site of head mesenchyme formation.

preoral gut figs. 126, 132–145
The projecting tip of the foregut anterior to the pharyngeal membrane; gradually atrophies. *Syn.:* Seessel's pocket.

primary bronchi *see* lung buds

primary follicle figs. 4–6
A small follicle of the mammalian ovary with but one layer of follicle cells surrounding an oocyte; the smallest or dormant primary follicles were formed during fetal life; they grow in response to follicle-stimulating hormone of the hypophysis.

primary mesenchyme figs. 28–30
A loose cluster of cells near the vegetal pole and within the blastocoel of sea urchins; derives from micromeres; contributes to the skeleton of the pluteus.

primary oocyte fig. 21
also see oocyte
The immature egg prior to completion of the 1st maturation division.

primary spermatocytes figs. 1, 2, 7, 36, 74
Large germ cells of the testis formed by growth of spermatogonia; undergo the first maturation division to form secondary spermatocytes.

primitive folds figs. 80, 83, 84
also see primitive streak
The thickened ridges of the primitive streak; formed by convergent flow of epiblast.

primitive groove figs. 80, 83, 84
also see primitive streak
A depressed trough between the primitive folds; a region of invagination of epiblast cells into the mesoderm.

primitive knot figs. 80, 82, 85
also see primitive streak
The anterior thickened end of the primitive streak. *Syn.:* Hensen's node.

primitive streak figs. 85, 92, 93, 103
A longitudinal thickening in the epiblast of early amniote embryos; formed by convergent flow of epiblast toward the caudal midline; site of invagination of epiblast cells into the mesoderm; consists of parallel longitudinal ridges (primitive folds), separated by a primitive groove, and an anterior thickening, the primitive knot.

primordial follicle
see dormant primary follicles

proamnion figs. 80, 85, 96
A crescent-shaped area lacking mesoderm around the head of early bird embryos; initially delimits the anterior end of the embryo; later drawn under the head by the head fold, invaded by the mesoderm, and contributes to the amnion.

prochromosome stage figs. 8, 9
also see leptotene stage
The earliest prophase stage of the first maturation division of spermatogenesis of some insects; chromosomes contract into discrete bodies of which there are a diploid number; unraveling of the prochromosomes leads to the next or leptotene stage.

proctodeum figs. 51, 61, 62, 69
An ectodermal invagination on the ventral side of the trunk at the base of the tail; breaks into the hindgut to form the anus. *Syn.:* anal pit.

pronephric duct fig. 72
also see mesonephric duct; pronephros
A tubule connecting the pronephros with the cloaca; arises from caudal growth of pronephric buds; subsequently becomes the mesonephric duct.

pronephros figs. 51, 68, 72, 105, 137
also see nephrotome
The first and most anterior kidney to form; derives from buds of nephrotomes which hollow out to form tubules—one pair per body segment; one end of each tubule opens as a nephrostome into coelom; the tubules link together to form the pronephric duct which grows posteriorly along the somites to the cloaca; is vestigial in amniotes but large in lower vertebrates and functions in the larval stage; the pronephric duct is appropriated by the mesonephros in amniotes.

pronucleus fig. 17
A haploid nucleus found in fertilized eggs, one pronucleus derives from the sperm and a second one from the egg; may fuse or enter prophase of the 1st cleavage division separately.

prosencephalon figs. 55, 59–64, 70, 71, 92–96, 106
also see diencephalon; telencephalon
The anterior primary brain vesicle; forms two lateral evaginations (optic vesicles) and a ventral evagination (the infundibulum); then differentiates into an anterior telencephalon and a posterior diencephalon. *Syn.:* forebrain.

pulmonary aorta fig. 164
The trunk of the pulmonary arteries; connects with the right ventricle; derives from the bulbus cordis by longitudinal division of the latter.

pulmonary arteries figs. 144, 149, 150, 166
Connect the pulmonary aorta with the lungs; basal sections derive from the 6th aortic arches.

pulmonary vein fig. 167
The vessel carrying blood from the lungs to the left atrium; arises as an outgrowth of the left atrium or in birds from sinus venosus and connects with the pulmonary plexus.

Rathke's pouch figs. 105, 106, 108, 111, 123, 126, 127, 133, 145, 151, 160
also see hypophysis
A dorsal evagination of the stomodeum extending under the diencephalon to the infundibulum in amniotes; becomes isolated from the stomodeum and forms the pars distalis (anterior lobe), the pars intermedia (intermediate lobe), and pars tuberalis of the hypophysis.

rectum figs. 173, 174
The posterior segment of the large intestine; formed by splitting off from the dorsal side of the cloaca.

reduction division *see* maturation division

residual bodies figs. 2, 74
Granules of degenerating cytoplasm sloughed off differentiating spermatids.

residual spermatogonium fig. 36
Large reserve germ cells; may proliferate mitotically to replace cells which have matured into sperm; each is enclosed by a follicle cell.

rhombencephalon figs. 61, 62, 65–67, 70, 71, 98–100
also see metencephalon; myelencephalon
The third and posterior primary brain vesicle extending from the mesencephalon to the spinal cord; divides into an anterior metencephalon and a posterior myelencephalon. *Syn.:* hindbrain.

right atrium figs. 149, 151, 165, 166
also see atrium
The right division of the primitive atrium separated from the left atrium by the atrial septum; receives blood from the sinus venosus and delivers it through the interatrial foramen to the left atrium and through the right atrioventricular canal to the right ventricle; after breathing begins, the interatrial foramen closes. *Syn.:* right auricle.

right auricle *see* right atrium

right horn of sinus venosus fig. 166
also see sinus venosus
The part receiving blood from the right common cardinal, right vitelline, right umbilical veins, and, later, inferior vena cava; eventually incorporated into the right atrium with its veins.

right ventricle figs. 149, 165–168
also see ventricle
A thick-walled heart chamber formed from the partitioning of the primitive ventricle by the ventricular septum; receives blood from the right atrium and delivers it to the pulmonary aorta.

right vitelline artery fig. 107
also see vitelline arteries
The arterial supply for the right half of the yolk sac.

right vitelline vein fig. 107
also see vitelline veins
The venous return for the right half of the yolk sac.

root of cranial nerve 5 fig. 130
also see cranial nerve 5
The part of the fifth cranial nerve connecting the semilunar ganglion to the metencephalon.

sclerotome figs. 118, 134, 161, 164, 171, 177, 178
also see somites
The medial, mesenchymatous division of the somite; arises from cells of the medioventral wall of the early somite; envelops the notochord and spinal cord; sclerotomes split transversely and adjacent halves fuse to form the rudiments of the vertebrae and ribs.

secondary mesenchyme figs. 29, 30
also see primary mesenchyme
Cells which migrate into the blastocoel from the wall of the archenteron during gastrulation in sea urchins; occupies the animal part of the blastocoel forming skeleton and muscle.

secondary spermatocytes figs. 2, 7, 10
Male germ cells formed from primary spermatocytes by the 1st maturation division; undergo at once the 2nd maturation division to form spermatids; distinguished from both primary spermatocytes and spermatids by intermediate size.

Seessel's pocket *see* preoral gut

segmental mesoderm figs. 85, 91–93, 102, 104, 121
Paraxial mesoderm extending posteriorly from the last somite; will form somites by segmentation.

segmentation cavity *see* blastocoel

semilunar ganglion figs. 148, 149, 156–159
see ganglion of cranial nerve 5.

seminiferous tubules figs. 1, 35, 73
Tubules within the testis; bounded by a thin basement membrane of connective tissue and containing the male germ cells and Sertoli cells.

sensory plate fig. 50
A thickened region of the ectoderm anterior to the neural plate of amphibian embryos; stomodeum, hypophysis, olfactory pits, lens placodes, mandibular arch, and neural crest differentiate from it.

sensory root *see* dorsal root of spinal nerve

septum fig. 35
In the frog testis, connective tissue membranes enclosing the seminiferous tubules.

serosa *see* chorion

Sertoli cell figs. 2, 36, 74
The sperm nurse cell of vertebrates; in mammals, a tall, columnar cell extending from the basement membrane to the lumen of the seminiferous tubule; the outline of the cell is irregular and obscure; the nucleus is light-staining with a prominent nucleolus; differentiating spermatids become embedded in

cytoplasmic pockets in Sertoli cells and withdraw at maturity. *Syn.:* sustentacular cells.

sinus venosus figs. 72, 104–108, 116, 122, 123, 125–128, 134, 143–146, 151–153
also see left horn of sinus venosus; right horn of sinus venosus; transverse sinus venosus
Initially the most posterior chamber of the heart, receiving the venous return and delivering it to the atrium; after the partitioning of the atrium, empties into the right atrium; disappears as a heart chamber by atrophy and by incorporation into the atria; originates the heart beat and later transfers that function to the atrium by forming the sinoatrial node.

small intestine figs. 171–174
also see duodenum; intestinal loop; intestine
The segment of gut following the stomach; arises from foregut and midgut in amphibians, from foregut and hindgut in amniotes.

somatic mesoderm figs. 68, 88, 95, 138
also see lateral plate mesoderm; somatopleure
The layer immediately outside the coelom; arises by splitting from lateral plate mesoderm; forms parietal peritoneum and, by fusion with myotomes, dermatomes, and epidermis, forms body wall and limbs; in the extraembryonic area, fuses with ectoderm to form the somatopleure of the amnion and chorion

somatopleure figs. 99, 100, 106
also see amnion; chorion
A double-layered membrane composed of ectoderm and somatic mesoderm; contributes to the body wall and extends into the extraembryonic area; forms the amniotic folds which by enveloping the embryo transform extraembryonic somatopleure into amnion and chorion.

somites figs. 49, 51, 54, 57, 61, 62, 67–71, 85, 90, 92, 93, 95, 101, 104–106, 108, 114, 117, 120, 122, 123, 126–128, 130, 131, 139, 145, 147, 148, 152, 153, 157, 160
also see dermatome; myotome; sclerotome
The segments of paraxial mesoderm; form first at the posterior end of the myelencephalon and extend progressively as a series of paired blocks posteriorly into the tail; are separated by intersomitic grooves and attach laterally to the nephrotomes; are primary segments of the body which establish all other segmental patterns; differentiate into a lateral dermatome, a middle myotome, and medial sclerotome. *Syn.:* epimere.

sperm figs. 1, 2, 7, 12, 14–16, 35, 36
also see differentiating spermatid
The mature male germ cell; in vertebrates, a small, haploid, highly specialized, flagellated cell which can attach to and penetrate egg membranes to activate the egg; formed from a spermatid through a complex differentiation called spermiogenesis. *Syn.:* spermatozoön.

spermatid figs. 1, 2, 7, 10, 11, 12, 13, 35, 36, 74
also see differentiating spermatid
A small haploid germ cell of the testis; formed from a secondary spermatocyte by the 2nd maturation division; embeds in a pocket within the Sertoli cell and differentiates into a sperm.

spermatocyte fig. 35
see primary spermatocytes; secondary spermatocytes

spermatogonia figs. 1, 2, 7, 8, 74
The "stem" germ cells of the testis; divide mitotically to regenerate the germinal epithelium against the loss of mature sperm; located near the basement membrane of the seminiferous tubule; may enter a prolonged growth phase forming primary spermatocytes.

sperm heads fig. 36
also see sperm; sperm tails
That part of the sperm containing the nucleus and acrosome.

sperm tails figs. 12, 73, 74
also see sperm; differentiating spermatid; immature sperm; centrioles
The flagellum of the sperm; derived from the cytoplasm of the spermatid; composed of an axial filament arising

S

from centrioles near the head, a mitochondrial sheath, fibers, and a plasma membrane.

spinal cord figs. 57–59, 61, 62, 68–72, 92, 101, 102, 104, 108, 115, 116, 121, 122, 123, 131, 132, 142, 145, 146, 151, 152, 161, 162, 167, 172, 176–178
also see neural tube
The central nervous system posterior to the brain; derives from posterior neural tube; bears a pair of spinal nerves for each body segment; wall differentiates into an inner ependymal layer, a middle mantle layer, and an outer marginal layer; the latter two layers are rudiments of gray matter and white matter, respectively.

spinal ganglia figs. 72, 116, 131, 132, 135, 149, 151, 153, 160, 161, 162, 167, 170, 177, 178
Ganglia borne on dorsal roots of spinal nerves; derive from neural crest and supply sensory nerve fibers of the spinal nerve. *Syn.:* dorsal root ganglia.

spinal nerves fig. 177
also see dorsal root of spinal nerve; spinal ganglia; ventral root of spinal nerve
Paired nerves emerging from spinal cord at each body segment; each is connected to the spinal cord by a dorsal root bearing a spinal ganglion and by a ventral root; the spinal nerve trunk divides immediately into a dorsal ramus and a ventral ramus; a ramus communicans connects to autonomic ganglia.

splanchnic mesoderm figs. 68, 88, 95, 138
also see lateral plate mesoderm; splanchnopleure
The layer between the coelom and endoderm; arises by splitting from lateral plate mesoderm; fuses with endoderm to form the wall of the gut and respiratory tract; forms mesenteries, visceral peritoneum, heart, and germinal epithelium; in the extraembryonic area, fuses with endoderm to form the splanchnopleure of the yolk sac and allantois.

splanchnopleure figs. 94, 106, 129
also see allantois; yolk sac
A double membrane composed of splanchnic mesoderm and endoderm; forms the gut wall and extends into the extraembryonic area to form the yolk sac and allantois.

stem bronchus fig. 169
The most posterior secondary bronchus of the developing lung; forms the lower lobe of the lung.

stem placental villus fig. 179
also see placental villi
In the placenta, the trunk or large branch of the placental villus; contains arteries and veins with fetal blood which arise from the umbilical arteries and veins; supports terminal villi.

stomach figs. 32–34, 72, 116, 135, 149, 153, 170–171
An enlarged segment of the foregut posterior to the esophagus; derives from archenteron in the sea urchin; the lining epithelium forms from gut endoderm but the muscle, blood vessels, and connective tissue develop from splanchnic mesoderm; in birds, the stomach differentiates into a proventriculus and a gizzard.

stomodeum figs. 31–55, 61, 62, 70, 71, 72, 95, 97, 104, 105, 106, 108, 112, 122, 123, 126, 127–129, 132, 133
also see mouth
The ectodermal rudiment of the mouth; an invagination in the anterioventral ectoderm of the head which contacts the anterior wall of the foregut; its floor is the pharyngeal membrane; rupture of the membrane opens the mouth into the pharynx; a rudiment of the hypophysis, called Rathke's pouch in amniotes, evaginates from the dorsal wall of the stomodeum.

stratum granulosa figs. 4, 5, 75
also see cumulus oophorus
The inner stratified epithelium of large ovarian follicles; derives from follicle cells of primary follicles; in mammals, contributes to the corpus luteum after ovulation.

subcardinal anastomosis fig. 176
also see subcardinal veins
A medial interconnection between the right and left subcardinal veins;

contributes to the prerenal segment of the posterior vena cava. *Syn.:* subcardinal sinus.

subcardinal veins figs. 144, 151, 173
Primitive paired veins of the trunk; they lie ventral to the mesonephroi and parallel to the posterior cardinals which they mostly replace;they subsequently contribute to the inferior vena cava and its branches.

subchorda *see* subnotochordal rod

subclavian artery figs. 144, 149, 150
The artery of the shoulder and forelimb; arises by the enlargement of the 7th intersegmental artery; in mammals, the right subclavian also receives contributions from the right 4th aortic arch and right dorsal aorta.

subclavian vein figs. 144, 150, 166
The vein of the forelimb; connects at first to posterior cardinal but later shifts to the anterior cardinal.

subgerminal cavity figs. 77, 78, 81, 87
A space beneath the hypoblast of the area pellucida in birds; becomes the cavity of the yolk sac; communicates through the intestinal portals with the foregut and hindgut.

subintestinal vein fig. 144
A vein in pig embryos extending from the base of the tail along the ventral margin of the intestine to the vitelline veins; initially drains the allantois, posterior limb buds, and intestine; mostly replaced by the development of the allantoic veins.

subnotochordal rod figs. 62, 69
A strand of cells lying between the midgut and notochord in amphibians; of endodermal origin; degenerates. *Syn.:* hypochordal rod; subchorda.

superior ganglion figs. 149, 156
also see ganglion of cranial nerve 9
The dorsal ganglion of the 9th cranial nerve; with the petrosal ganglion, supplies sensory fibers to the nerve.

superior mesenteric artery figs. 172–175
also see vitelline arteries
The arterial supply of the small intestine; derived from the vitelline arteries.

superior mesenteric vein fig. 173
The main branch of the hepatic portal vein; drains the digestive tract.

sustentacular cells *see* Sertoli cell

sympathetic ganglia fig. 166
A series of paired ganglia dorsal to the aorta and connected to the spinal nerves by the rami communicans; is part of the autonomic nervous system; derives from neural crest.

systemic arch *see* aortic arch 4

tail fig. 171
also see tail bud
The extension of the body posterior to the anus; derives from the tail bud.

tail bud figs. 51, 59, 60, 62, 104, 108, 122, 126, 127, 143
The rudiment of the tail and posterior trunk; a mass of undifferentiated tissue projecting from the posterior end of the embryo; derives from primitive streak in amniotes; contributes to neural tube, somites, and notochord.

tail fin fig. 71
A blade-like extension of the border of the tail in amphibians; continuous anteriorly with the dorsal fin.

tail fold figs. 104, 121, 141
A depressed fold encircling the tail bud and connecting anteriorly with the body folds; forms part of the boundary between the embryonic and extraembryonic areas; undercuts the tail bud and posterior trunk forming hindgut.

tail gut *see* postanal gut

telencephalon figs. 104, 105, 108, 114, 115, 122, 123, 126, 127–129, 135, 136, 145, 153, 162, 163, 164
also see cerebral hemispheres
The anterior division of the prosencephalon; the greatly enlarged roof forms the cerebral hemispheres; the floor forms the olfactory bulbs, hippocampus, and corpus striatum; the cavities are the lateral ventricles of the brain.

testicular cyst fig. 7
also see testicular lobe wall
In the grasshopper testis, a compartment within a testicular lobe bounded by connective tissue septa and containing a group of germ cells at the same stage of maturation.

t u

testicular lobe wall figs. 7, 13
also see testicular cyst
In the grasshopper testis, the connective tissue capsule enclosing a lobe or major division of the testis; the lobe is divided into cysts, each containing a cluster of germ cells; the apical end of the lobe contains proliferating spermatogonia with more mature germ cells extending toward the opposite end which opens into a vas deferens.

testis figs. 1, 2
The male gonad; the organ in which sperm differentiate; secretes the male sex hormone testosterone in vertebrates.

tetrad *see* bivalent

theca externa figs. 5, 37
The outer connective tissue layer of Graafian follicles; arises from the stroma of the ovary; is the outer wall of the ovary in amphibians.

theca interna figs. 5, 37
A vascular layer between the theca externa and the stratum granulosa of large ovarian follicles; contains endocrine gland cells, connective tissue, and blood vessels; contributes to the corpus luteum after ovulation or to the interstitial tissue after follicular atresia in mammals.

thyroid figs. 62, 65, 72, 113, 126, 127, 132, 145, 161
An endocrine gland in the throat region; forms as a ventral diverticulum of the pharynx at the level of the 2nd branchial arch; the rudiment bifurcates and migrates posteriorly, becoming isolated from the pharynx.

tongue figs. 148, 151–153
also see copula; lateral swellings
An organ that arises by fusion of several elevations on the floor of the mouth and pharynx; these elevations include two lateral swellings and a median tuberculum impar on the mandible, the copula on the hyoid arch, and a contribution from the 3rd branchial arch.

trachea figs. 145, 149, 151, 165, 166
The part of the respiratory tract connecting the laryngotracheal groove with the lung buds or, later, the larynx with the primary bronchi; arises with the lung buds as a ventrocaudal diverticulum of the pharynx; muscle and connective tissues develop from the splanchnic mesoderm of the ventral mesoesophagus.

transverse septum figs. 72, 153, 168, 169
A mass of mesenchyme posterior to the heart, incompletely separating the pericardial cavity from the peritoneal cavity; encloses the veins that enter the heart; the liver is attached to its caudal face; contributes to the diaphragm in mammals

transverse sinus venosus fig. 168
also see sinus venosus
A narrow middle part carrying blood from the left horn to the sinoatrial opening; eventually forms the coronary sinus.

trigeminal nerve *see* cranial nerve 5

truncus arteriosus *see* ventral aorta

trunk organizer fig. 49
Inductor of trunk parts; consists of middle and posterior notochord and the somites; follows the head organizer over the dorsal lip during gastrulation.
Syn.: trunk inductor

tuberculum impar fig. 160
also see tongue
A median elevation on the mandible in the floor of the mouth; fuses with the lateral swellings to form the body of the tongue.

tuberculum posterius fig. 62
An elevation on the floor of the diencephalon marking its posterior boundary.

tunica albuginea figs. 4, 5, 36
A fibrous connective tissue capsule or membrane enveloping the ovary and testis.

ultimobranchial body fig. 163
An evagination from the caudal surface of the fourth pharyngeal pouch; may represent the fifth pouch; fuses with the thyroid rudiment probably forming the parafollicular cells of the thyroid gland.
Syn.: postbranchial body

umbilical arteries figs. 144, 149–152, 171–175
The arterial blood supply to the chorioallantois of birds and the placenta of mammals; a pair of vessels arising from the posterior end of the aorta; forms

the common iliac and hypogastric arteries in mammals, and, after birth, the lateral umbilical ligaments. *Syn.:* allantoic arteries.

umbilical cord figs. 147, 148, 153, 171, 172
The narrowed connection of the embryo to the extraembryonic membranes; the outer wall is amnion and may contain the yolk stalk, allantoic stalk, vitelline blood vessels, umbilical blood vessels, and a gelatinous connective tissue; in birds, separates from the umbilicus just before hatching; in mammals, is bitten in two after birth; the stump dropping off in a few days.

umbilical vein figs. 135, 144, 146, 149–153, 171–175
Initially, paired embryonic vessels draining the allantois of birds or the placenta of mammals; the left vein atrophies early, the right atrophies after hatching or birth, forming, in mammals, the ligamentum teres. *Syn.:* allantoic vein.

ureter fig. 175
The excretory duct of the metanephros; derives from the stalk of the metanephric diverticulum connecting at first with the mesonephric duct, its site of origin; later shifts to the cloaca in birds or to the urinary bladder in mammals. *Syn.:* metanephric duct.

ureteric bud *see* metanephric diverticulum

urogenital sinus figs. 148, 152, 174
also see cloaca
A chamber split from the ventral part of the cloaca of mammals; receives the mesonephric ducts, Mullerian ducts, and allantoic stalk; contributes to the bladder; forms the urethra and, in females, the vestibule of the vagina as well.

vagus nerve *see* cranial nerve 10

valve of sinus venosus fig. 166
Valve of the sinoatrial opening. *Syn.:* valvulae venosae.

valvulae venosae *see* valve of sinus venosus

vegetal pole fig. 43
The end of the embryonic axis centered in the yolky region of the egg; opposite the animal pole.

vein fig. 73
In the testis, veins, arteries, and capillaries branch in the interstitial connective tissue between the seminiferous tubules; blood vessels do not penetrate the tubules.

vena capitis figs. 144, 150
The principal embryonic vein draining the venous plexi of the brain and later the dural sinuses; derived from the anterior segment of the anterior cardinal vein.

ventral aorta figs. 94, 97, 106, 107, 113, 125, 127, 133, 145, 161, 162
The outlet of the embryonic heart; lies in the floor of the pharynx and conducts blood from the bulbus cordis to the aortic arches; forms the innominate arteries and the ascending aorta. *Syn.:* aortic sac, truncus arteriosus.

ventral hindgut diverticulum fig. 51
also see hindgut
A ventral extension of the hindgut toward the proctodeum; contributes to the cloaca.

ventral lip figs. 45–48
The margin of the blastopore toward the animal pole and at the ventral side of the embryo; derives from the ventral marginal zone and forms ventral mesoderm in amphibians.

ventral liver bud *see* posterior liver diverticulum

ventral mesentery fig. 135
also see dorsal mesocardium; lesser omentum; ventral mesoesophagus
A double layer of splanchnic mesoderm attaching parts of the foregut to the ventral body wall; forms the transient mesocardia, roots of the lungs, lesser omentum, and falciform ligament of the liver.

ventral mesoderm figs. 49, 51, 62
also see lateral plate mesoderm
The extension of the lateral plate into the ventral body region; split by the coelom into somatic and splanchnic mesoderm.

V

ventral mesoesophagus figs. 168, 169
also see dorsal mesocardium
The ventral mesentery of the esophagus; forms the roots of the lungs and the transient mesocardia.

ventral pancreas figs. 149, 172
also see dorsal pancreas
A ventral evagination of the liver diverticulum (two in birds and amphibians) which grows, branches, and fuses with the dorsal pancreas to form one glandular mass of the adult pancreas.

ventral ramus of spinal nerve fig. 170
The main ventral branch of the spinal nerve trunk; innervates the viscera, body wall, and limbs.

ventral root of spinal nerve figs. 170, 177, 178
The ventral division of a spinal nerve connecting the trunk of the nerve to the basal plate of the spinal cord; composed of motor nerve fibers arising from neuroblasts in the mantle layer of the basal plate. *Syn.:* motor root.

ventral vein fig. 150
Transient veins extending along the ventral margin of the mesonephros in pig embryos and draining into the posterior cardinal veins; subsequently replaced by the subcardinal veins.

ventricle figs. 72, 92, 94, 99, 104–108, 116–118, 122, 123, 125–129, 136, 143, 145, 148, 151–153, 170
also see heart; left ventricle; right ventricle
The thick-walled heart chamber that, in the embryo, receives blood from the atrium and delivers it under high pressure to the bulbus cordis; in amniotes, is partitioned into right and left ventricles delivering blood to the pulmonary aorta and ascending aorta, respectively.

ventricular septum figs. 166, 167, 168
A muscular partition arising from the posterior wall of the primitive ventricle; grows anteriorly, fusing with the endocardial cushion and bulbar septum; divides the ventricle into right and left ventricles.

vertebral arteries figs. 144, 150, 160
A pair of longitudinal vessels extending anteriorly from the subclavian arteries to the basilar artery under the myelencephalon; with the internal carotids, provides the arterial supply to the brain; arises from anastomosis of anterior intersegmental arteries.

vesicular follicle *see* Graafian follicle

villus cytotrophoblast fig. 180
also see placental villi
The inner epithelial layer covering placental villi; cells are separate and may show mitotic figures; probably forms the overlying syntrophoblast.
Syn.: Langhans cells

villus syntrophoblast figs. 179, 180
also see peripheral syntrophoblast; placental barrier; villus cytotrophoblast
The outer epithelial covering of placental villi, the part of the chorionic surface and placental barrier in contact with maternal blood; a true syncytium; possesses a brush border on its free surface; partly underlaid by and derived from the cytotrophoblast; probable site of synthesis of placental hormones.

visceral cleft *see* branchial cleft

visceral groove *see* branchial groove

visceral pouch *see* pharyngeal pouch

vitelline arteries figs. 104, 106, 107, 125, 126, 127, 129, 139, 143, 145, 160
The arterial supply of the yolk sac; arise as ventral branches of the dorsal aortae; form the superior mesenteric artery and, in mammals, the coeliac and inferior mesenteric arteries as well. *Syn.:* omphalomesenteric arteries.

vitelline membrane fig. 14
A membrane enveloping the egg or oocyte; lies immediately outside the plasmalemma; formed while the oocyte is in the ovary and, in some species, after fertilization, separates from the egg to form the fertilization membrane.

vitelline plexus figs. 92, 101-103, 107
A network of small vessels in the yolk sac; some enlarge to form the vitelline veins and arteries.

vitelline veins figs. 72, 92-95, 100, 106-108, 117, 118, 125-127, 136-138, 143-146, 150, 173
also see anterior vitelline veins; posterior vitelline vein

Vessels that provide initially the venous return from the yolk sac; proximal ends fuse, forming successively the atrium, and sinus venosus of the heart, and the ductus venosus; also form the hepatic vein, hepatic sinusoids, and the hepatic portal vein; distal branches degenerate with the yolk sac; in amphibians, form around the yolk endoderm of the midgut. *Syn.:* omphalomesenteric veins.

vitelline vessels figs. 96, 109, 111, 112, 130, 140
also see vitelline arteries; vitelline plexus; vitelline veins
The blood vessels of the yolk sac; arise from the blood islands.

wing bud figs. 122, 126-129, 137, 138, 143-146
The rudiment of the wing; arises as a thickening of somatic mesoderm of the body wall; later bears an ectodermal thickening, the apical ridge; homologous to the foreleg bud and arm bud.

Wolffian body *see* mesonephros

Wolffian duct *see* mesonephric duct

X-chromosome figs. 9, 10
also see leptotene stage
The sex chromosome which usually occurs double in females and single in males, where it may be associated with a Y-chromosome; often exists in a contracted or heteropyknotic state.

yolk figs. 51, 75, 77, 78
A reserve food mixture within the ovum; in birds formed into yolk spheres up to 100 microns in diameter; composed mainly of lipids and proteins.

yolk endoderm figs. 54, 59-62, 68, 70, 71
A mass of large yolky cells in the floor of the midgut in amphibians; derives from the vegetal hemisphere; subsequently disintegrates and the yolk is absorbed.

yolk plug figs. 45-49
A mass of large yolky cells filling the blastopore of the amphibian gastrula; derives from the vegetal hemisphere of the blastula; invaginates to form the yolk endoderm of the neurula.

yolk sac figs. 96-100, 108-111, 115, 116, 118-121, 127-130, 134-142, 145, 146, 148, 151
A bag-like extraembryonic membrane formed as an extension of the midgut; in vertebrates with large eggs, encloses and absorbs the yolk; in mammals, is filled with fluid; arises from splanchnopleure and contains vitelline blood vessels; the earliest blood-forming organ; the source of the primordial germ cells; forms a placenta in some elasmobranchs and mammals (pig); degenerates eventually.

yolk stalk fig. 149
The narrow connection of the yolk sac to the midgut; contains the vitelline arteries and veins.

zona pellucida figs. 4,5
A thick membrane containing mucopolysaccharide surrounding the eggs of mammals; called a zona radiata in the ovary when perforated by cytoplasmic processes of the oocyte and follicle cells; penetated by sperm during fertilization and encloses the embryo during cleavage.

Photographic Data

EMBRYOLOGICAL PREPARATIONS

All prepared slides photographed for this atlas were selected from the embryological collection of the Department of Biology, Wayne State University. The slides were commercially produced by several biological supply houses. Only the opaque mounts of chick embryos (figs. 86 and 124) were specially prepared for the atlas. For these two figures, chick embryos were fixed in a saturated solution of mercuric chloride and transferred to 70% alcohol. The embryos were photographed in alcohol. The pig embryo opaque mount (fig. 147) was fixed in Bouin's fluid and preserved in 70% alcohol for photography. All amphibian material was prepared from *Rana pipiens* embryos; the *Ascaris* was *A. megalocephala.*

EQUIPMENT

Microscope—Bausch and Lomb Dynoptic microscope, Model CBR–9, with achromatic condenser.

Camera—Bausch and Lomb Photomicrographic Camera, Model L.

Illuminators—Bausch and Lomb Research Illuminator with a ribbon filament lamp was used for all photomicrographs. The substage Fluorescent Illuminator was used with the Micro Tessar lens.

Filters—Set of Wratten color filters used to increase or decrease the contrast of the photomicrographs as required.

Enlarger—Omega D2 with a Wollensak lens, 135 mm, *f*/4.5.

PHOTOGRAPHIC SUPPLIES

Films—Kodak 4″ × 5″ cut film, LS Pan, Pantomic X, Plus X, Ektachrome.

Developers—Kodak D19 was used for most plates, but D76 was used to reduce contrast of some whole mount photomicrographs.

Photographic Paper—Kodabromide in contrast surfaces from F1 to F5 as needed.

LIST OF FIGURES WITH LENSES USED FOR EACH

Figure	Objective	Ocular
Color Plate	16 mm achromat	5× Huygenian
1	16 mm achromat	5× Huygenian
2	4 mm achromat	7.5× Hyperplane
3	4 mm achromat	10× Hyperplane
4	16 mm achromat	5× Huygenian
5	22.7 mm achromat	5× Huygenian
6	48 mm achromat	5× Huygenian
7	16 mm achromat	5× Huygenian
8	4 mm achromat	7.5× Hyperplane
9	4.3 mm fluorite	10× Hyperplane
10–13	4 mm achromat	7.5× Hyperplane
14–19	4.3 mm fluorite	7.5× Hyperplane
20	16 mm achromat	10× Huygenian
21–31	16 mm apochromat	7.5× compensating
32	16 mm achromat	5× Huygenian
33, 34	22.7 mm achromat	10× Huygenian
35	16 mm achromat	5× Huygenian
36	4 mm achromat	7.5× Hyperplane
37	16 mm achromat	5× Huygenian
41–45, 52–58	22.7 achromat	5× Huygenian
59–62	30 mm achromat	5× Huygenian
63–69	22.7 mm achromat	5× Huygenian

70	48 mm achromat	5× Huygenian
71	30 mm achromat	5× Huygenian
73	16 mm achromat	5× Huygenian
74	4 mm achromat	7.5× Hyperplane
75–77	30 mm achromat	5× Huygenian
78	16 mm achromat	5× Huygenian
79, 80	30 mm achromat	5× Huygenian
81–84	22.7 mm achromat	5× Huygenian
85	30 mm achromat	5× Huygenian
86	48 mm Micro Tessar	—
87–91	22.7 mm achromat	5× Huygenian
92, 93	30 mm achromat	5× Huygenian
96–103	22.7 mm achromat	5× Huygenian
104	48 mm achromat	5× Huygenian
105	30 mm achromat	5× Huygenian
107, 108	48 mm achromat	5× Huygenian
109–121	30 mm achromat	5× Huygenian
122	48 mm achromat	5× Huygenian
123	30 mm achromat	5× Huygenian
124, 125	48 mm Micro Tessar	—
127–129	48 mm achromat	5× Huygenian
130–137	30 mm achromat	5× Huygenian
138	16 mm achromat	5× Huygenian
139–142	30 mm achromat	5× Huygenian
143	48 mm Micro Tessar	—
145, 146	48 mm achromat	5× Huygenian
147, 148, 151–178	48 mm Micro Tessar	—
179	16 mm achromat	5× Huygenian
180	4 mm achromat	7.5× Hyperplane